Andrzej Pękalski (Ed.)

Diffusion Processes: Experiment, Theory, Simulations

Proceedings of the Vth Max Born Symposium
Held at Kudowa, Poland, 1-4 June 1994

Springer-Verlag Berlin Heidelberg GmbH

Editor

Andrzej Pękalski
Institute of Theoretical Physics
University of Wrocław
Pl. M. Borna 9
PL-50-205 Wrocław, Poland

ISBN 978-3-662-13960-8 ISBN 978-3-540-49038-8 (eBook)
DOI 10.1007/978-3-540-49038-8

CIP data applied for

Originally published by Springer-Verlag Berlin Heidelberg in 1994
Softcover reprint of the hardcover 1st edition 1994

This book was processed using the LATEX macro package
SPIN: 10127074 55/3140-543210 - Printed on acid-free paper

Lecture Notes in Physics

The Editorial Policy for Proceedings

The series Lecture Notes in Physics reports new developments in physical research and teaching – quickly, informally, and at a high level. The proceedings to be considered for publication in this series should be limited to only a few areas of research, and these should be closely related to each other. The contributions should be of a high standard and should avoid lengthy redraftings of papers already published or about to be published elsewhere. As a whole, the proceedings should aim for a balanced presentation of the theme of the conference including a description of the techniques used and enough motivation for a broad readership. It should not be assumed that the published proceedings must reflect the conference in its entirety. (A listing or abstracts of papers presented at the meeting but not included in the proceedings could be added as an appendix.)
When applying for publication in the series Lecture Notes in Physics the volume's editor(s) should submit sufficient material to enable the series editors and their referees to make a fairly accurate evaluation (e.g. a complete list of speakers and titles of papers to be presented and abstracts). If, based on this information, the proceedings are (tentatively) accepted, the volume's editor(s), whose name(s) will appear on the title pages, should select the papers suitable for publication and have them refereed (as for a journal) when appropriate. As a rule discussions will not be accepted. The series editors and Springer-Verlag will normally not interfere with the detailed editing except in fairly obvious cases or on technical matters.
Final acceptance is expressed by the series editor in charge, in consultation with Springer-Verlag only after receiving the complete manuscript. It might help to send a copy of the authors' manuscripts in advance to the editor in charge to discuss possible revisions with him. As a general rule, the series editor will confirm his tentative acceptance if the final manuscript corresponds to the original concept discussed, if the quality of the contribution meets the requirements of the series, and if the final size of the manuscript does not greatly exceed the number of pages originally agreed upon. The manuscript should be forwarded to Springer-Verlag shortly after the meeting. In cases of extreme delay (more than six months after the conference) the series editors will check once more the timeliness of the papers. Therefore, the volume's editor(s) should establish strict deadlines, or collect the articles during the conference and have them revised on the spot. If a delay is unavoidable, one should encourage the authors to update their contributions if appropriate. The editors of proceedings are strongly advised to inform contributors about these points at an early stage.
The final manuscript should contain a table of contents and an informative introduction accessible also to readers not particularly familiar with the topic of the conference. The contributions should be in English. The volume's editor(s) should check the contributions for the correct use of language. At Springer-Verlag only the prefaces will be checked by a copy-editor for language and style. Grave linguistic or technical shortcomings may lead to the rejection of contributions by the series editors. A conference report should not exceed a total of 500 pages. Keeping the size within this bound should be achieved by a stricter selection of articles and not by imposing an upper limit to the length of the individual papers. Editors receive jointly 30 complimentary copies of their book. They are entitled to purchase further copies of their book at a reduced rate. As a rule no reprints of individual contributions can be supplied. No royalty is paid on Lecture Notes in Physics volumes. Commitment to publish is made by letter of interest rather than by signing a formal contract. Springer-Verlag secures the copyright for each volume.

The Production Process

The books are hardbound, and the publisher will select quality paper appropriate to the needs of the author(s). Publication time is about ten weeks. More than twenty years of experience guarantee authors the best possible service. To reach the goal of rapid publication at a low price the technique of photographic reproduction from a camera-ready manuscript was chosen. This process shifts the main responsibility for the technical quality considerably from the publisher to the authors. We therefore urge all authors and editors of proceedings to observe very carefully the essentials for the preparation of camera-ready manuscripts, which we will supply on request. This applies especially to the quality of figures and halftones submitted for publication. In addition, it might be useful to look at some of the volumes already published. As a special service, we offer free of charge LATEX and TEX macro packages to format the text according to Springer-Verlag's quality requirements. We strongly recommend that you make use of this offer, since the result will be a book of considerably improved technical quality. To avoid mistakes and time-consuming correspondence during the production period the conference editors should request special instructions from the publisher well before the beginning of the conference. Manuscripts not meeting the technical standard of the series will have to be returned for improvement.

For further information please contact Springer-Verlag, Physics Editorial Department II, Tiergartenstrasse 17, D-69121 Heidelberg, FRG

PREFACE

The 1994 Max Born Symposium was the fifth organised by the Institute of Theoretical Physics of the University of Wrocław. Its main topic was, as suggested by the title, diffusion. On looking more closely at the phenomenon of diffusion we have discovered that it is omnipresent. Not only do many branches of physics use its concepts, but also it plays a vital role in many problems in astronomy, biology, geology, and even the social sciences! Hence we decided to invite lecturers representing a wide spectrum of points of view on the topic. We have asked physicists from different specializations - experimentalists, theoreticians, and "gamblers", i.e. the Monte Carlo people. As for non-physicists we have invited researchers in the domains of biology, astronomy, and geology. We hoped that such interdisciplinary talks would prove fruitful for all sides. As is the usual fate of such meetings, not all the speakers who had agreed to come were able to show up in Kudowa and therefore, e.g., we had no lecture on the role of diffusion in geology. On the other hand, Michel Droz, who was unable to attend the meeting, was kind enough to send us his lecture.

The lectures presented in this volume are organised only in two categories - invited and contributed talks. Otherwise the order follows the timetable of the symposium with courses on physics intermixed with those on biophysics or astronomy. Such an order reflects our belief that more and more we are coming to treating science as one body.

It is a pleasure to thank all the lecturers, some of them coming a long way, for their contribution not only to the scientific program, but also for creating a truly warm and friendly atmosphere at the symposium.

Numerous duties connected with the organization of the symposium have been carried out by friends and colleagues from the Institute of Theoretical Physics of the University of Wrocław - Dr. Cz. Oleksy, Dr. M. Dudek, Dr. M. Wolf, W. Hann, Z. Koza, K. Skwarek, M. Bordzań, and G. Czarnecki. Their help is gratefully acknowledged.

Our sponsors were first of all the University of Wrocław, then the Polish Academy of Sciences and the Batory Foundation. We thank them all.

At the end of the book we include the list of the lectures delivered at the fourth Max Born Symposium. The proceedings of that meeting have not been published.

Wrocław
September 1994

Andrzej Pękalski

Director of the Symposium

Contents*

* The names of the speakers are underlined

Selfdiffusion of polymer chains in solutions and melts

K. Binder

Institut für Physik, Johannes Gutenberg Universität Mainz
D–55099 Mainz, Staudinger Weg 7, Germany

Abstract: Anomalous diffusion of monomers of polymer chains, as well as motion of these chains as a whole, is discussed with an emphasis on Monte Carlo simulations and simple scaling concepts. While the behavior of isolated chains in good solvents or Theta–solvents without excluded volume interactions is fully accounted for by the Rouse model, the behavior is less clear both for isolated chains in bad solvents and for chains in dense melts. Collapsed chains are shown to diffuse as $g_3(t) \equiv \langle[\mathbf{r}_{CM}(t) - \mathbf{r}_{CM}(0)]^2\rangle \propto t^{x_3}$ where the (effective?) exponent x_3 simply seems to be linearly temperature-dependent for temperatures T lower than the Θ–temperature, $x_3 \approx T/\Theta$. A relaxation time $\tau \propto N^3$ is found, and scaling scenarios which possibly can explain these results are developed.

Short (not entangled!) chains in dense melts also are found to exhibit anomalous center of mass–diffusion, $g_3(t) \propto t^{x_3}$ with $x_3 \approx 0.8 - 0.85$, contrary to expectations from the Rouse model. Therefore also the crossover from the Rouse–like behavior for chain length N less than the entanglement chain length N_e to reptation–like behavior for long chains shows some unexpected features.

Finally we briefly discuss the motion of chains in constrained geometry, such as chains constrained in straight tubes and chains end–grafted on a wall in a polymer brush.

1 Introduction and overview

The Brownian motion of long flexible polymer chains in solution or melts has been of long–standing interest and still poses challenging questions [1, 4]. There is an interesting interplay between the constraint of chain connectivity and the interactions between monomers (and solvent molecules, if present). The geometrical size of the polymer coil (its gyration radius R_{gyr}, for instance) scales with the number N of subunits ("effective" monomers) of linear dimension b like fractal objects [5]

$$R_{gyr} \propto bN^{\nu} \ . \tag{1}$$

Here ν is a "critical exponent" (if the coil has on the average a spherical shape, the volume in d dimensions is $\propto R_{gyr}^d \propto N^{\nu d}$ and the density $\rho \propto N/$volume

$\propto N^{1-\nu d} \neq N^o$ in general). The time needed for a coil to renew its configuration then also exhibits a law (critical slowing down [6])

$$\tau_N \propto R_{gyr}^{z'} \propto N^z \quad, z = \nu z' , \tag{2}$$

where z, z' are dynamic exponents [6, 7]. For times t less than the chain relaxation time τ_N inner monomers of the chain exhibits anomalous diffusion, i.e. a subdiffusive increase of the mean square displacement with time,

$$g_1(t) \equiv \langle [\mathbf{r}_i(t) - \mathbf{r}_i(0)]^2 \rangle \propto t^{x_1} \quad, x_1 < 1 \text{ for } t < \tau_N , \tag{3}$$

where $\mathbf{r}_i(t)$ is the position of an inner monomer (far from the free ends of the chain) in the d–dimensional space. Sometimes the center of mass of the chain exhibits anomalous diffusion as well [9, 10]

$$g_3(t) \equiv \langle [\mathbf{r}_{cm}(t) - \mathbf{r}_{cm}(0)]^2 \rangle \propto t^{x_3} , \quad x_3 < 1 \text{ for } t < \tau_N . \tag{4}$$

In this paper, we shall consider only the simplest case, which for dilute chains with ideal gaussian configurations, for which $\nu = 1/2$ in $d = 3$, reduces to the well known Rouse model [4, 11]. I.e., solvent molecules only act as a heat bath inducing random conformational changes of the chain, and do not transmit hydrodynamic forces (which would yield the Zimm model [4, 12]). While the extension of the Rouse model for isolated chains in good solvents (for which $\nu \approx 0.588$ in $d = 3$ [13] and $\nu = 3/4$ in $d = 2$ [14] is rather well understood [7]-[9], [15, 16], the dynamics of collapsed chains in poor solvents [2], [17]-[19] (Fig. 1) still remains a puzzle [9, 10], [20]-[22]. Such problems are not surprising, since the collapsed chains (at temperatures T less than the Theta–temperature [2, 17] Θ) are very dense objects with strong interactions among the beads.

Another case where the dynamics of chains is difficult to understand are dense melts, where another length scale d_τ (the "tube diameter", if the reptation model [2]-[4],[23]-[28] is invoked) enters, and several regimes of subdiffusive monomer motion with different exponents x_1 must be distinguished. But the reptation picture, where a chain relaxes by a curvilinear diffusion along its own contour, has been challenged by various alternative approaches, such as the mode coupling theory of entangled polymeric liquids [29, 30].

Finally such additional length scales also must be considered when one discusses the motions of chains in constrained geometry, e.g. diffusion of chains in straight tubes (Fig. 2) [31, 32] or in polymer brushes (where chain ends are restricted to a binding wall, Fig. 3) [34]-[44]. In the latter case the situation is particularly complicated, since one must distinguish cases where the end-grafted monomer is strictly fixed at the wall and where it is still mobile in lateral direction [41], and also the spatial structure in the z–direction is inhomogeneous, due to the gradual decrease of monomer density in the outer region of the brush. So several length scales occur (blob size $\xi(0) = \sigma^{-1/2}$ at the wall and $\xi(h - d_{fl}) = d_{fl}$ near the end of the brush (h is the brush height, $h \propto R_{gz} \propto \sigma^{1/(2\nu)-1/2}N$), chain linear dimension in the directions perpendicular to the z–direction, $R_{g\perp} \propto \sigma^{1/(4\nu)-1/2}N^{1/2}$ [43, 44]). Understanding the dynamics of chains in this nonuniform situation is a particular challenge but will remain out of detailed consideration here.

In Sec. 2 we shall now recall more details about the Rouse model for gaussian coils and its extension including excluded volume. Sec. 3 describes some simulation results for the dynamics of collapsed chains, and scaling speculations to account for them. Sec. 4 recalls the reptation predictions for meansquare displacements in melts and the corresponding evidence from simulations. Alternative concepts on chain dynamics in this entangled geometry will only be briefly mentioned, while Sec. 5 contains some conclusions and comments on the dynamics in the constrained geometries (Fig. 2,3).

2 The Rouse model for gaussian chains and in good solvents

Let us model a polymer chain as a sequence of beads at position $\{\mathbf{R}_1, \mathbf{R}_2, ..., \mathbf{R}_N\}$ connected by springs (Fig. 4, [45]). The effective potential energy U is $U = (k/2)\sum_{n=2}^{N}(\mathbf{R}_n - \mathbf{R}_{n-1})^2$, where $k = 3k_BT/b^2$ because on the coarse-grained length scale b entropic forces dominate.

Assuming that the solvent acts as a heat bath, we obtain the Langevin equation [1, 4, 11]

$$\zeta\frac{d\mathbf{R}_n(t)}{dt} = -k(2\mathbf{R}_n - \mathbf{R}_{n+1} - \mathbf{R}_{n-1}) + \mathbf{f}_n(t), \;\; n = 2, \cdots, N-1 \tag{5}$$

where ζ is a friction coefficient and $\mathbf{f}_n(t)$ a random force describing Gaussian uncorrelated noise,

$$\langle \mathbf{f}_n(t)\rangle = 0, \; \langle f_{n\alpha}(t) f_{m\beta}(t')\rangle = 2\zeta k_BT\delta_{nm}\delta_{\alpha\beta}\delta(t-t') \, . \tag{6}$$

For large N we may neglect effects due to the free ends by using periodic boundary conditions along the chain, $\mathbf{R}_{N+1} \equiv \mathbf{R}_1, \mathbf{R}_0 \equiv \mathbf{R}_N$, and then Eq. (5) holds for all n. Due to the translational invariance in the coordinate system along the chain we have

$$\mathbf{R}_n(t) \propto \exp(iQn)\exp(-t/\tau_Q) \tag{7}$$

where Q is a momentum–type coordinate in the coordinate system along the chain, conjugate to the index n labeling the monomers, and the time τ_Q can be written in terms of the rate $W \equiv k/\zeta = 3k_BT/b^2\zeta$ as

$$(W\tau_Q)^{-1} = 2(1-\cos Q) \approx Q^2 \quad (Q \to 0) \tag{8}$$

The periodic boundary condition requires $QN = \pi p, p = 1, 2 \ldots$ and hence we obtain a spectrum of Rouse modes

$$\tau_{(p)}^{-1} = W\pi^2p^2/N^2 = (3k_BT\pi^2/N^2\zeta)p^2 \; , \; p = 1, 2, \cdots \tag{9}$$

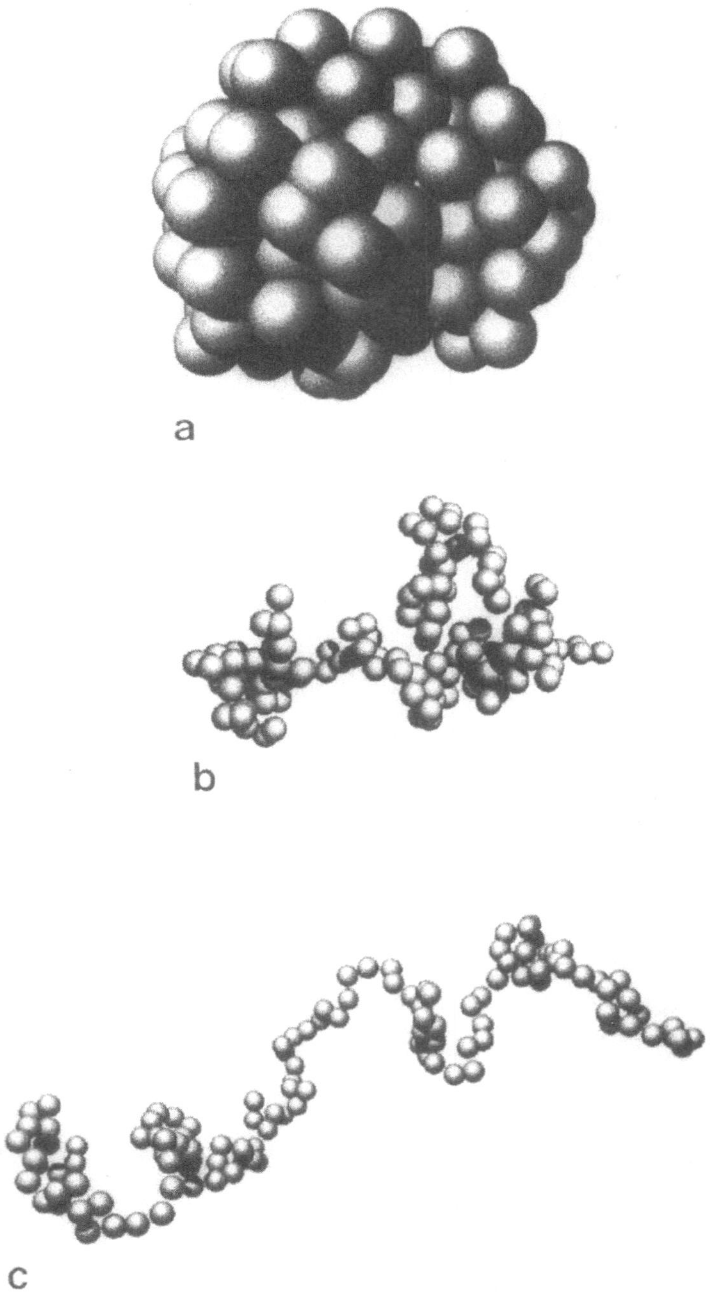

Fig. 1. Snapshot pictures of a well–equilibrated polymer chain with $N = 128$ beads, using the off–lattice bead–spring model of Sec. 3, after 524000 Monte Carlo steps (MCS) per bead, at $T = 0.30(a)$, $T = \Theta = 0.58(b)$ and $T = 1.20(c)$. For each bead, a sphere with radius l_{min} is drawn (cf. Sec. 3). From Ref. 22.

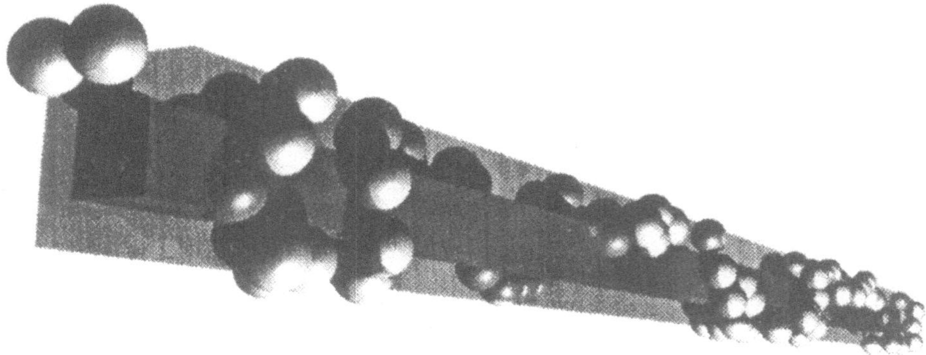

Fig. 2. Snapshot picture of a chain with $N = 128$ effective monomers confined in a tube with quadratic cross section $D_T \times D_T$ with $D_T = 2\, l_{\max}$ (cf. Sec. 3 below) for strong wall attraction ($\epsilon/k_BT = -3$). The inner concentric tube with cross section $D_T \times D_T$ with $D_T = l_{\max}$ has only been drawn to enhance the spatial view of the chain, it has no physical effect whatsoever. One can see that the chain winds itself around in an irregular fashion along the attractive walls (outer tube). From Ref. 33.

Now a distance n along the chain corresponds to a distance $b\sqrt{n}$ in real space, for gaussian random walks. Therefore Q is not conjugate to a distance in real space but rather to (distance)^2, and hence Q corresponds to $(ql)^2$ when we consider scattering under wave vector $\mathbf{q}$ [46]. This sketchy argument can be substantiated by a detailed calculation of the intermediate dynamic scattering function

$$S(\mathbf{q},t) \equiv \frac{1}{N} \sum_{i=1}^{N} \sum_{j=1}^{N} \langle \exp\{i\mathbf{q} \cdot [\mathbf{R}_i(t) - \mathbf{R}_j(0)]\} \rangle \ , \tag{10}$$

and one can show that for $N \to \infty$ [46]

$$S(\mathbf{q},t)/S(q,0) \propto \exp\{-\omega_c(\mathbf{q})t\}\ ,\ \omega_c(\mathbf{q})t \gg 1 \tag{11}$$

with

$$\omega_c(\mathbf{q}) \propto W(qb)^{z'}\ ,\ z' = 4. \tag{12}$$

While for normal diffusion $w_c(\mathbf{q}) = Dq^2$ where D is the diffusion constant, the dynamic exponent $z' = 4$ implies subdiffusive behavior of the monomer displacements, Eq. (3), with

$$x_1 = 1/2\ ,\ \text{Rouse model without excluded volume}\ . \tag{13}$$

Of course, for finite N Eq. (11) only holds for $t < \tau_{(1)} \equiv \tau_N$, while for $t > \tau_N$ ordinary diffusion takes over [4]

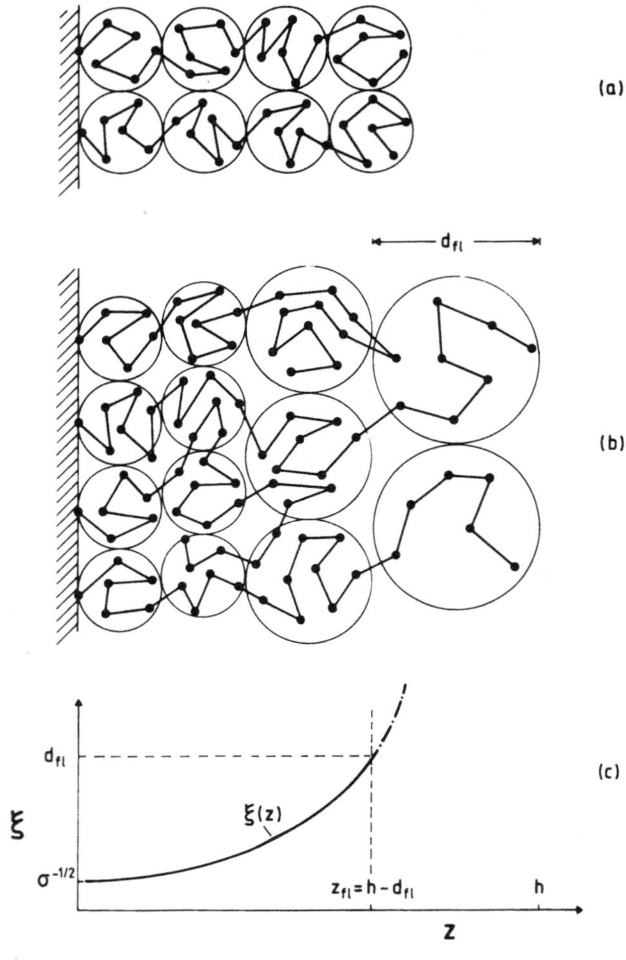

Fig. 3. Schematic illustration of the structure of polymer brushes, where chains are grafted with one end at a repulsive wall (shaded) at a surface coverage σ. In a good solvent, the chains can be modeled as a sequence of blobs of uniform diameter $d_{blob} = \sigma^{-1/2}$ (Alexander [36] picture, part (a)) or as a sequence of blobs, whose diameter $\xi(z)$[(b), (c)], also called "screening length", smoothly increases until the size $d_{fl} \propto \sigma^{1/(2\nu-1)} N^{1-\nu}$ of the last blob [43]. The gyration radius is very anisotropic, the component in z-direction scales as $R_{gz} \propto \sigma^{\frac{1}{2\nu}-\frac{1}{2}} N$, while the components perpendicular to the z-direction scale as $R_{g\perp} \propto \sigma^{\frac{1}{4\nu}-\frac{1}{2}} N^{1/2}$. From Ref. 44.

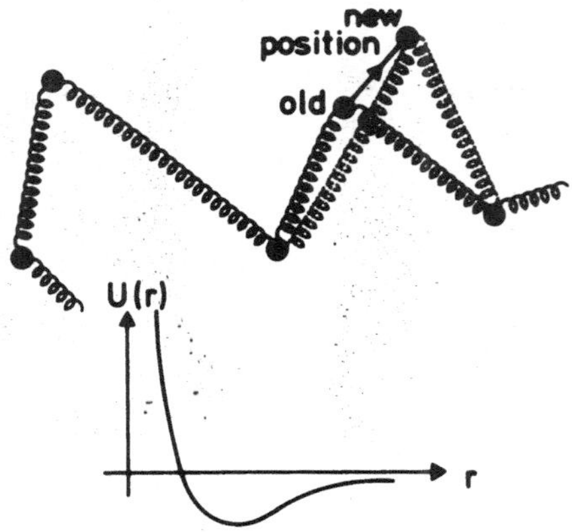

Fig. 4. Off–lattice bead–spring model for polymer chains. This model has no excluded volume and leads to gaussian chain conformations, $R^2_{gyr} = b^2 N/6$. Excluded volume is incorporated by using the repulsive part of the Lennard–Jones potential between the beads. Using a potential with both repulsive and attractive parts, as sketched, one can have swollen coils (good solvent) at high temperature and collapsed coils (bad solvent) at low temperature. The harmonic or anharmonic forces represent the covalent bonds along the chain. From Ref. [45].

$$g_1(t) = g_3(t) = 6D_N t \, , t > \tau_N \propto W^{-1} N^2, D_N = k_B T/N\zeta. \tag{14}$$

Note that Eq. (10) can be understood directly by the simple argument that a random displacement of an effective monomer by a distance comparable to b (Fig. 4) will lead to a center of gravity displacement b/N. Now a sensible choice of time units is that every monomer attempts such a reorientation at a rate W per unit time, and hence we have to add up N such random displacement squares $N(b/N)^2 W \propto k_B T/(N\zeta)$ which is Eq. (14) if there is no anomalous diffusion,

$$x_3 = 1 \, , \text{ Rouse model} \, , \tag{15}$$

cf. Eqs. (4),(14). On the other hand, we may invoke the scaling argument that

$$g_1(\tau_N) = g_3(\tau_N) \propto R^2_{gyr} \propto b^2 N \; ; \tag{16}$$

Eq. (16) says that during the relaxation time τ_N both an individual monomer and the center of mass must have travelled a distance of the order of R_{gyr}. Since we must have $g_3(t > \tau_N) = 6D_N t$ a smooth matching condition requires also $g_3(\tau_N) \propto D_N \tau_N \propto D_N W^{-1} N^2$, which – together with Eq. (16) again yields $D_N \propto b^2 W/N$, i.e. Eq. (14). This scaling argument corroborates the fact that in the Rouse model without excluded volume there is no anomalous diffusion of the center of mass, i.e. Eq. (15).

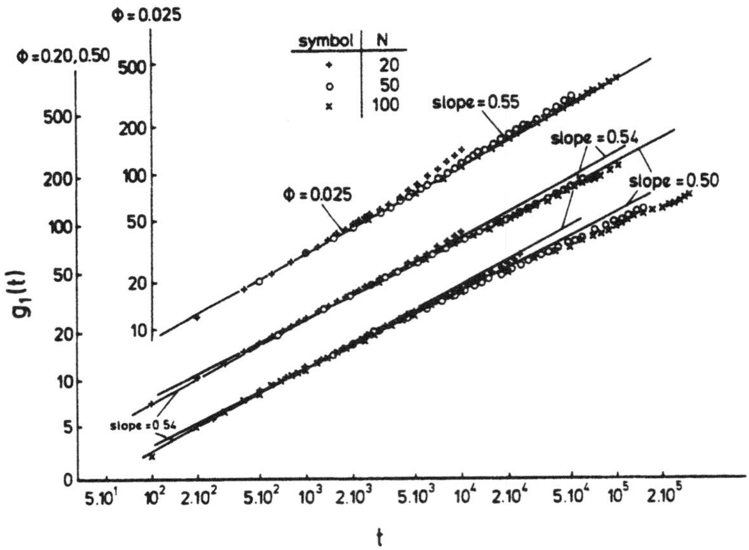

Fig. 5. Log–log plot $g_1(t)$ vs. t for the three–dimensional athermal bond fluctuation model on the simple cubic lattice, at three volume fractions of occupied lattice sites: $\Phi = 0.025$ (topmost data set with offset ordinate axis), $\Phi = 0.20$ (data points in the middle) and $\Phi = 0.50$ (lower part). Time is measured in units of attempted moves per monomer. Here $\Phi = 0.50$ (lower part). Time is measured in units of attempted moves per monomer. Here $\Phi = 0.025$ corresponds to a dilute and $\Phi = 0.20$ to a semidilute solution, while $\Phi = 0.5$ corresponds to a concentrated solution or melt. From Ref. [15]

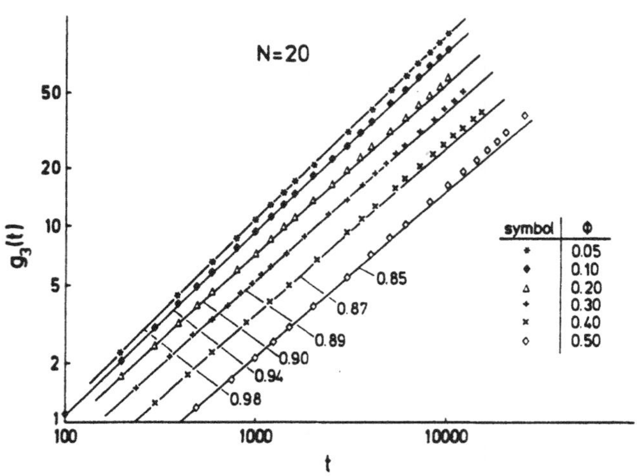

Fig. 6. Log–log plot of the center of gravity displacement $g_3(t)$ versus time, for the three–dimensional bond fluctuation model at various volume fractions Φ as indicated. Numbers at straight lines indicate $x_3(\Phi)$. From Ref. 47.

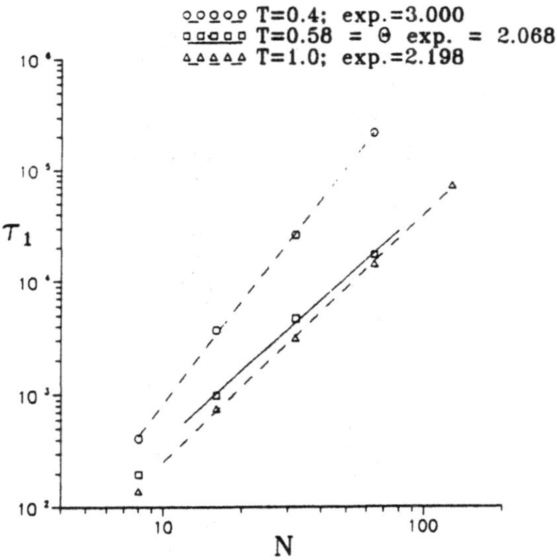

Fig. 7. Log–log plot of the chain relaxation time τ_1 defined from the monomer mean square displacement $(g_1(\tau_1) \equiv \langle R^2_{gyr} \rangle)$ versus chain length N, for three temperatures as indicated, for the bead–spring model (Fig. 4c,d) using a Morse potential with $\epsilon/k_B \equiv 1$. Note that $T = 0.58$ is the Θ–temperature for this model. From Ref.9.

If we assume that Eq. (15) holds also in the good solvent case with excluded volume, where $R^2_{gyr} \propto b^2 N^{2\nu}$, Eq. (16) yields

$$R^2_{gyr} = b^2 N^{2\nu} = D_N \tau_N \propto (k_B T/N\zeta)\tau_N, \; \tau_N \propto N^{2\nu+1}\,, \tag{17}$$

i.e. the dynamic exponents z, z' in Eq. (2) are

$$z = 2\nu + 1, z' = 2 + 1/\nu\,. \tag{18}$$

Dynamic scaling [6] then also implies [32]

$$x_1 = 2/z' = (1 + 1/2\nu)^{-1} \approx 0.54\,. \tag{19}$$

This prediction was tested by simulations [15], see Fig. 5. The data for the dilute case are indeed independent of chain length N for $t \ll \tau_N$, as it should be, and within statistical accuracy compatible with Eq. (19). For the "semidilute" concentration (where coils have started to overlap and excluded volume is screened at distances larger than the screening length $\xi(\Phi) \propto \Phi^{-\nu/(d\nu-1)}$ [2]) one finds a crossover from $t^{0.54}$ at very short times { where $g_1(t) \ll \xi^2$} to a behavior $g_1(t) \propto t^{1/2}$ at intermediate times, as expected from the gaussian coil statistics that takes over at distances exceeding $\xi(\Phi)$. For the densest system (and long enough chains) the data for large times fall even systematically below the ideal Rouse-like behavior $g_1(t) \propto t^{1/2}$, due to the gradual onset of entanglement effects, see Sec. 4. However, over the same time range where $g_1(t) \propto t^{1/2}$ for short chains ($N = 20$) in dense systems there does occur anomalous diffusion, with an effective exponent $x_3(\Phi)$ that depends on volume fraction [47], see Fig. 6. Only for $\Phi \to 0$ does $x_3(\Phi)$ tend to Eq. (15). The implication is that the Rouse model is not strictly valid even for short chains in a melt.

3 Dynamics of collapsed chains

In order to study the dynamics of chains at and below the Θ–temperature, an off-lattice bead–spring model of the type of Fig. 4 is used, with a Morse potential, $U_M = \epsilon\{\exp[-2\alpha(r - r_{\min})] - 2\exp[-\alpha(r - r_{\min})]\}$, where the constants are chosen $\epsilon/k_B \equiv 1, r_{\min} = 0.8$ and $\alpha = 24$. The springs are described by a harmonic potential in the range from $l_{\min} = 0.4 < l < l_{\max} = 1.0, U(l) = K(l - l_o)^2$, with $l_o = 0.7, k = 10$, while $U(l < l_{\min}) = U(l > l_{\max}) = \infty$. An analysis of the chain length dependence of the radii shows that for this model the Θ–temperature occurs at $\Theta = 0.58 \pm 0.01$ (typical configurations of the chains are shown in Fig. 1).

Fig. 7 shows a log–log plot of the chain relaxation time τ_1 defined by $g_1(\tau_1) \equiv \langle R^2_{gyr}\rangle$, which should be of the same order as the Rouse time τ_N discussed in Sec. 2.

Within statistical errors the exponent $z \approx 2.20$ in the good solvent regime ($T = 1.0$) and at the Θ–temperature ($z \approx 2.07$) is in agreement with the respective theoretical predictions, $z = 2\nu + 1$ {Eq. (18)}, using $\nu \approx 0.59$ and $\nu = 1/2$, respectively. However, for $T < \Theta$ (in the regime where the chains are collapsed, cf. Fig. 1) a much larger exponent is found, $z \approx 3$; the scaling relation $z = 2\nu + 1$ does not hold for collapsed chains, at least not in the framework of our model, since $\nu = 1/3$ for collapsed chains would yield $z = 5/3 \approx 1.67$ [21].

The dramatic failure of the simple Rouse–type scaling {Eq. (18)} is corroborated by the observation of a much faster decrease of the diffusion constant with chain length (Fig. 8a) and of anomalous diffusion (Fig. 8b). Strangely enough, for inner monomers the Rouse behavior {Eqs. (3),(13) = $g_1 \propto t^{1/2}$} still seems to hold, while the center of mass motion shows clearly subdiffusive behavior, with an (effective?) exponent x_3 that approximately satisfies the relation

$$x_3 \approx (T/\Theta) \; , T < \Theta \, . \tag{20}$$

Thus the regime of anomalous diffusion disappears at the Θ–point!

A possible explanation of this behavior is found studying the rates for the various displacements at fixed temperature as function of chain length [10]. These rates seem to decrease with increasing N as power laws,

$$g_1(t) = (W_1 t)^{x_1} \propto N^{-y_1} t^{x_1} \; , t < \tau_N \tag{21}$$

$$g_3(t) = (W_3 t)^{x_3} \propto N^{-y_3} t^{x_3}, t < \tau_N \, , \tag{22}$$

and to ensure a smooth crossover we assume

$$g_1(\tau_N) \approx g_3(\tau_N) \approx \langle R_g^2\rangle \propto N^{2/3} \, . \tag{23}$$

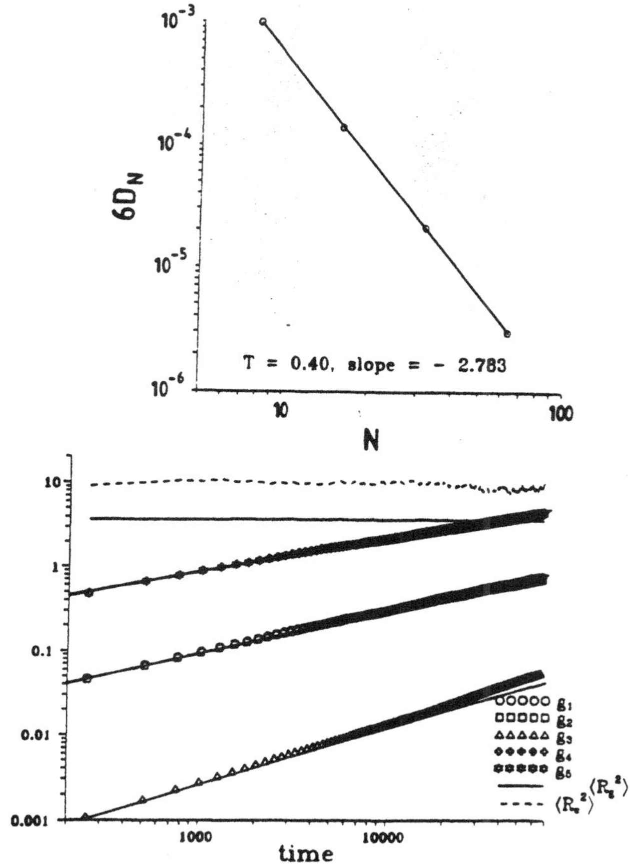

Fig. 8. a) Log–log plot of the chain diffusion constant D_N vs. N for the bead–spring model of Figs. 1,4c,7 at a temperature $T = 0.4$ in the region where the chains are collapsed. Straight line indicates an exponent $y \approx 2.8$ in the relation $D_N \propto N^{-y}$. From Ref. 9. b) Log–log plot of mean square displacements $g_1, g_2 \equiv \langle [\mathbf{r}_i(t) - \mathbf{r}_{CM}(t) - (\mathbf{r}_i(0) - \mathbf{r}_{CM}(0))]^2 \rangle, g_3, g_4 \equiv \langle [\mathbf{r}_{end}(t) - \mathbf{r}_{end}(0)]^2 \rangle, g_5 \equiv \langle [\mathbf{r}_{end}(t) - \mathbf{r}_{CM}(t) - (\mathbf{r}_{end}(0) - \mathbf{r}_{CM}(0))]^2 \rangle$ vs. time, for $T = 0.4$. Horizontal straight line indicates $\langle R^2_{gyr} \rangle$, broken line the mean-square end–to end distance $\langle \mathbf{R}^2_{end} \rangle$. Slopes of straight lines are $x_1 \approx 0.5$ (for g_1, g_2) but $x_3 \approx 0.72$ (for g_3) and $x_e \approx 0.4$ (for g_4, g_5). From Ref. 10.

For $t \gtrsim \tau_N$ we must have ordinary diffusion, $g_1(t) \approx g_3(t) = 6D_N t$ with $D_N \propto N^{-y}$.

These assumptions yield the scaling laws ($\tau_N \propto N^2$, Eq. (2))

$$z = (y_1 + 2/3)/x_1 = (y_3 + 2/3)/x_3 = y + 2/3 \tag{24}$$

Noting $x_1 = 1/2, y_1 \approx 0.7$ [10] would imply $z \approx 2.8$, while $x_3 \approx 0.74$, $y_3 \approx 1.7$ [10] would imply $z \approx 3.2$, and thus $y \approx 2.13 - 2.53$. These estimates agree with the direct observation (Figs. 7,8a) only very roughly.

A speculative explanation [9, 10] of these results assumes that either in a collapsed chain a net displacement is effected only if a monomer at a surface is involved, yielding $W_1 \propto$ (surface/volume) $\propto N^{-1/3}$, or – alternatively – that an end monomer must move at the surface $\{W_1 \propto (\text{surface/volume})/N \propto N^{-4/3}\}$. Obviously, the latter assumption is in rough agreement with observation. The associated center of mass rate then should be $W_3 \propto W_1/N \propto N^{-7/3}$, which would yield $y_3 = 7x_3/3 \approx 1.87$, again roughly compatible with the observation. But it is possible that Eqs. (21)–(24) are too simple, since they assume a single regime of subdiffusive motion characterized by a single exponent x_3 for the center of mass motion for all times $t < \tau_N$. The data do not really rule out several regimes with different exponents, however, and such a behavior with several different exponents x_3, as it occurs for the motion of entangled chains in melts (see Sec. 4). Also it is posssible that the N–dependence described by power–laws in Eqs. (21),(22) is spurious, reflecting the fact that for small N in collapsed chains there is not yet a clear division into "surface" and "bulk", and one should note that there are indications that the bulk of collapsed chains is in a glass–like frozen state [10]. Also the role of knots (self–entanglements) [20] remains to be clarified. More work clearly is required to understand the dynamics of collapsed coils!

4 Dynamics of chains in dense melts

In a dense melt, chains have gaussian random walk–type configurations and since in $d = 3$ the density of one chain inside its volume $\{\rho = N/(\text{volume}) \propto N/R^3_{gyr} \propto N/(bN^{1/2})^3 = b^{-3}N^{-1/2}\}$ is much smaller than the melt density ($\rho_{melt} = b^{-3}$), the volume of a chain contains monomers of $N^{1/2}$ other chains: this strong entangledness of one chain with many other chains produces an anomalously slow relaxation of long chains in melts [1]-[4]. Experimentally this shows up in a diffusion constand $D_N \propto N^{-2}$ and a melt viscosity (proportional to the chain relaxation time τ_N) $\eta_N \propto N^{3.4}$ [3] if $N \gg N_e$ (the "entanglement" chain length). The explanation of these findings has stimulated a longstanding debate [4], [23]-[30], [47]-[53]. According to the "tube model" of de Gennes [23] and Edwards [4], the net effect of the entanglements among the chains is that on a coarse grained scale the motion of the chains is essentially restricted to a diffusion along its own contour ("reptation"). The proper length scale for this curvilinear motion is the tube diameter d_T, related to N_e via the standard random walk relation, $d_T^2 \propto b^2 N_e$: as long as one considers monomer mean square displacements $g_1(t) \ll d_T^2$, the entanglements are not yet effective, one observes simple Rouse–like motion. Fig. 9a gives a schematic description of the reptation predictions for the mean square displacements $g_1(t)$ and $g_3(t)$, where prefactors of order unity are ignored throughout. When $g_1(t) = b^2(Wt)^{1/2} = d_T^2 = b^2 N_e$, i.e. for $Wt = W\tau_e = (d_T/b)^4 = N_e^2$, one crosses over to a diffusion of the chain along its own contour. Since this is again a random walk, a path of length L along this walk corresponds only to a distance $\sqrt{L}$ in real space. Thus in this regime the displacement of monomers increases as $g_1(t) = bd_T(Wt)^{1/4}$, and the center of gravity shows anomalous diffusion,

$$g_3(t) = \frac{b^2 d_T^2}{R_{gyr}^2}(Wt)^{1/2} = \frac{N_e}{N} b^2 (Wt)^{1/2}. \tag{25}$$

At the Rouse time $\tau_R = N^2/W = (R_{gyr}/b)^4$ the monomer density inside the tube is equilibrated, and a normal Rouse–type motion occurs again, but with reduced prefactors,

$$g_3(t) = \frac{b^4 d_\tau^2}{R_{gyr}^4}(Wt)^1 = \frac{N_e}{N^2} b^2 (Wt)^1 , \tag{26}$$

$$g_1(t) = \frac{b^2 d_T}{R_{gyr}}(Wt)^{1/2} = \sqrt{\frac{Ne}{N}} b^2 (Wt)^{1/2} . \tag{27}$$

Of course, both the crossover at $\tau_e = N_e^2/W$ and the crossover at $\tau = \tau_R$ are smooth and spread out over a broad time interval, one does not expect such sharp crossovers as drawn in the schematic Fig. 9a for the sake of clarity. It is then seen that Eqs. (26),(28) merge for

$$g_1(t) = g_3(t) = R_{gyr}^2 \quad \text{if } t = \tau_N = N^3/(N_e W) = R_{gyr}^6/(b^4 d_\tau^2) \ . \tag{28}$$

Eqs. (26),(28) thus imply $D_N \propto N^{-2}, \tau_N \propto N^3$ (the experimental result $\eta \propto N^{3.4}$ usually is interpreted as an effect of insufficient approach to the asymptotic limit $N \to \infty$: for N not much larger than N_e additional effects such as "tube leakage", fluctuations in tube length etc. are important [4, 26]. However, despite considerable efforts (e.g. [25]) a fully convincing microscopic derivation of the reptation mechanism does not exist, and possibly different alternative theories [29, 30, 50]-[53] could also account for the experimental data. E.g., in the mode coupling theory [29, 30] the interpretation of d_τ is that of a length scale where crossover from simple Rouse dynamics to mode–coupled dynamics occurs. As also shown in Fig. 9a, this theory also implies $D_N \propto N^{-2}$, $\tau_N \propto N^3$ and anomalous diffusion is implied at intermediate scales, though with exponents slightly different from that of the reptation theory $\{g_3(t) \propto t^{9/16}$ instead of $t^{1/2} = t^{8/16}$, and at $Wt = cN^{5/2}$ an additional crossover occurs, where c is a constant$\}$. Despite this similarity of the results, the mode–coupling approach [29, 30] is based on isotropic rather than curvilinear motions of the monomers being dominating.

Simulations [47] do show evidence of two crossovers for both $g_1(t)$ and $g_3(t)$ {Fig. 9b} but clearly cannot distinguish between the two alternative scenarios shown in Fig. 9a. And there is another difficulty, that $g_3(t) \propto t^{0.8}$ already in the Rouse regime and there is no crossover in $g_3(t)$ at $t = \tau_e$, unlike the predictions of either reptation or mode coupling theory {Fig. 9a}. This difficulty is not related primarily to reptation, since this anomalous diffusion occurs already for short nonentangled chains in dense systems (Fig. 6). This implies that there is already a difficulty with the Rouse description for the small scale motion in polymer melts. Since both reptation and mode coupling are built on Rouse relaxation on small scales, Figs. 6,9b cast doubt on both theories! On the other hand, one

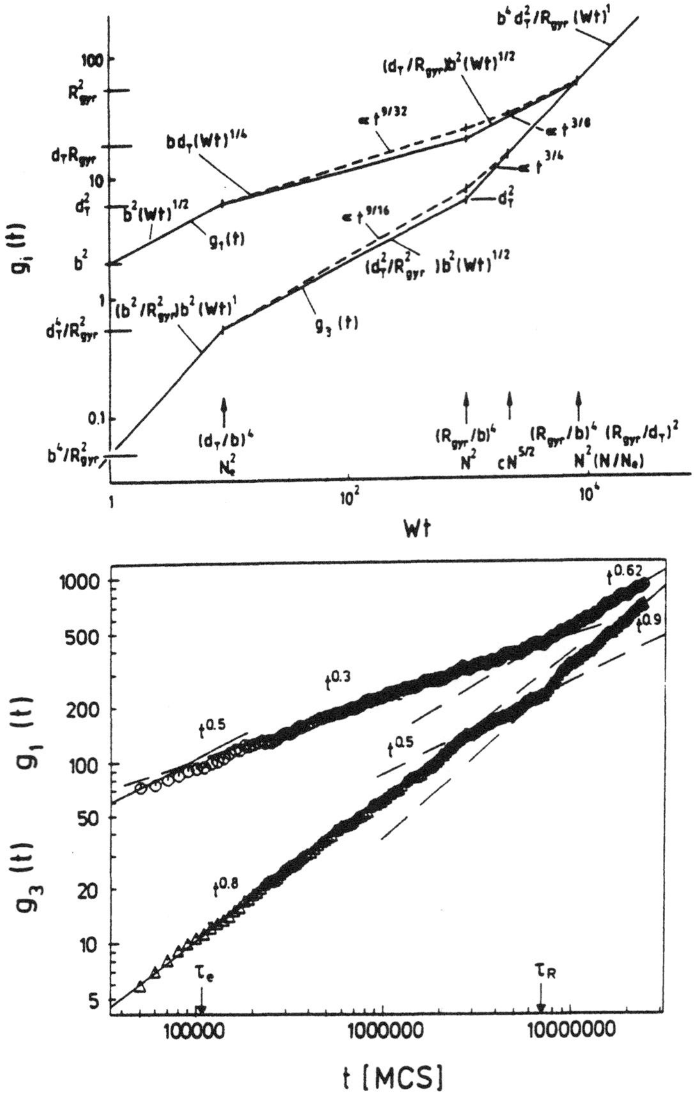

Fig. 9.

a) Schematic description of the time–dependent monomer displacement $g_1(t)$ and the center of mass displacement $g_3(t)$ according to reptation theory (full curves) and the mode coupling theory (broken curves). For explanations cf. text.
b) Monte–Carlo results for the mean square displacements $g_1(t), g_3(t)$, obtained from the athermal bond fluctuation model for $N = 200$ at a volume fraction $\Phi = 0.5$ at the simple cubic lattice. From Ref. [47].

must be careful and not abandon any of these theories too quickly – numerical evaluation of the mode coupling theory [30] shows that even for $N = 10^5$ one has smooth crossovers spread out over about 8 decades in time, and clearcut power laws cannot yet be distinguished.

5 Discussion

Evidence from computer simulation suggests that the Rouse–like description can be carried over from gaussian chains (Θ–solvent) to swollen chains (good solvents), but not to collapsed chains (bad solvent). The latter seems to show anomalous diffusion of the center of gravity over a broad range of times, and the scaling of the diffusion constant and the relaxation time with chain length are not understood. While in real experiments, of course, hydrodynamics implies

Stokes law $\{D_N \propto (\eta_s R_{gyr})^{-1} \propto N^{-1/3}$ where η_s is the solvent viscosity$\}$, at the same time one can expect that solvent–mediated hydrodynamic forces are screened inside a collapsed coil; so the present treatment should be of interest for the actual chain relaxation time τ_N. Many biological molecules occur in collapsed states, and we hope that experimental studies of this problem will come.

Computer simulations of chain dynamics in melts now have produced clear evidence of an intermediate length scale d_T [47, 48], which can be interpreted as the tube diameter if reptation theory is invoked. This length scale can also be seen in the intermediate neutron scattering function [28, 54]-[57], but probably neither experiment [54, 55] nor simulation [56, 57] can rule out the alternative mode coupling description [29, 30]. A fundamental difficulty seen in the simulation, but not predicted by any theory, is an anomalous diffusion of the center of gravity of chains in dense melts, under conditions where the Rouse model is expected to hold {Figs. 6, 9b}. We are not aware of any theory explaining this behavior, and thus a quantitative comparison between simulation and theory is hampered. This strange behavior is neither seen in inner monomer displacement {Fig. 9b} nor in the scattering function [54]-[58]: the latter does agree with the Rouse model at short times! Thus experimental probes which are sensitive to the short–time behavior of the center of mass–displacement $g_3(t)$ in melts would be very desirable.

The different length scales (b, d_T, R_{gyr}) then allow for several characteristic times where crossover from one regime to another one occurs: e.g. {Fig. 9a} $\tau_e \propto N_e^2/W$, $\tau_R \propto N^2/W$, $\tau_N \propto N^3/(WN_e)$. Simulations now can see two of these crossovers {Fig. 9b} but clearly one would need more decades in time (and still longer chains!) for a quantitative analysis: this is hardly possible on present day computers.

Different length scales also occur when we study the dynamics in pores (Fig. 2), where the chain is a sequence of "blobs" with a diameter comparable to the pore diameter D_T, and the longitudinal chain linear dimension (parallel to such a straight tube) then is $R_{gl} \propto bN(b/D_T)^{1/\nu-1}$. The time for a chain to diffuse its own size then is

$$\tau_N \sim R_{gl}^2/D_N \propto R_{gl}^2 N/W \propto b^2 N^3 (b/D_T)^{2/\nu-2}/W \tag{29}$$

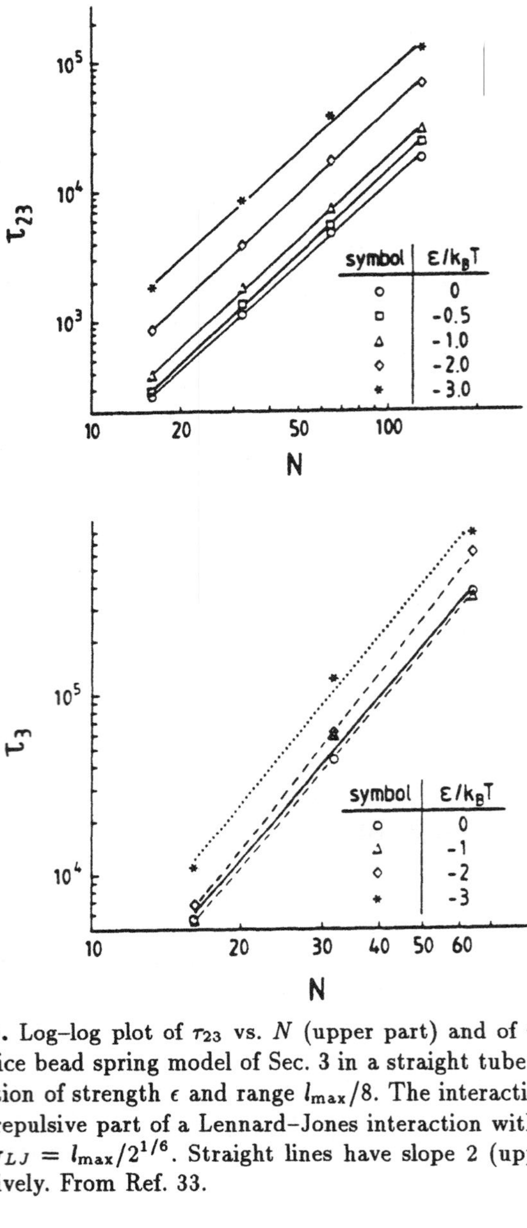

Fig. 10. Log–log plot of τ_{23} vs. N (upper part) and of τ_3 vs. N (lower part), for the off–lattice bead spring model of Sec. 3 in a straight tube with a square–well attractive interaction of strength ϵ and range $l_{\max}/8$. The interaction between beads was chosen as the repulsive part of a Lennard–Jones interaction with strenght $\epsilon_{LJ} = k_BT/10$ and range $\sigma_{LJ} = l_{\max}/2^{1/6}$. Straight lines have slope 2 (upper part) and 3 (lower part), respectively. From Ref. 33.

while the time to equilibrate density fluctuations along the chain is again a Rouse time $\tau_R \propto N^2$. These times can be deduced from the time–dependent displacements $g_{2l}(t)$, $g_{3l}(t)$ in longitudinal direction by defining characteristic times τ_{23}^l, τ_3^l via ($\tau_{23}^l \propto \tau_R$, $\tau_3^l \propto \tau_N$)

$$g_{2l}(t = \tau_{23}^l) = g_{3l}(t = \tau_{23}^l),\; g_{3l}(t = \tau_3^l) = 2R_{gl}^2/3\,. \tag{30}$$

Fig. 10 shows that the expected behavior $\tau_3^l \propto N^2$ is indeed verified [33], irrespective of the energy of interaction ϵ with the wall of the pore. Also in polymer brushes (Fig. 3) two such times $\tau_{\|} \propto N^3, \tau_{\perp} \propto N^2$ can be identified [44], related to fluctuations in the longitudinal and perpendicular dimensions of the chains in the brush.

Thus, despite the inhomogeneity of chain configurations in pores (Fig. 2) or brushes (Fig. 3), the extension of the Rouse model to these situations seems to be rather well understood. Note, however, that only brushes with rather short chains where entanglements among the chains are negligible have been considered so far [40, 44].

Acknowledgements

This work is based on a fruitful collaboration with A. Milchev, W. Paul, J. Wittmer, D.W. Heermann, K. Kremer, and P.Y. Lai. It is a great pleasure to thank them.

References

[1] Stockmayer W.H.: in *Molecular Fluids*, ed by R. Balian and G. Weill (Gordon and Breach, New York 1976)

[2] De Gennes P.G.: *Scaling Concepts in Polymer Physics* (Cornell University Press, Ithaca, New York 1979)

[3] Ferry J.D.: *Viscoelastic Properties of Polymers*, third edition (J. Wiley, New York 1980)

[4] Doi M. and Edwards S.F.: *The Theory of Polymer Dynamics* (Clarendon Press, Oxford, 1986)

[5] Bunde A and Havlin S.: (eds.), *Fractals and Disorderd Systems* (Springer, Berlin 1991)

[6] Hohenberg P.C. and Halperin B.I.: Rev. Mod. Phys. **49**, 435 (1977)

[7] Beretti A. and Sokal A.D.: J. Statist. Phys. **40**, 483 (1985)

[8] Sokal A.D.: in *Monte Carlo and Molecular Dynamics Simulations in Polymer Science*, ed. by K. Binder (Oxford University Press, Oxford 1994)

[9] Milchev A., Paul W., and Binder K.: J. Chem. Phys. **99**, 4786 (1993)

[10] Milchev A., and Binder K.: Europhys. Lett. (in press)

[11] Rouse P.E.: J. Chem. Phys. **21**, 1272 (1953)

[12] Zimm B.H.: J. Chem. Phys. **24**, 269 (1956)

[13] Le Guillou J.C., and Zinn–Justin J.: Phys. Rev. **B21**, 3976 (1980)

[14] Nienhuis B.: Phys. Rev. Lett. **49**, 1062 (1982)

[15] Paul W., Binder K., Heermann D.W., and Kremer K.: J. Phys. (Paris) **II1**, 37 (1991)

[16] Carmesin I. and Kremer K.: Macromolecules **21**, 2819 (1988)

[17] Des Cloizeaux J. and Jannink G.: *Polymers in Solution: Their Modelling and Structure* (Oxford University Press, Oxford 1990)

[18] Lifshitz I.M., Grosberg A.Yu., and Khokhlov A.R.: Rev. Mod. Phys. **50**, 683 (1978)

[19] Grosberg A.Yu, and Kuznetsov D.V.: Macromolecules **25**, 1970, 1980, 1991, 1996 (1992)
[20] Grosberg A.Yu., Nechaev S.K., and Shakhnovich E.I.: J. Phys. (Paris) **49**, 2095 (1988); Grosberg A.Yu., Rabin Y., Havlin S., and Neer A.: Europhys. Lett. **23**, 373 (1993)
[21] Naghizadeh J. and Kovac J.: Phys. Rev. Lett. **59**, 1710 (1987)
[22] Milchev A. and Binder K.: Macromol. Theory & Simul., submitted
[23] De Gennes P.G.: J. Chem. Phys. **69**, 572 (1972)
[24] De Gennes P.G.: Macromolecules **9**, 587, 594 (1976)
[25] Hess W.: Macromolecules **19**, 1395 (1986); **20**, 2589 (1987); **21**, 2620 (1988)
[26] Lodge T.P., Rotstein N.A., and Prager S.: Adv. Chem. Phys. **79**, 1 (1990)
[27] Des Cloizeaux J.: Europhys. Lett. **5**, 437 (1988); **6**, 475 (1988) [E]; Macromol. **23**, 4678 (1990); **25**, 835 (1992)
[28] Des Cloizeaux J.: J. Phys. (Paris) **II3**, 1523 (1993)
[29] Schweizer K.S.: J. Chem. Phys. **91**, 5802, 5822 (1989); J. Non-Cryst. Solids **131-133**, 634 (1991); Physica Scripta **T49**, 99 (1993)
[30] Szamel G.: Phys. Rev. Lett. **70**, 3744 (1993), Szamel G. and Schweizer K.S., Phil. Mag. in press; J. Chem. Phys. **100**, ... (1994)
[31] Daoud M. and De Gennes, P.G.: J. Phys. (Paris) **38**, 85 (1977)
[32] Kremer K. and Binder K.: J. Chem. Phys. **81**, 6381 (1984)
[33] Milchev A., Paul W., and Binder K.: Macromol. Theory Simul. **3**, 305 (1994)
[34] Halperin A., Tirrell M., and Lodge T.P.: Adv. Polym. Sci. **100**, 31 (1991)
[35] Milner S.T.: Science **251**, 905 (1991)
[36] Alexander S.: J. Phys. (Paris) **38**, 983 (1977)
[37] Milner S., Witten T., and Cates M.: Macromolecules **21**, 2610 (1988)
[38] Halperin A. and Alexander S.: Europhys. Lett. **6**, 329 (1988); Macromolecules **21**, 2403 (1989)
[39] Klushin L.I. and Skvortsov A.M.: Macromolecules **24**, 1549 (1991)
[40] Murat M. and Grest G.S.: Macromolecules **22**, 4054 (1989)
[41] Lai P.-Y. and Binder K.: J. Chem. Phys. **95**, 9288 (1991)
[42] Lai P.-Y. and Binder K.: J. Chem. Phys. **97**, 586 (1992)
[43] Wittmer J., Johner A., Joanny J.F., and Binder K.: J. Chem. Phys. **101** (1994, in press)
[44] Binder K., Lai P.-Y., and Wittmer J.: Faraday Disc., subm.
[45] Binder K.: in *Computational Modeling of Polymers*, ed. by J. Bicerano (M. Dekker, New York 1992)
[46] De Gennes P.G.: Physics (N.Y.) **3**, 97 (1967)
[47] Paul W., Binder K., Heermann D.W., and Kremer K.: J. Chem. Phys. **95**, 7726 (1991)
[48] Kremer K. and Grest G.S.: J. Chem. Phys. **92**, 5057 (1990); J. Chem. Soc. Faraday Trans. **88**, 1707 (1992)
[49] Skolnick J. and Kolinski A.: Adv. Chem. Phys. **78**, 223 (1990)
[50] Fixman M.: J. Chem. Phys. **89**, 3892 (1988)
[51] Rostiashvili V.G.: Sov. Phys. JETP **70**, 563 (1990)
[52] Douglas J.F. and Hubbard J.B.: Macromolecules **24**, 3163 (1991)
[53] Wilson J.D. and Loring R.F.: J. Chem. Phys. **99**, 7150 (1993)
[54] Richter D., Farago B., Fetters L.J., Huang J.S., Ewen B., and Lartigue C.: Phys. Rev. Lett. **64**, 1389 (1990)
[55] Richter D., Butera R., Fetters L.J., Huang J.S., Farago B., and Ewen B.: Macromolecules **25**, 6156 (1992)

[56] Wittmer J., Paul W., and Binder K.: Macromolecules **25**, 7211 (1992)
[57] Wittmer J., Paul W., and Binder K.: J. Phys. (Paris), **II4**, 873 (1994)
[58] Richter D., Baumgärtner A., Binder K., Ewen B., and Hayter J.B.: Phys. Rev. Lett. **47**, 109 (1981); **48**, 1695 (1982)

Diffusion on Fractal Substrates

Raf Dekeyser,[1] *Amos Maritan*[2] *and Attilio L. Stella*[2]*

[1] Instituut voor Theoretische Fysica
Katholieke Universiteit Leuven, B-3030 Leuven, Belgium
[2] Dipartimento di Fisica e Sezione I.N.F.N.,
Universitá di Padova, I-35131 Padova, Italy

1 Introduction

The study of static (geometric) properties of fractals has by now been widely accepted as the proper tool for describing physical objects that exhibit some kind of self-similarity [1, 2]. Lately, a lot of attention has also been devoted to the dynamical properties of fractals. This was triggered by some experiments in which the energy density of the vibrational states of some supposedly fractal polymers played an important role. These excitations were called "fractons" by Alexander and Orbach [3], who also pointed out the relation between these fractons and other dynamical properties like diffusion and resistivity [4].

In this contribution we first briefly review the basic concepts in this domain and the relation between the various dynamical phenomena. Thereafter, we discuss two different deterministic fractal structures for which the dynamical properties can be studied exactly by an iterative calculation. The first model, inspired by numerical studies of diffusion on random walk substrates, will give rise to a power-law dynamical behaviour, with fractal dynamical exponents that depend in a continuous way on a free parameter of the model. The second model leads to a logarithmic time-dependence of the diffusion, instead of a power-law behaviour.

Both examples constitute, in a sense, frontier applications of two important paradigms, which, from the start, played a very important role in the field. The first one is the use of deterministic hierarchical or fractal lattices to model, at least qualitatively, the complicated random self-similarity found in nature [5]. The simplifications inherent in such modelling make the application of renormalization group (RG) methods, our second paradigm, natural and feasible. We speak here of frontier applications in the sense that the model calculations that we deal with aim at an explanation of quite complicated issues, whose control – even numerically – is sometimes at the border of our present computational possibilities. In one case, the difficulty of the problem leads us to the necessity of reconsidering the standard strategy for extracting information on critical behaviour from the RG flow.

* With acknowledgement for financial support by the I.N.F.M.

2 Fractal Diffusion and Other Dynamical Properties

Let us concentrate on a d-dimensional lattice, on which a fractal structure with Hausdorff-dimension D is defined (e.g., by deleting from the lattice some specific sites and bonds). The diffusion equation describes the time evolution of the probability $P_i(t)$ to find a diffusing particle at site i at time t; it is given by

$$\Delta' P_i(t) = \frac{\partial}{\partial t} P_i(t) , \tag{1}$$

where the quasi-Laplacian Δ' is defined as

$$\Delta' P_i = \sum_j (P_j - P_i) , \tag{2}$$

with j running over the nearest neighbours of site i in the fractal structure. This equation is usually studied with the initial condition

$$P_i(0) = \delta_{i,0} . \tag{3}$$

Equation (1) can formally be solved in terms of the eigenfunctions of the Δ'-operator, obtained (e.g. in an obvious continuous description) from

$$(\Delta' + E)\Psi_E(\mathbf{r}) = 0 . \tag{4}$$

(The presence of some suitable boundary conditions has to be assumed.) The solution to the diffusion equation can then be written as

$$P(\mathbf{r}, t) = \int dE \; \rho(E)\Psi_E^*(0)\Psi_E(\mathbf{r})\, \mathrm{e}^{-Et} . \tag{5}$$

$\rho(E)$ represents the spectral density of the Δ'-operator. This is closely related to the vibrational spectrum of the fractal, since the vibrations have in a similar way to obey the equation[3]

$$\Delta' u(\mathbf{r}, t) = \frac{\partial^2}{\partial t^2} u(\mathbf{r}, t) . \tag{6}$$

The vibrational modes are described by the same functions $\Psi_E(\mathbf{r})$ and they correspond to frequencies $\omega = \sqrt{E}$. Since the low frequency spectral density $\tilde{\rho}(\omega)$ of phonons behaves on a regular d-dimensional lattice like ω^{d-1}, the spectral or fracton dimension $\tilde{d}$ was defined [3] through the relation

$$\tilde{\rho}(\omega) \sim \omega^{\tilde{d}-1} \qquad (\omega \to 0) , \tag{7}$$

which is equivalent to

$$\rho(E) \sim E^{-1+\tilde{d}/2} \qquad (E \to 0) . \tag{8}$$

---

[3] The equation actually describes a rather peculiar form of scalar vibrations.

Using this behaviour in (5) shows us immediately that the solution to our diffusion equation has the asymptotic property

$$P(0,t) \sim t^{-\tilde{d}/2} \qquad (t \to \infty) . \tag{9}$$

The diffusion of a particle on a fractal structure is usually described in terms of the effective dimensionality d_{w} of the random walk problem, through the relation

$$t \sim R^{d_{\mathrm{w}}} \qquad \text{with} \qquad R^2 = \langle r^2(t) \rangle , \tag{10}$$

which links the mean square displacement to the diffusion time. Alternatively, one can define an exponent ν with

$$\langle r^2(t) \rangle \sim t^{2\nu} \qquad (\nu = 1/d_{\mathrm{w}}) . \tag{11}$$

The exponent d_{w} is not independent from the spectral dimension $\tilde{d}$; it can be argued that the diffusion probability should behave for large times as

$$P(\mathbf{r},t) \approx r^{-D} f(r^{d_{\mathrm{w}}}/t) , \tag{12}$$

where the factor r^{-D} appears for reasons of normalization. With

$$g(x) = x^{-D/d_{\mathrm{w}}} f(x) ,$$

this can also be written as

$$P(\mathbf{r},t) \approx t^{-D/d_{\mathrm{w}}} g(r^{d_{\mathrm{w}}}/t) . \tag{13}$$

Comparison with (9) then tells us that

$$\tilde{d} = \frac{2D}{d_{\mathrm{w}}} . \tag{14}$$

This equation allows us to determine the low frequency scaling of the phonon spectral density through the asymptotic behaviour of the diffusion for $t \to \infty$.

It was soon discovered [5] that the previously defined exponents are also related to the resistivity exponent ζ, defined through

$$\Omega(r) \sim r^{\zeta} \qquad (r \to \infty) , \tag{15}$$

where $\Omega(r)$ is the electrical resistance between two points at a distance r on the fractal, if the bonds of the fractal structure are replaced by identical resistances. On a regular d-dimensional lattice we can estimate $\Omega(r)$ from the picture that between two points A and B at a distance r there exist roughly r^{d-1} parallel conducting paths of length (and thus of resistivity) r. This leads to the formula

$$\Omega(r) \sim \frac{1}{\sigma} \frac{r}{r^{d-1}} = \frac{1}{\sigma} r^{2-d} . \tag{16}$$

σ is the local conductivity, and this should contain the information about the fractal structure. This quantity can be calculated from the Einstein relation as a function of the density n and the diffusion coefficient C:

$$\sigma = \frac{e^2}{k_B T} n C \ . \tag{17}$$

The diffusion coefficient at a distance r on the fractal can be estimated as follows:

$$C = \frac{d}{dt} \langle r^2(t) \rangle \sim t^{2/d_w - 1} \sim r^{2-d_w} \ . \tag{18}$$

The density n obviously behaves like

$$n \sim r^{D-d} \tag{19}$$

and thus

$$\sigma \sim r^{D-d_w+2-d} \ . \tag{20}$$

Using this result in (16) gives us

$$\zeta = d_w - D \ . \tag{21}$$

This relation, which can also be derived from the Green function of the Δ'-operator [6], provides us with an independent way of determining the dynamical exponents through a direct calculation of the exponent ζ.

3 Diffusion on Two-dimensional Random Walks

The diffusion problem on random two-dimensional substrates generated by random walks has led in the literature to some contradictory statements. It has been claimed [7] that in this case, since $D = d = 2$, the random walk is homogeneous in space, and we should therefore have $\zeta = 0$, as in the regular two-dimensional lattice. This would lead to $d_w = 2$ or $\nu = 1/2$ and hence $\tilde{d} = 2$. Resistivity calculations [8], on the other hand, seemed to favour $\zeta = 1$, hence $d_w = 3$ and $\tilde{d} = 4/3$. This has led the present authors to an extensive numerical study of the diffusion of a particle during 100 time steps on random walks of 20,000 steps on the square lattice [9], which strongly supported the results of the resistivity calculations with $\nu = 1/3$. These results were contradicted by a calculation [10] of the diffusion on N-step random walks in the limit $N \to \infty$, which led to the result

$$\nu - \frac{1}{2} \sim 1/\ln N \ . \tag{22}$$

In order to understand what is going on, one should remember that the *infinite* two-dimensional random walk is known to visit all sites of the lattice and diffusion on a regular lattice is always described by $\nu = 1/2$. The asymptotic $t \to \infty$ behaviour of the diffusion on a *finite* random walk, on the other hand, can only be described by $\nu = 0$. This means that all finite, but eventually large, random walks can be described by an effective $\nu_{\text{eff}}(t)$, for which we should thus expect that it varies from 1/2 for $t = 0$ to 0 for $t \to \infty$. We believe that the dynamical behaviour of a large two-dimensional random walk can be characterized by three different regimes (either in time or in diffusion length): an initial regime, where the normal two-dimensional character is manifested (e.g. $\nu = 1/2$), an intermediate regime where the true fractal behaviour of the walk manifests itself, and a final regime where the limiting boundary of the walk starts hindering further diffusion. In this sense, the random walk situation could have some resemblance with that of realistic systems like Silica aerogels[11], where the fractal behaviour is found for lengths L satisfying $a \ll L \ll \xi$, where a is an atomic length scale and ξ some correlation length. In our case, a would rather be the average size of densely filled clusters and instead of ξ we should have the average diameter of the random walk.

It is illuminating to compare the finite random walk with, e.g., finite realisations of Sierpinsky gaskets. We calculated exactly $< r^2(t) >$ on respectively the 4-th, 5-th, 6-th and 7-th generation of simple Sierpinsky triangular gaskets for times up to $t = 1200$, starting from an arbitrary interior point. Whereas the infinite gasket corresponds to $\nu = \ln 2/\ln 5 \approx 0.43$, our numerical study for the finite gaskets showed an effective ν-exponent fluctuating around the value 0.43 up to a crossover time t_x, which scales with the size R of the system like $t_x \sim R^{1/\nu}$.

The situation is quite different in the random walk case. The diameter of an N-step walk is proportional to $\sqrt{N}$; if the random walk would be comparable to a finite-size fractal with diffusion exponent $\nu = 1/2$, we should expect to see a rather flat $\nu_{\text{eff}}(t) = 1/2$ up to $t \approx N$. We always see, however, that ν drops rather sharply to a much lower value on a much shorter time scale. The actual value seems to saturate around 1/3 with a rather long flat plateau, although a simple log-log analysis is too crude to give an accurate estimate (an alternative estimate was given elsewhere[9]). If this is indeed the correct value for ν in the fractal regime, as we suppose, we should see the crossover to $\nu = 0$ appear only at times of the order $t_x \approx N^{3/2}$. For $N = 2,000$, e.g., this already means $t_x \approx 100,000$, which is unfortunately too high to study numerically.

In conclusion, we want to stress that the results obtained by Manna et al.[10] are only relevant for the initial time regime. Indeed, their 100-step blind ant walks correspond perhaps to 60-step myopic ant walks, as studied by us; these short diffusion times are implemented on very long random walks, up to 640,000 steps in their case, to be compared with our 20,000 steps. It is clear that they are in a completely different regime for the t/N-ratio, i.e. they are testing compact regions of the random walk structure. Furthermore, we have extended our Monte Carlo diffusions on a 20,000-step random walk up to $t = 10,000$, confirming in a crude log-log analysis the picture of a long flat plateau around 1/3 for the ν-values.

In order to obtain an independent confirmation of our result, we have also investigated $P(0,t)$, the probability for return to the origin during a diffusion process, whose behaviour is described by (9). We have therefore generated large samples of random walks of varying lengths, and we solved exactly the diffusion equation on their substrates for the occupation probabilities $P(\mathbf{r},t)$ for $t \leq 2000$. Through log-log calculations we obtain the following results: $\tilde{d} = 1.08$ for $N = 5,000$ and $\tilde{d} = 1.18$ for $N = 20,000$ (sample size: 1380). A linear extrapolation in $1/N$ yields $\tilde{d} = 1.22$, whereas an extrapolation in $1/\ln N$, in analogy with the calculations of Ref. [10], gives $\tilde{d} = 1.30$. These results are compatible with $\tilde{d} = 4/3$ (or $d_{\rm w} = 3$) and they rather clearly exclude the value $\tilde{d} = 2$ (or $d_{\rm w} = 2$).

4 Hierarchical Model with Random Walk Similarities

In order to understand better the results from our simulations, we tried to construct a deterministic model of which one could hope that it possesses the essential properties that lead to the anomalous dynamical behaviour as was observed in the two-dimensional random walks. We therefore looked at the patterns that were formed by some large, but finite, random walks on a regular lattice. These patterns seem to be composed of many densely visited regions of various sizes, interconnected by narrow pathways. This scheme, moreover, seems to repeat itself over many length scales. Led by these considerations, we have constructed in a hierarchical way a fractal structure that incorporates the same succession of dense regions and narrow links, hoping to be able to calculate exactly some dynamical properties, like diffusion rate or resistivity, on these fractals.

In Fig. 1 we present the iterative construction of such a model, in which at each step four blocks of the previous step are connected with one another into a larger block. Since there is still some variety in the way in which one performs these connections, we actually have a class of models.

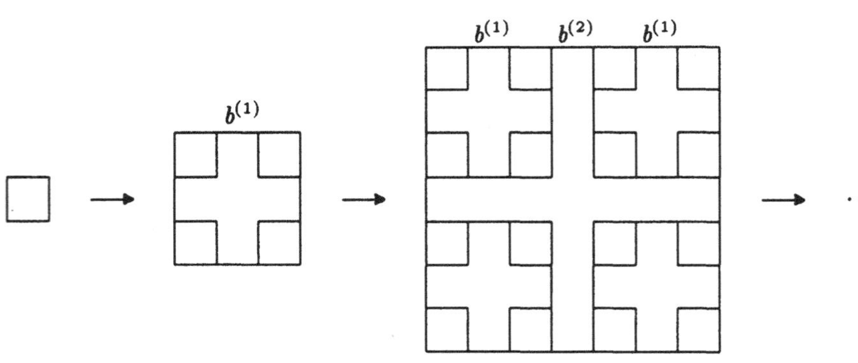

Fig. 1. Iterative construction of the class of fractal models, depending on a family of parameters $\{b^{(n)}\}$.

The links on the structure represent paths along which a particle may diffuse. The elementary units of the zeroth step of the structure are squares with some initial transition rate a along the sides and zero transition rate along the diagonal. The links connecting these elementary units may be classified according to the order in which they appear in the iterative construction of the fractal. We may thus introduce transition rates $b^{(n)}$ with $n = 1, 2, 3, \ldots$ for this class of links. The most simple choice might seem to take $b^{(n)} = b = a$ for all n, which would correspond to the problem of a regular lattice with missing links. Keeping in mind the pictures of our random walk samples, however, we want to leave open the possibility to have longer links between the blocks at higher steps of the construction. The essential property of these links should be that they have a smaller transition rate than the bonds of the elementary squares.

We have solved the diffusion problem on this class of models, characterized by the transition rates a and $b^{(n)}$, through a renormalization procedure by decimation. By this procedure a block, constructed from four unit squares linked together, is mapped onto a single square (see Fig. 2), by decimating out 12 from the original 16 sites of the block. By repeating this decimation n times, the linear size of the blocks is reduced after the n-th step from $(2^n - 1)$ units to $(2^{n-1} - 1)$ units, which means that asymptotically the lengths are rescaled by a factor 2.

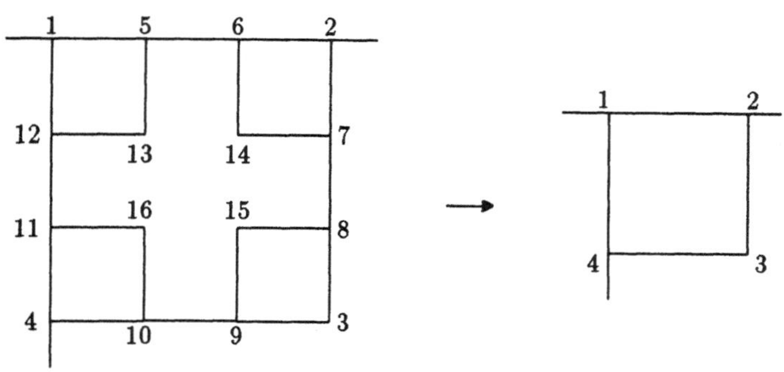

Fig. 2. Scheme for the renormalizatio procedure by decimation for the diffusion equations.

We have numbered the sites of our original block from 1 to 16 in such a way that sites numbered from 1 to 4 are saved by the decimation procedure (see Fig. 2). We suppose that the transition rates for the site occupation probabilities inside this block are given by a, b and c (See Fig. 3), where a and b are regular nearest neighbour transition rates (the b being equal to $b^{(1)}$ in the first step), and c is a diagonal transition rate. On the initial fractal we suppose $c = 0$, but since it acquires a nonzero value through the renormalization procedure, we include it in our calculation. Furthermore, sites 1 through 4 may or may not have an outgoing link with a transition rate $b^{(n)}$ (with $n > 1$), depending on their position in the fractal. We indicate this transition rate by b_1, which must be zero if the link is missing.

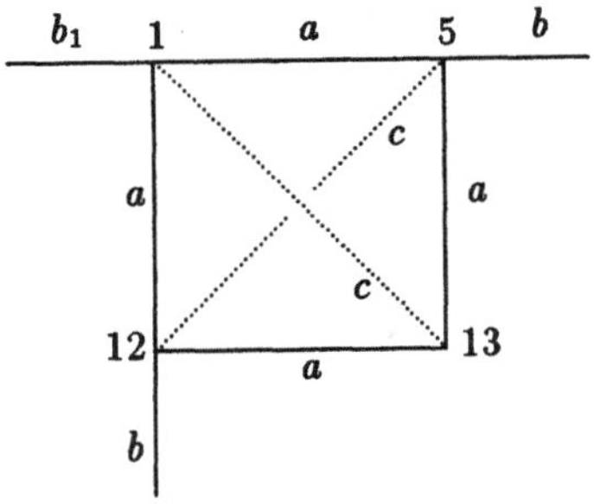

Fig. 3. Transition rates for the diffusion.

Let $P_k(t)$ indicate the probability for finding the particle at site k at time t. These quantities obey the master equation

$$\frac{dP_k}{dt} = \sum_j W_{kj}(P_j - P_k) \ , \tag{23}$$

where $W_{ij} = W_{ji}$ denote the already mentioned transition rates. We define $X_k(\lambda)$ as the Laplace transform of the $P_k(t)$. From (3) these $X_k(\lambda)$ obey a set of 16 equations, which may easily be constructed from Fig. 2. Using the definition

$$p = 2a + c \ , \tag{24}$$

and denoting with $X_k^0(\lambda)$ the Laplace transform of the occupation probability of an external neighbour to site k ($k = 1, \ldots, 4$), we have, e.g.,

$$(\lambda + p + b_1)X_1 = a(X_5 + X_{12}) + cX_{13} + b_1 X_1^0 \tag{25}$$

$$\ldots \quad \ldots$$

$$(\lambda + p + b_1)X_4 = a(X_{10} + X_{11}) + cX_{16} + b_1 X_4^0 \tag{26}$$

$$(\lambda + p + b)X_5 = a(X_1 + X_{13}) + cX_{12} + bX_6 \tag{27}$$

$$\ldots \quad \ldots$$

$$(\lambda + p + b)X_{12} = a(X_1 + X_{13}) + cX_5 + bX_{11} \tag{28}$$

$$(\lambda + p)X_{13} = a(X_5 + X_{12}) + cX_1 \tag{29}$$

$$\ldots \quad \ldots$$

$$(\lambda + p)X_{16} = a(X_{10} + X_{11}) + cX_4 \tag{30}$$

A straightforward elimination procedure of the functions X_5 through X_{16} can always be normalized in such a way that we end up with the original value of a in equations such as

$$(\lambda' + p' + b_1')X_1 = a(X_2 + X_4) + c'X_3 + b_1' X_1^0 \tag{31}$$

and analogous equations for X_2, X_3 and X_4. We defined

$$p' = 2a + c' \ . \tag{32}$$

Since a remains constant, we may arbitrarily choose it to be equal to unity from now on: $a = 1$.

Since we want to study the long time behaviour of these equations, we can restrict ourselves to small values of λ, and hence we expand the obtained equations in ascending powers of λ. This leads to

$$b^{(n)\prime} = \gamma b^{(n+1)} \ , \tag{33}$$

$$c' = \frac{b(1+c)}{2(1+c)+b(3+c)} + \mathcal{O}(\lambda) \ , \tag{34}$$

$$\lambda' = 4\gamma\lambda \ , \tag{35}$$

$$\gamma = \frac{(1+c)+b(2+c)}{b(1+c)} + \mathcal{O}(\lambda) \ . \tag{36}$$

One should remark that, together with their transformation, the b's shift down one step in their eventual hierarchy of the $\{b^{(n)}\}$-values.

It is immediately clear from (33) that, if initially all $b^{(n)}$ are equal, they will finally be renormalized to infinity, since $\gamma > 1$. In this case, we obtain a fixed point

$$b^\star = \infty \tag{37}$$

$$c^\star = \sqrt{2} - 1 \ , \tag{38}$$

and from (35) follows in this case

$$\left(\frac{\lambda'}{\lambda}\right)^\star = 4 + 2\sqrt{2} \ . \tag{39}$$

Since each step in the decimation procedure corresponds to a length rescaling by a factor 2, we expect from (10) that this implies a rescaling of the time by a factor 2^{d_w}, from which we may infer that

$$d_w = \frac{\ln(\lambda'/\lambda)}{\ln 2} \ . \tag{40}$$

Our infinite-b fixed point corresponds then to $d_w = 2.77$ or $\tilde{d} = 1.44$.

Another possibility is to start from a model, in which the $b^{(n)}$ are not equal to one another, but become smaller with growing n, indicating in this way some lengthening of the effective diffusion path along these links. If the hierarchy of $b^{(n)}$-values is chosen properly, one might obtain any desired finite fixed point value for $b^\star$. In such cases we have:

$$c^\star = \frac{\left[2(b^\star)^2 + 4b^\star + 1\right]^{1/2} - b^\star - 1}{b^\star + 2} \ , \tag{41}$$

$$\left(\frac{\lambda'}{\lambda}\right)^\star = \frac{2(c^\star + 1)}{c^\star} \ . \tag{42}$$

In order to obtain these values we should, at least asymptotically for large n, have started with a situation where

$$b^{(n)} \simeq b f^{-n} \tag{43}$$

with

$$f = (c^\star + 1)/2c^\star \ . \tag{44}$$

From these formulas, it is clear that, when $b^\star$ goes from 0 to ∞, $c^\star$ increases monotonically from 0 to $\sqrt{2} - 1$ and f from ∞ to $1 + \sqrt{2}/2$.

If in (43) f is smaller than $f_c = 1 + \sqrt{2}/2$, then b always iterates to $b^\star = \infty$ and c to $c^\star = \sqrt{2} - 1$.

Since our renormalization steps correspond to a rescaling of the lengths by a factor 2, we have in general (see Fig. 4)

$$d_w = \begin{cases} 2 + \ln f / \ln 2 & (f \geq f_c) \\ 2 + \ln f_c / \ln 2 & (f < f_c) \ . \end{cases} \tag{45}$$

This means that we have a whole class of models with nonuniversal diffusion dimensions $2.77 \leq d_{\mathrm{w}} \leq \infty$. One special case immediately draws our attention. The $b^{(n)}$ are diffusion strengths on links between clusters with a linear dimension 2^n. In realistic fractal structures grown by some random procedures, a good guess would be that links between such clusters would themselves have lengths of the order 2^n. This can be modeled in our scheme by making the diffusion strengths $b^{(n)}$ shrink by factors $f = 2$. This special and obvious choice leads to the values $b^\star = 4$, $c^\star = 1/3$, $(\lambda'/\lambda)^\star = 8$, $d_{\mathrm{w}} = 3$, and it gives us the Alexander-Orbach value $\tilde{d} = 4/3$ [3].

Completely independent of the previous renormalization calculation, we have also performed [12] an iterative calculation of the electrical resistivity between opposing diagonal points of the fractal structure in increasing order of the construction. For all f, these resistivities Ω_n scale like $2^{n\zeta}$, where $\zeta = d_{\mathrm{w}} - 2$ with d_{w} exactly given by (45). This calculation thus proves that relation (21) is valid for our whole class of models.

5 A Model with Logarithmic Time Dependence

It has been known for many years that some special models may give rise to extremely slow diffusion, such that (5) has to be replaced with a logarithmic time dependence

$$R \sim (\ln t)^k \ , \tag{46}$$

for some value of k. Such behaviour has been demonstrated for a particle on a one-dimensional chain, subject at each site to an independent random bias [13]. Another mechanism leading to the same behaviour is that arising in a comb-like structure with teeth of variable length, along which a bias field pushes the particle towards the tip. A suitable power law distribution of teeth lengths can be shown to imply delay times, which make diffusion along the comb basis logarithmic [14]. There is some numerical evidence [15, 16] that the same combination

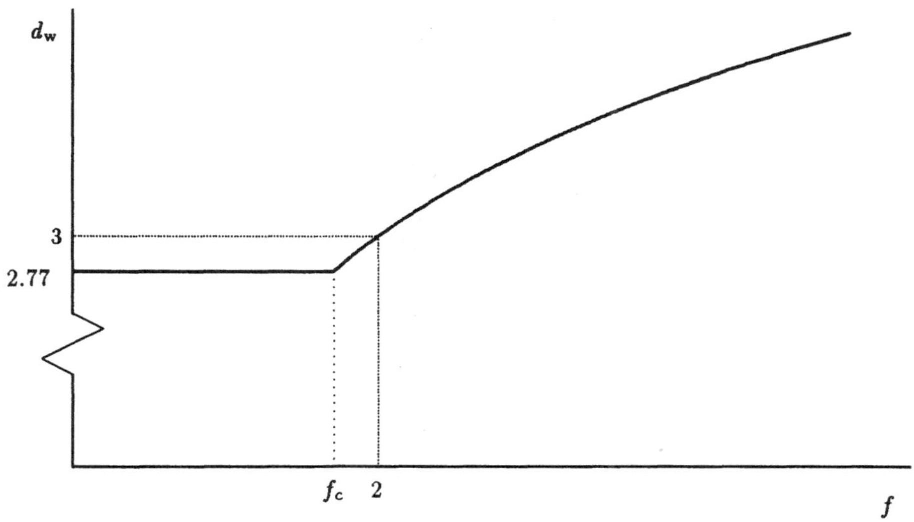

Fig. 4. Exponent d_w versus rescaling factor f for the diffusion family $\{b^{(n)}\}$.

of dangling ends and a bias field can give rise to a logarithmic behaviour also in higher-dimensional models like the infinite incipient cluster of percolation. This was recently confirmed by a renormalization group (RG) calculation on the hierarchical T-fractal [17], and we will briefly report on these results. The calculation is of some methodological interest, since the RG recursion equations are singular at the fixed point, such that the standard analysis for extracting the asymptotic behaviour has to be reconsidered.

The iterative construction of the T-fractal, which has a central site $j = 0$, is illustrated in Fig. 5. It is a ramified structure with fractal dimension $D = \ln 3/\ln 2$. The resulting structure, after two iterations, is shown in Fig. 6.

A diffusing particle hops between nearest neighbour sites of the T-structure according to the master equation, equivalent to (1):

$$P_i(t+\tau) = P_i(t) + \sum_{i \neq j} [W_{ij} P_j(t) - W_{ji} P_i(t)] \ , \tag{47}$$

where the W_{ij} represent the hopping probabilities during time τ. It is assumed

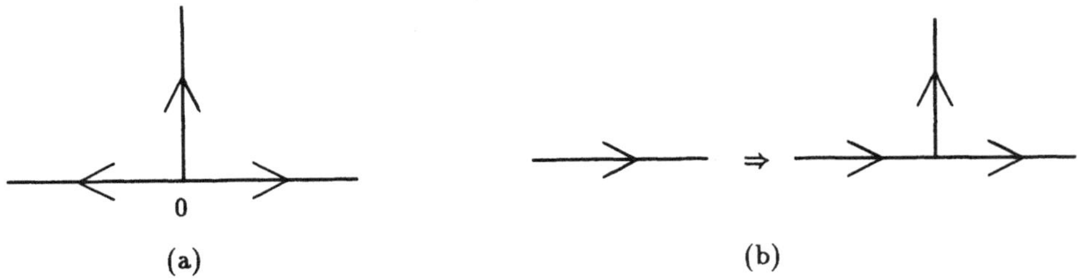

Fig. 5. (a) Starting configuration and (b) iterative step for the construction of the T-fractal.

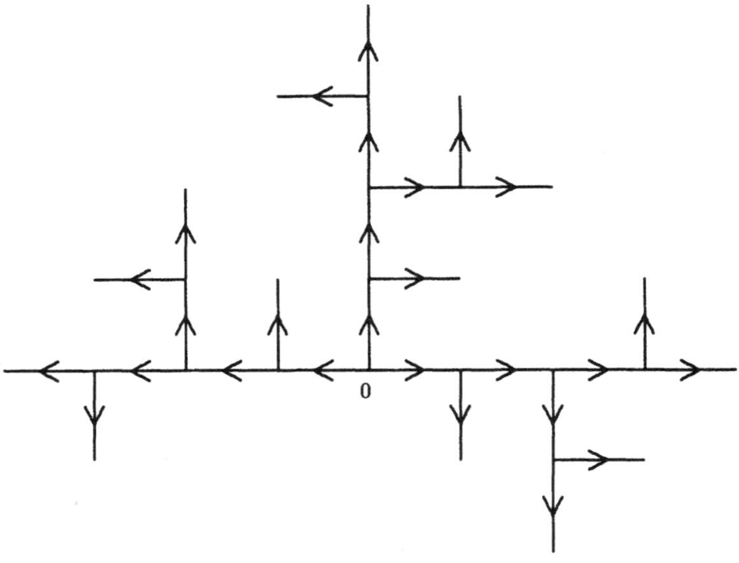

Fig. 6. T-fractal after two steps in the iterative construction.

that $W_{ij} = W_+$ or W_-, according to whether one goes from j to i following the direction of the bond arrows of Fig. 4 or against them. In order to preserve positive probabilities, the following relations should be satisfied:

$$2W_+ + W_- \leq 1$$
$$3W_+ \leq 1 \, .$$

We will furthermore assume that

$$W = \frac{W_-}{W_+} \leq 1 \, , \tag{48}$$

which means, as a consequence of the direction of the arrows, that there is a bias, pushing the particle to higher "chemical" distance from the origin. We put $\tau = 1$, and introduce discrete Laplace transforms [18]:

$$\tilde{P}_i^0(z) = \sum_{n=0}^{\infty} P_i(n)(1+z)^{-1-n} \ . \tag{49}$$

This transforms (47) into

$$(1+z)\tilde{P}_i^0(z) - P_i(0) = \tilde{P}_i^0(z) + \sum_{j \neq i} \left[W_{ij}\tilde{P}_j^0(z) - W_{ji}\tilde{P}_i^0(z) \right] \ . \tag{50}$$

With the definitions $\omega = z/W_+$ and $\tilde{P}_k(\omega) = W_+\tilde{P}_k^0(z)$, and using the initial condition $P_i(0) = \delta_{i,0}$, this can also be written as

$$\left(\alpha(i)\omega + \sum_{j \neq i} \frac{W_{ji}}{W_+} \right) \tilde{P}_i(\omega) = \sum_{j \neq i} \frac{W_{ij}}{W_+} \tilde{P}_j(\omega) + \delta_{i,0} \ . \tag{51}$$

Equation (50) actually gives us $\alpha(i) = 1$ for all i, but since the RG procedure will generate three different values α_0, α_1 or α_3, according to whether i is the origin or a site with coordination number 1 or 3, we introduce from the beginning such a variable $\alpha(i)$. The only parameters entering (51) will thus be $\boldsymbol{\alpha} = (\alpha_0, \alpha_1, \alpha_3)$ and W.

The RG transformation is now performed by eliminating from the equations (51) the $\tilde{P}_j$ corresponding to sites j introduced at the last step in the iterative construction of the T-fractal. This corresponds to performing the inverse of the operation illustrated in Fig. 5(b). The sites of the old structure which are not decimated have distances reduced by a factor 2, if measured in terms of the new lattice spacing. The decimated equations can be expressed in a form similar to that of the original equations, if we use the following renormalized parameters:

$$W' = W^2 \tag{52}$$

$$\alpha_0'\omega' = (\alpha_0\omega + 3)\left(\alpha_3\omega + 2 + W - \frac{W}{\alpha_1\omega + W} \right) - 3W - 3 \tag{53}$$

$$\alpha_1'\omega' = (\alpha_1\omega + W)(\alpha_3\omega + 2 + W) - 2W - W^2 \tag{54}$$

$$\alpha_3'\omega' = (\alpha_3\omega + 2 + W)\left(\alpha_3\omega + 2 + W - \frac{W}{\alpha_1\omega + W} \right) - 2 - 3W - W^2 \tag{55}$$

$$\tilde{P}_i' = \frac{\alpha_1\omega + W}{(\alpha_1\omega + W)(\alpha_3\omega + 2 + W) - W} \tilde{P}_i \ . \tag{56}$$

In order to see how the further analysis should be performed, we first consider the unbiased case, corresponding to the (unstable!) fixed point $W^* = 1$ of (52).

Large t-values correspond to small ω, and in the limit $\omega \to 0$ and $W = 1$ (53-55) become:

$$\alpha_0'\omega' = (2\alpha_0 + 3\alpha_1 + 3\alpha_3)\omega \tag{57}$$

$$\alpha_1'\omega' = (3\alpha_1 + \alpha_3)\omega \tag{58}$$

$$\alpha_3'\omega' = (3\alpha_1 + 5\alpha_3)\omega \ . \tag{59}$$

Under repeated action of such a transformation, the value of ω'/ω and the vector $\boldsymbol{\alpha}$ will tend respectively to the largest eigenvalue and the corresponding eigenvector of the matrix

$$\begin{pmatrix} 2 & 3 & 3 \\ 0 & 3 & 1 \\ 0 & 3 & 5 \end{pmatrix} \ . \tag{60}$$

Since this largest eigenvalue is 6, and since ω plays here the same role as the Laplace transform variable λ introduced for the study of the model (23) in the previous section, we can now write in analogy with (40):

$$d_{\mathrm{w}} = \frac{\ln(\omega'/\omega)^*}{\ln 2} = \frac{\ln 6}{\ln 2} \ , \tag{61}$$

which describes indeed the unbiased diffusion on the T-fractal [18].

If we try to use the same method in the neighbourhood of the second fixed point $W^* = 0$, which controls the biased diffusion, we get into trouble since the transformation matrix for $\boldsymbol{\alpha}$ now becomes to leading order in W:

$$\boldsymbol{\alpha}'\omega' = \begin{pmatrix} 1 & 3/W & 3 \\ 0 & 2 & W \\ 0 & 2/W & 3 \end{pmatrix} \boldsymbol{\alpha}\omega \ . \tag{62}$$

Although there is a regular largest eigenvalue 4, we can not repeat the previous analysis since we will never reach a finite $\boldsymbol{\alpha}^*$: under the renormalization the α_0 and α_3 will for small W at every step be multiplied with larger and larger factors $(W^{2^n})^{-1}$. After n iterations, the leading contribution to $(\boldsymbol{\alpha}\omega)^{(n)}$ can easily be estimated from (62) to be of the form

$$(\boldsymbol{\alpha}\omega)^{(n)} \approx \left(a_0 2^n/W^{2^n}, a_1 2^n, a_3 2^n/W^{2^n}\right) \ , \tag{63}$$

the a_i's being suitable constants depending on initial conditions. This means that, if one wants to retain the original form (51) for the diffusion with a fixed point value for the vector $\boldsymbol{\alpha}$, one should renormalize the variable $\omega^{(n)}$ not by multiplying it with a constant factor at each step as $\Lambda^n\omega$, but in the following form:

$$\omega^{(n)} = \omega 2^n/W^{2^n} \ , \tag{64}$$

or

$$\ln\omega = \ln(\omega^{(n)}2^{-n}) + 2^n \ln W \ . \tag{65}$$

Simultaneously, the average distance travelled by a diffusing particle renormalizes as

$$\left\langle R^{(n)} \right\rangle \approx 2^{-n} \langle R \rangle \ , \tag{66}$$

or

$$\langle R \rangle \sim 2^{n} \sim \frac{|\ln \omega|}{|\ln W|} \sim \frac{\ln t}{|\ln W|} \ . \tag{67}$$

(Remark that the main contribution to the Laplace transform comes from the region of values where $\omega t \approx 1$.) For a more detailed derivation of this result, we refer to Reference [17].

As anticipated, the singular character of the RG mapping gives rise to logarithmic diffusion. An extra result is the logarithmic dependence on W, which has been conjectured on the basis of numerical evidence or on heuristic arguments [16].

6 Conclusions

As anticipated in the introduction, the applications of the dynamical RG discussed here have shown how far reaching this method can be for the qualitative understanding of very complicated, and sometimes controversial, mechanisms of diffusion on fractal structures. The key to success is of course an appropriate choice of the hierarchical model on which the RG should be applie d. Even if the relevance of such results to the corresponding problems on real random fractal structures can always be questioned, there is no doubt that the exercise is suggestive and instructive on its own. For the diffusion phenomena discussed here, more realistic models can only be treated by numerical methods, and even then with much difficulty. Certainly the analytical insight provided by our toy models is of help to our general understanding, by providing simplified but clear mechanisms of explanation. In addition, considering borderline applications can lead to interesting methodological spinoffs for the RG strategy itself. The relevant example here is our singular RG recursion for logarithmic diffusion. A singular transformation, at first sight, seems to contrast with the common wisdom about RG strategies. Here we showed that such singular transformation can be handled within an exact treatment and is the natural mechanism for explaining the phenomenon.

References

[1] B.B.Mandelbrot, *The Fractal Geometry of Nature* (Freeman, San Francisco, 1982).

[2] P.G. de Gennes, *Scaling Concepts in Polymer Physics* (Cornell University Press, Ithaca, 1979).

[3] S.Alexander and R.Orbach, J.Phys. (Paris) Lett. **43**, L 625 (1982).

[4] S.Alexander, J.Bernasconi, W.R.Schneider and R.Orbach, Rev.Mod. Phys. **53**, 175 (1981).

[5] Y.Gefen, A.Aharony and S.Alexander, Phys.Rev.Lett. **50**, 77 (1983).

[6] M.E.Cates, J.Phys. (Paris) **46**, 1059 (1985).
[7] S.Havlin, G.H.Weiss, D.Ben-Avraham and D.Movshovitz, J.Phys. A **17**, L849 (1984).
[8] J.R.Banavar, A.B.Harris and J.Koplik, Phys.Rev.Lett. **51**, 1115 (1983).
[9] R.Dekeyser, A.Maritan, and A.L.Stella, Phys.Rev.Lett. **58**, 1758 (1987) and Phys.Rev. A **36**, 2338 (1987).
[10] S.S.Manna, A.J.Guttmann and B.D.Hughes, Phys.Rev. A **39**, 4337 (1989).
[11] D.W.Schaefer and K.D.Keefer, Phys.Rev.Lett. **56**, 2199 (1986).
[12] R.Dekeyser, A.Maritan and A.Stella, Phys.Rev. A **40**, 5299 (1989).
[13] Y.G.Sinai, Theory of Prob. and its Appl. **27**, 256 (1982).
[14] S.Havlin, A.Bunde, Y.Glaser and H.E.Stanley, Phys.Rev. A **34**, 3492 (1986).
[15] D.Stauffer, J.Phys. A **18**, 1827 (1985).
[16] A.Bunde, H.Harder, S.Havlin and H.E.Stanley, J.Phys. A **20**, L865 (1987).
[17] A.Maritan, G.Sartoni and A.L.Stella, Phys.Rev.Lett. **71**, 1027 (1993).
[18] A.Giacometti, A.Maritan and A.L.Stella, Int.J.Mod.Phys. B **5**, 709 (1991).

Stellar Diffusion

Henryk Cugier

Astronomical Institute of the Wroclaw University, ul. Kopernika 11, 51-622 Wroclaw, Poland

1 Introduction

The astrophysicists are faced with extremaly broad region of physical conditions realized in various objects of the Universe. Even if we limit ourselves to stars, physics of the stellar material is not always well understood. The best situation is for Main Sequence (MS) stars, where plasma effects due to high (but not very high) density and temperature are relatively well elaborated. With the exception of free electrons in stellar cores, the velocity distribution functions are not far from the maxwellian ones. More precisely, the solution of the Boltzmann equation leads to the maxwellian distribution functions if the temperature, pressure, and concentration of chemical elements is exactly uniform in the gas mixture, and if there is no external force acting selectively on the gas components. Clearly, such conditions are never realized within stars and deviations of the velocity functions from the maxwellian distribution should be derived in principle. Basically, these deviations are not the same for the various components of the gas mixture which leads to diffuse different elements relative to one another.

The first attempt to recognize the importance of diffusion of chemical elements within stars was made by Chapman [5]. In his computation of diffusion processes, Chapman considered the effects of the pressure and temperature gradients. He believed that thermal diffusion oposed gravitational diffusion in stars. Eddington [15] pointed out that this was not the case for ionized particles. He found that heavy particles diffuse towards hotter regions, viz. thermal diffusion acts in the same direction as gravitational diffusion. However, at that time the general conclusion was that these results were in contradiction with the observations of light and heavy elements in the same stars.

Later, when large chemical anomalies were discovered in some stars, astronomers came to the idea that these chemical peculiarities could be the result of diffusion processes appearing at the surface, cf. Michaud [22], Watson [46] and Vauclair and Vauclair [42]. As a basic physical process the atomic diffusion must be included in stellar model calculations unless there are more efficient transport processes that compete with it and wipe out its effects. Diffusion within stars competes with radiative acceleration acting selectively on different elements, meridional circulation, accretion of matter and mass loss by stellar wind. Perturbations due to other physical effects like turbulence, magnetic fields, etc. have also been discussed in literature. In addition to studies of chemical anomalies in stars, helio- and astero-seismology offer new powerful methods to test predictions of the diffusion and mixing processes in stars.

The plan of this paper is the following. First, in Sect. 2, we briefly review characteristic time scales for a few processes mentioned above. The diffusion theory, which is the primary goal of this Conference, is desribed in Sect. 3 in a more detailed manner. Next, we confront the model calculations with the observations of Main Sequence (MS) stars (Sect. 4), White Dwarf (WD) stars (Sect. 5) and Close Binary stars (Sect. 6). Finaly, in Sect. 7 the astrophysical context of diffusion is summarised.

2 Order of Magnitude Estimations

A diffusion velocity in stars may be estimated by considering a test particle with charge e_1 scaterred on field particles with charge e_2. The field particle should be regarded as an ensemble of (say) electrons and protons. For further simplicity we assume that the average mass per particle of the whole mixture is not different from the mass of the test particle, m. As a characteristic collision cross section in a fully ionized plasma we can adopt

$$\sigma = l^2 = \left(\frac{e_1 e_2}{kT}\right)^2. \tag{1}$$

The mean time between collisions is given by

$$\tau = (\sigma N v_{th})^{-1}, \tag{2}$$

where N is the number density of scattering centers and v_{th} the mean thermal velocity $(kT/m)^{1/2}$. The test particle of mass m is accelerated by gravity g, between randomizing collisions. The mean downward velocity is then

$$v_{drift} = g\tau = g(kT)^{3/2} m^{1/2} (e_1 e_2)^{-2}. \tag{3}$$

For quasi static models $dp/dr = \rho g$ and $\rho = Nm$ so

$$v_{drift} = \left(\frac{dp}{dr}\right) \frac{(kT)^{3/2}}{N^2 m^{1/2}} (e_1 e_2)^{-2} \tag{4}$$

Applying the last equation to the case of helium in a typical MS star model with $1 M_\odot$ leads to diffusion velocities only of the order of 10^{-10} cm s^{-1} at the base of the convective zone; thus in 4.5 10^9 yr (the Sun age) the travelled distance would only be of the order of 2 10^7 cm, a small (0.0003) fraction of the solar radius. There are a few factors which make the picture more interesting: thermal diffusion enhances the above pressure term, diffusion speads up as the star ages, and in certain regions, density and geometric factors combine to yield somewhat greater compositional changes than one would estimate from simple scaling laws.

Diffusion may operate faster in WD stars due to steeper pressure gradient in these objects in comparison with MS stars, cf. Schatzman [34]. Nevertheless, the expected diffusion velocity within stars is so small that the question whether stellar regions are quiet enough to develope separation of elements immediately arises.

There are several possibilities of mixing of material in stars. Among them convection is the most efficient mechanism providing a mixing time scale

$$\tau = (l/v)\,(R_c/l)^2, \tag{5}$$

where l and v are mixing length and turbulent velocity of convective blobs, R_c is the size of the convective zone. For example, in the solar case the velocity of convective blobs can reach about 2 km s^{-1} near the surface but decreases fast with depth. Stellar convective zones are highly turbulent and therefore would be well mixed. The mixing time scale consists of several years only. Thus the atomic diffusion is not able to separate the chemical elements within convective zones.

Von Zeipel [44] demonstrated, however, that a star cannot be in radiative equilibrium (it is the case when the convection does not occur and the energy is transported to the surface by radiation only) and rotates as a solid body, because of the elipsoidal shape of the equipotentials of gravity. Vogt [45], Eddington [16] and Sweet [38] suggested that there should be a circulation of matter from the hotter pole to the cooler equator, creating loops of meridional circulations. The time scale of this process according to Fricke and Kippenhahn [17] is

$$\tau = 2\pi G < \rho > \omega^{-2}\tau_{KH}, \tag{6}$$

where τ_{KH} is the thermal time given by 3.1 $10^7\frac{M^2}{RL}$ yr for a star with mass M, radius R and luminosity L expressed in solar units, ω is the angular velocity, $< \rho >$ is the average density of the star and G - gravitational constant. In the envelopes of A and B-type stars the circulation velocity is about $v_c = 10^{-5}{V_{100}}^2$ cm s^{-1}, where V_{100} is the equatorial velocity in units of 100 km s^{-1}. Typical diffusion velocity in atmospheres of these stars is $v_d = 10^{-5}$ cm s^{-1}. The two above velocities became comparable for $V_{100} = 1$.

Furthermore, a star with mass M moving through interstellar gas should accrete the latter with the rate which can be estimated as:

$$dM/dt = 10^{-15}(M/M_\odot)^2 {v_6}^{-3} n_H \qquad [M_\odot\ \mathrm{yr}^{-1}], \tag{7}$$

where v_6 is a relative velocity in units 10^6 cm s^{-1} and n_H is the number density of interstellar gas. Taking into account the fact that the mass of the stellar photosphere is of the order of 0.5 10^{-10} $M_\odot$ a star during $\tau = 5\ 10^4 {v_6}^3 (M/M_\odot)^{-2} {n_H}^{-1}$ yr can get a new atmosphere from interstellar gas with its chemistry. Such possibility is very real for WD stars since for $n_H = 1$ cm^{-3} and $v_6 = 3$ we have $\tau = 100$ yr only.

Finaly, in the absence of calculations from first principles, various phenomenological parametrizations of turbulent diffusion coefficients have been proposed to evaluate the potential effects of turbulence (Schatzman [35], Vauclair [41]). It is often assumed that turbulence wipes out the effects of atomic diffusion.

3 The Diffusion Theory

An application of kinetic theory to stellar diffusion problem is described in detail in the text book by Chapman and Cowling [6]. This approach is based on the Chapman-Enckog solution of the Boltzmann equation. Another approach arises from the use of velocity moments of the Boltzmann equations, cf. Burgers [4]. The second theory (briefly described here) yield a more useful system of multicomponent flow equations than the earlier approach. Noerdlinger [28] first applied this theory for diffusion of elements within stars investigating the case of helium diffusion for the Sun.

The standard assumptions made in the theory of stellar diffusion are the following:

1. the particles in the gas have approximately Maxwellian velocity distributions,
2. the temperatures are the same for all species of particles,
3. the mean thermal velocities are much greater than the diffusion velocities,
4. magnetic fields are unimportant.

We would like to introduce some definitions and basic equations which are essential for our purposes. Each species of particle has a distribution function $f_s(\mathbf{v}, \mathbf{x}, t)$, mean number density N_s, ionic charge $e_s = z_s e$, mass m_s, temperature T (here assumed the same for all species s), and mean velocity

$$\mathbf{u}_s = \int \mathbf{v} f_s d\mathbf{v}. \tag{8}$$

If ρ is the total mass density and ρ_s is the partial density ascribed to species s then the mean fluid velocity is defined as

$$\rho \mathbf{u} = \Sigma_s \rho_s \mathbf{u}_s, \tag{9}$$

The peculiar velocity $\mathbf{c}_s$ and drift velocity $\mathbf{w}_s$ for species s are defined by

$$\mathbf{c}_s = \mathbf{v} - \mathbf{u}; \quad \mathbf{w}_s = \int \mathbf{c}_s f_s d\mathbf{v}. \tag{10}$$

According to Burgers (cf. also Muchmore [27]) the diffusion equation for species s is

$$\nabla p_s - \frac{\rho_s}{\rho} \nabla p - N_s q_s \mathbf{E} = \Sigma_t K_{st} \left[(\mathbf{w}_t - \mathbf{w}_s) + z_{st} \left(\frac{m_t}{m_s + m_t} \mathbf{r}_s - \frac{m_s}{m_s + m_t} \mathbf{r}_t \right) \right]. \tag{11}$$

The heat flow equation for species s, under the same set of assumptions, is

$$\frac{5}{2} N_s k \nabla T = -\frac{5}{2} \Sigma_t K_{st} z_{st} \frac{m_t}{m_s + m_t} (\mathbf{w}_s - \mathbf{w}_t) - \frac{2}{5} K_{ss} z''_{st} \mathbf{r}_s$$

$$-\Sigma_t K_{st} \frac{3 m_s{}^2 + m_t{}^2 z'_{st} + 0.8 m_s m_t z''_{st}}{(m_t + m_r)^2} \mathbf{r}_s$$

$$+ \Sigma_t K_{st} \frac{m_s m_t (3 + z''_{st})}{(m_s + m_t)^2} \mathbf{r}_t. \tag{12}$$

The subscripts here denote individual species of particles (p_s is the partial pressure for species s). The quantity $\mathbf{E}$ is the electric field; k is Boltzmann's constant; T is the temperature. The $\mathbf{r}$'s are "residual heat flow vectors". They are defined by

$$\mathbf{r}_s = \frac{1}{2}\left[m_s (kT)^{-1} \int f_s \mathbf{c}_s c_s{}^2 d\mathbf{v} - 5\mathbf{w}_s \right]. \tag{13}$$

Transport coefficients appear in the form of the resistance coefficients, K_{st}, z_{st}, z'_{st} and z''_{rs} The resistance coefficients K are inversely related to the usual diffusion coefficients. In the low-density - weak coupling - limit where the Debye-Huckel theory applies, the resistance coefficients become

$$K_{st} = K_{ts} = \frac{2}{3}(2kT\mu_{st})^{1/2} N_s N_t \sigma_{st}, \tag{14}$$

$$z = 0.6, \tag{15}$$

$$z' = 1.3, \tag{16}$$

and

$$z'' = 2. \tag{17}$$

In Eq. (11) μ_{st} is the reduced mass for species s and t; σ_{st} is the collision cross-section:

$$\sigma_{st} = 2\pi^{1/2} e_s{}^2 e_t{}^2 (kT)^{-2} ln\,(\Lambda_{st}). \tag{18}$$

A collision process between a pair of particles of type s and t is controlled by the "plasma parameter":

$$\Lambda_{st} = \mathrm{abs}\,(3\mathrm{kTr_D}/\,(\mathrm{e_s e_t}))\,, \tag{19}$$

where

$$r_\mathrm{D} = \left[kT/\left(4\pi\Sigma_s N_s e_s{}^2\right)\right]^{1/2} \tag{20}$$

is the Debye length. In a pure hydrogen plasma, the number of particles in a sphere of radius r_D is $\Lambda/9$.

The Eqs. (1) and (2) with the steady state conditions of no net mass flow with respect to the center of mass

$$\nabla \Sigma \rho_s \mathbf{w}_s = 0 \tag{21}$$

and no electrical current

$$\nabla \Sigma e_s \mathbf{w}_s = 0 \tag{22}$$

form the system of linear equations which can be solved for $\mathbf{w}$'s, $\mathbf{r}$'s and $\mathbf{E}$. In the stellar core Eq. (18) should be modified to allow changes of ρ_s due to nuclear burning of elements.

4 Main Sequence Stars

The chemical composition of most stars is uniform and close to the solar composition. However, about one percent of stars show significant anomalies. Such objects are known as Chemically Peculiar (CP) stars. The theory of the chemical evolution of stars could explain it. The problem of uniform chemical composition for the majority of stars has been solved in the following way. All hydrogen and almost all helium have been formed by Big Bang. The rest of the elements are produced by massive stars and are returning into interstellar medium by supernovae and asymptotic giant branch stars. The interstellar gas is well mixed which provides for uniformity of stellar chemistry.

Knowledge of the helium abundance in stars is of fundamental importance to both cosmology and stellar structure theory. The atomic diffusion can alter the results of canonical stellar evolution theory, cf. Vauclair et al. [43], Noerdlinger [28], Richer at al. [33], Proffitt and Michaud [31] and Proffitt [30].

4.1 Helio- and Asteroseismology

The relationship between pulsations and diffusion has been long debated in the case of δ Scuti stars which are located in the well known instability strip on MS and slightly above MS band. The kappa-mechanism acting in the zone with the He^{++} partial ionization drives pulsations in these stars. However, in the instability strip there are also nonpulsating metalic stars (Am) belonging to a broad class of Chemically Peculiar (CP) objects. The exclusion between pulsation and metallicism is well established. Breger [3] first suggested that this dichotomy could be explained by helium diffusion. A star crossing the instability strip in the H-R diagram along its evolutionary path, will or will not appear as a variable star, according to its helium content in the appropriate layers. According to Vauclair et al. [43] after a time equal to the helium diffusion time scale the second helium convection zone may disappear, which allows characteristic Am anomalies to develope. If no helium is left, there is no instability strip anymore. Model calculations indicate that the rotation with $V_{rot} > 100\ \mathrm{km\,s^{-1}}$ prevents the diffusion of helium down (cf. Sect. 2) and such stars can pulsate. This can be an explanation of the mentioned above dichotomy among A-type stars.

Until recently, only a few parameters (age, mass, luminosity, radius, and surface heavy-element abundances) were measured for the Sun. Helioseismic observations have recently provided a great deal of information about the solar interior and place new constrains on the models. Solar p-modes are closely related to simple acoustic waves and as such their frequencies depend strongly on the sound speed as a function of radius in the interior. This allows the observed p-mode frequencies to be inverted to estimate the solar interior sound speed. Once the sound speed profile in the solar interior is known, a variety of other information can be inferred. The transition between the nearly adiabatic gradient within the solar convection zone, and the subadiabatic gradient below it, leads to a distinct feature in the sound speed profile. It can be used to determine the base of the adiabatically stratified region in the Sun, viz. the extend of the convective zone. The variation of the adiabatic exponent in the second He ionization zone also leads to a feature in the sound speed that can in principle be used to determine the He abundance of the solar convection zone.

It is interesting to note that while the gravitational setling of He can have a large effect on the surface of a star, the He abundance cannot be directly measured in F and G stars. Dziembowski, Pamyatnykh and Sienkiewicz [14] report that a study of solar oscillations suggests a value of the solar surface He abundance of about 0.23 or 0.24. This is significantly lower than the value of the initial He abundance needed to match the solar luminosity (Y= 0.28 or 0.29 with most current models).

Christensen-Dalsgaard, Proffitt and Thompson [8], Guzik and Cox [18] and Proffitt [30] have made detailed studies of the effects of diffusion on p-mode oscillation frequencies. All of these papers show that including diffusion results in improved agreement with observations. The surface He abundance suggested by Dziembowski et al. [14] is most consistent with the case of pure diffusion without turbulence. Effects of heavy-element settling on internal structure of the Sun was initiated by Aller and Chapmman [1] and recently reinvestigated by Proffitt [30]. Proffitt's model that includes both He and heavy-element settling requires an initial He mass fraction $Y = 0.280$ and has a surface He abundance of $Y = 0.251$ at the solar age. Unfortunately, the numbers shown here depend on the thermodynamic quantities, cf. Kosovichev et al. [21]. The papers cited above used the new equation of state recently elaborated by Mihalas, Dappen and Hummer [26], which together with the new opacity data (Iglesias et al. [19] and Seaton et al. [36]) form the base of the present understanding of plasma physics at the densities $\rho < 10^{-2}$ g cm^{-3}. In particular, these new data result in a breakthough in the understanding of β Cephei star variability.

The observed modes in β Cephei stars occur in narrow frequency range (the periods are about a few hours) and the oscillating spectra are scarce. However, in addition to precise frequency measurements, the nonadiabatic observables should be regarded as important data for asteroseismology, cf. Cugier et al. [10], [11]. The name nonadiabatic observables denotes amplitudes ratios and phase difference for any pair of oscillating parameters such as light in selected filter, colour or radial velocity. A comparison of the diagnostic diagrams in Fig. 1 reveals a strong dependence on the abundance of chemical elements, as well as the choice of the opacity data. A significant difference between the plots obtained with the use of the OPAL (Iglesias et al. [19]) and the OP (Seaton et al. [36]) opacities demonstrates the usefulness of the two-colour photometry of β Cephei stars for testing stellar opacities.

4.2 Chemically Peculiar Stars

At the region on Main Sequence from about $T_{eff} = 7000$ K to about $T_{eff} = 20000$ K various kinds of chemically peculiar stars are observed, viz. berylium-deficient stars (Be-d), metalic stars of spectral types A and F (Am-Fm), mercury-manganese stars (Hg-Mn), magnetic stars of spectral types B and A (Bp-Ap), helium weak stars (He-w), helium rich stars (He-R) and helium weak stars with

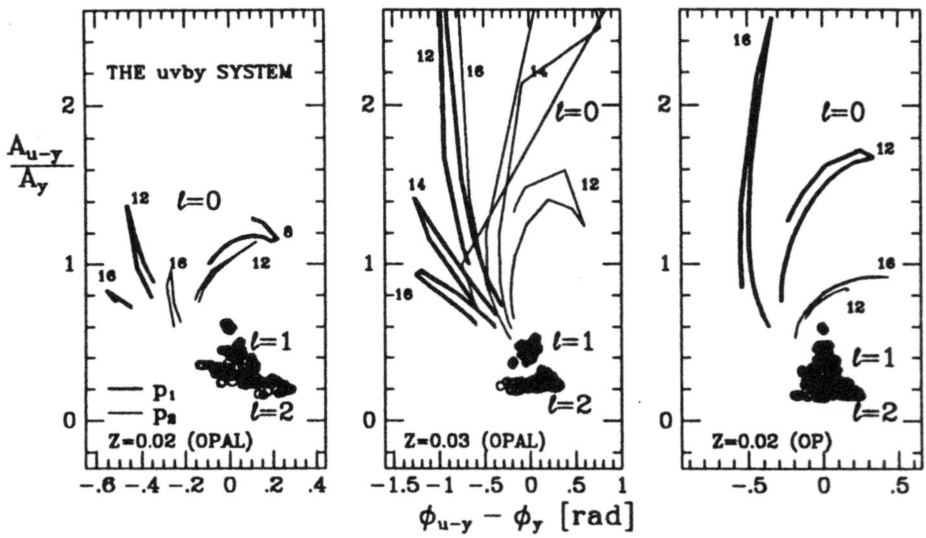

Fig. 1. The colour to light amplitude ratio *vs.* the phase difference diagrams for the Stromgren photometric system. For $l = 0$ modes the lines correspond to the Main Sequence part of the evolutionary track where fundamental (p_1) and first overtone (p_2) modes are unstable. The value of stellar mass is given in $M_\odot$. The data for unstable modes with $l = 1$ and 2 in corresponding models are plotted as dots and circles, respectively. The stellar models were calculated with the OPAL for two values of Z (0.02 and 0.03) and with the OP opacities for Z = 0.02.

abnormal He^3/He^4 isotopic ratio. A recent survey of the general predictions of the diffusion theory in MS stars has been given by Michaud [23].

The differential diffusion of heavy elements offers an explanation of the CP phenomenon. The competition between radiative acceleration (g_R) and gravity determines whether diffusion leads to over- or underabundances of a given element. Selective radiation forces arise through the process of nonisotropic absorption and isotropic reemision of photons by the atoms. Both continuum and the epectral lines contribute to the radiative acceleration. The contribution of the lines is generally much larger than the contribution of the continuum.

Surface abundances for He, Li and Mn predicted for non-rotating stars are shown in Fig. 2a. As one can see, there is settling by a factor of order two for Li and He and 1.5 for Mn around 6000 K. As the T_{eff} increases, the settling increases for both Li and He, the Li gap appearing within about 150 K of the T_{eff} near 6800 K. As the T_{eff} further increases, the Li abundance is the largest around 7050 K. It decreases again at higher T_{eff}. The Mn abundance increases considerably for $T_{eff} > 7000$ K, the Am star domain. It is even larger for

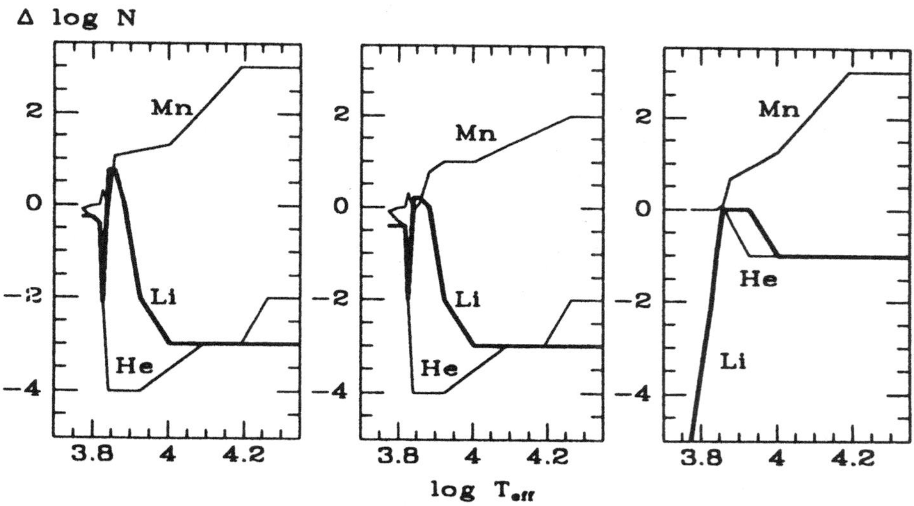

Fig. 2. (a). Expected abundances of He, Li and Mn as a function of T_{eff} in Main Sequence evolutionary models at half of the MS life time. The expected abundance is in the logarithmic scale relative to the original abundance. Atomic diffusion is included. (b). The same as panel (a) but in the presence of a mass loss rate of 10^{-15} $M_\odot\ yr^{-1}$. (c). The same as panel (a) but in the presence of meridional circulation, for an equatorial rotation velocity of 50 km s^{-1}.

$T_{eff} > 10000$ K, where HgMn stars are observed. Large overabundances, at least by two orders of magnitude are predicted for heaviest elements such as earth and iron peak elements. The anomalies are predicted to be much larger than observed for most elements; other phenomena must complete with diffusion to lead to the observed anomalies.

The effect of mass loss in A stars has been modeled by Michaud and Charland [25]. A mass loss rate of 3 10^{-16} to 10^{-15} $M_\odot yr^{-1}$ reduces the overabundances to observed values in AmFm stars for iron peak elements as well as rare earths without perturbing the agreement for underabundant elements. Expected abundances of He, Li and Mn in the presence of diffusion and mass loss are shown in Fig. 2b.

The mixing of stellar material by the meridional circulations can prevent the diffusional settling of elements. Only atmospheres of slow rotating A and B type stars are quiet enough to make diffusion efficient in agreement with observations of Am, Ap, HgMn stars. Figure 2c illustrates model predictions for the equatorial velocity of rotation equal to 50 km s^{-1}, cf. also Charbonneau and Michaud [7] where detailed model calculations are given.

Among observed objects, mercury-manganese stars probably represent the closer approach to the ideal situation of pure diffusion in stellar atmospheres. These stars are hot enough to have no outer convection zone, they have no observed magnetic field and a small rotational velocities. The results of detailed calculations generally agree with the observations. The *V sin i* at which the

HgMn and AmFm phenomena disappear give an argument in favour of the meridional circulation process.

The Am stars are CP stars located at the cool end of this zone, without strong organized magnetic fields. The chemical analysis of Am stars have revealed a number of anomalies. For instance Lithium is found either with the standard cosmic abundance, or underabundant. It is never strongly overabundant, contrary to what the diffusion theory predicts in absence of mass-loss or turbulent mixing. The three elements C, N and O are deficient in Am stars. The deficiency is clearly marked for the coolest Am stars and decreases to nothing for the hottest ones. The Aluminium and Iron abundances are only moderately enhanced with respect to normal late A stars. The overabundances found for the rare-earths are of the order of a factor 20. Quite interesting is the fact that the overabundance factor is very uniform, and preserves the odd-even effect present in the sun and is predicted by the theory of the s-process.

The helium deficiency observed in most chemically peculiar stars is naturally explained by diffusion: the cosmic helium abundance is so large that the radiation force is not able to support it in stellar atmospheres. The helium-deficient stars with large He^3/He^4 are also well explained by diffusion. During diffusion process He^3 slowly replaces He^4 as a result of their mass difference. MS helium-rich stars may be explained by mass loss or magnetic fields coupled with diffusion.

Barium stars exhibit overabundances of barium and of other *s*-process elements relative to Fe as well as a correlation between carbon and the *s*-process element abundances. Most of barium stars are giants and all or almost all of them have a degenerate dwarf companion. An attractive explanation for the barium syndrome origin is the accretion of the nuclear processed material from white dwarf companion.

4.3 Problem of Lithium in MS Stars

Observations of Li has promoted the development of a great variety of models invoking transport processes in the outermost layers of these stars (e.g., microscopic diffusion, turbulent diffusion, meridional circulation, mass loss). In stellar interiors Lthium (an Bryllium) nuclei ar destroyed at relatively low temperatures (about 2.5 10^6 and 3.5 10^6 K, respectively). In low MS stars, these temperatures are reached slightly below the base of the convective zone or inside the convective zone. Therefore, the processes which transport matter between the outermost layers and regions of the stellar interior may cause affect the contents of Li and Be in the atmospheric layers. Figure 3 compiles Li abundances measurements for F and later stars in the Pleiades (circles) and Hyades (asterisks) stars. These two clusters are different in age: $6 - 7\ 10^7$ yr and 7.6 10^8 yr = 0.76 Gyr for Pleiades and Hyades, respectively. A comparision of these two clusters reveals important differences. For stars with $T_{eff} > 6300$ K the abundance of Li in Pleiades is virtually constant and close to the value of log N(Li) = 3.3. In the Hyades, however, note the strong absence of Li at the temperature range between 6500 and 6900 K. This effect ("Li gap") was discovered by Boesgaard and Tripicco [2]. For stars with $T_{eff} < 6300$ K, there is a trend to lower abundances with decreasing

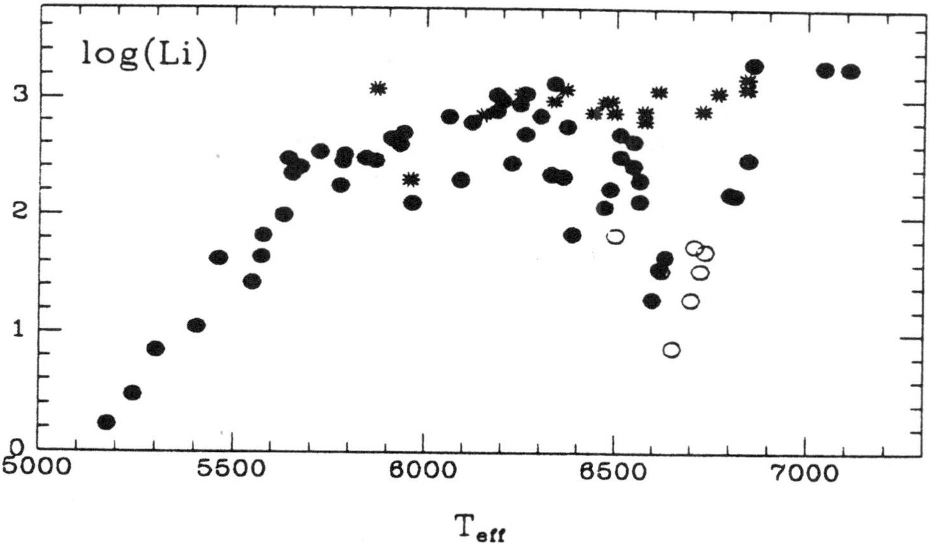

Fig. 3. Li measurements in Pleiades (circles) and Hyades (asterisks) stars. Open circles indicate upper limits.

the effective temperature. The Li gap was confirmed in other clustes, viz. NGC 752 (age 1.7 Gyr), Coma Berenices (0.5 Gyr), Ursa Major Group (0.3 Gyr) and Prasepe (0.7 Gyr). Clearly, Li abundance depends markedly on a stellar age but other stellar parameters may also be important including metalicity, rotation velocity and others. Measurements in their present state do not allow to establish definitive conclusions on which process is the most important in controlling the observed behaviour of Li abundances in F-M stars. Recent reviews on the problem of Li in MS stars can be found in Michaud and Charbonneau [24] and Rebolo [32].

5 White Dwarf Stars

Relative to "normal stars" these objects are characterized by small radii, high densities and strong surface gravity. Typical values for white dwarfs (WD) are $R = 10^{-2}\ R_{\odot}$, $\rho = 10^{6}$ g cm^{-3} and $g = 10^{8}$ cm s^{-2}. Their configuration is supported against the large gravity by the pressure of highly degenerate electrons. In very interior of a WD, the degenerate electrons provide a high thermal conductivity, which (together with small luminosity) does not allow large temperature gradients. The situation is different when going to the outermost layers. With decreasing ρ we expect to find a non-degenerate outher layer with heat transfer controled by radiation or convection.

Most white dwarfs have surface convection zones. Velocities of convective elements may be as large as several km s^{-1}, while diffusion velocities are orders of magnitude smaller. Thus diffusion will only operate to separate material in deeper layers but the chemical composition changes produced in these layers will next propagate through the convection zone up to the stellar surface. The results of detailed model calculations (cf. Muchmore [27]) indicate that if a WD forms with the atmosphere consisting partly of hydrogen and partly of helium, the hydrogen and helium should rapidly separate. For instance helium is depleted from the atmosphere of a hydrogen-rich WD stars with $T_{eff} = 12000$ K in only 10 yr. Metals settle out of the atmosphere in 10^4 yr or less. In the case of helium WD stars hydrogen diffuses upward, while iron and other elements considered diffuse downward. Generally, model calculations indicate that iron diffuses most rapidly and carbon most slowly. The diffusion time scales for the separation of metals out of the atmospheres are $10^6 - 10^8$ yr depending on the star's mass.

There are two time scales for changes in the structure of a white dwarf: a time scale for cooling and a time scale for changes in the atmospheric opacity. In a typical case, diffusion time scales are much shorter than evolutionary time scales. Any metals originally in the atmospheres of white dwarfs must have sunk out of sight long ago. The metals presently seen in the atmospheres of some white dwarfs cannot be original to these stars. One hypothesis to account for their presence is that accretion from the interstellar medium continually supplies an influx of metals onto the star. The scenario of a competition between diffusion and accretion to explain the composition of white dwarfs makes a variety of predictions about the relative abundances of different elements.

In WD stars diffusion generally occurs in a sufficiently dense gas that the usual treatment of the diffusion coefficients is not valid. In particular the electrons may be weakly degenerate and pressure ionization effects are likely to be important. Neither the weak coupling approximation nor the strong coupling approximation is valid. Uncertain pressure ionization effects are likely to be important.

6 Close Binary Stars

In the case of a single star, it is necessary to mix the unevolved envelope matter with the matter from the deep interiors of stars or to lose the external envelope with unprocessed matter to get the chemically evolved matter from the deep interiors on the stellar surface. In the case of close binary systems, however, large-scale transfer of mass from one component to the other is the most basic phenomenon in their evolution, cf. e.g., Paczynski [29]. When a sufficiently large fraction of mass is lost by the originally more massive star, substantial changes in the surface relative abundances of H, He, C, N and O because of the CNO cycle are expected. A carbon (C^{12}) deficiency and nitrogen (N^{14}) enrichment may be observed in classical Algol-type stars even if the mass loss does not expose very deep layers with H, He and O anomalies. This is caused by the fact that the CN part of the CNO cycle in massive stars takes place above the convective core. In particular the mass-losing secondary in Algol itself (β Persei) should be stripped down to layers with $M_r < 0.36$ of the original mass. The chemical composition of these layers was radically changed prior to the start of mass transfer and carbon depletion as large as 2.0 dex. may be expected, cf. Cugier and Hardorp [12]. The mass transfer from the secondary component to the primary one means that any C poorness of the secondary's surface material will also be reflected in the primary star.

The carbon abundance analysis for the primaries of eight Algol-type stars was made by Cugier and Hardorp [12] and Cugier [9]. They found that observed abundances of C^{12} are in fact clearly lower than the solar value, but markedly larger than the expected ones. In order to explain this discrepancy Cugier and Hardorp suggested that a large-scale mixing of matter occurs in these stars. One of the possibilities is a thermohaline mixing of matter due to an inversion of the mean molecular weight developed in a mass accreting star receiving helium-rich matter from the companion, cf. De Greve and Cugier [13]. This behaviour of C^{12} in Algols was futher confirmed both observationaly by Tomkin et al. [39] and theoretically by Sarna [33].

It is interesting to note that the term termohaline convection was introduced by Stern [37] who considered the situation in which a layer of warm salt water is above a layer of fresh cold water of slightly higher density. When the hot salt water is cooling off then its density becomes higher than the fresh water's one and an istability developes. Experiments show that it leads to the formation of fingers with the salt water which sink down. A blob of matter with higher molecular weight, $\mu + \Delta\mu$, immersed in a region with molecular weight μ was considered by Ulrich [40] and Kippenhahn et al. [20]. Such a blob sinks with the velocity v_μ and after a path-length L the blob is destroyed and mixed with the surroundings. It can be described as a "diffusion process" with a "diffusion coefficient" $D = v_\mu L$.

7 Summary

This review is dedicated to the astrophysical context of diffusion processes. In particular, atomic diffusion as a basic physical process must be included in stellar model calculations unless there are more efficient transport processes that wipe out its effects. Diffusion within stars does not develope in convective zones. In radiative equlibrium regions diffusion competes with radiative acceleration acting selectively on different elements, meridional circulation, accretion of matter, mass loss by stellar wind, turbulence and magnetic fields.

Almost ideal situation of pure diffusion in stellar atmospheres is probably realized in mercury-manganese stars. The helium deficiency observed in most chemically peculiar stars is also naturally explained by diffusion. Highly monoelemental atmosphere of white dwarf stars reflects the efficiency of diffusion in the case when steep pressure gradients are developed. On the other hand, the situation is not clear for a long searched problem of Li abundance in stars. Measurements in their present state do not allow to establish definitive conclusions on which process is the most important in controlling the observed behaviour of Li.

In addition to studies of chemical anomalies in stars, helio- and asteroseismology offer powerful methods to test theoretical predictions. The solar *p*-mode oscillations suit this purpose best. A recent theoretical discovery that the usual opacity mechanism acting in the zone with temperature close to 2 10^5 K due to a large number of Fe lines is responsible for oscillations observed in β Cephei stars should draw attention as a new constrain on stellar diffusion in early type stars.

Classical Algol-type binaries should also be regarded as stars where mixing and diffusion processes effectively take place.

References

[1] L. H. Aller, S. Chapman: Diffusion in the Sun. Astrophys. J. **132**, 461 – 472 (1960)

[2] A. M. Boesgaard, Tripicco: Lithium in the Hyades cluster. Astrophys. J. Lett. **302**, L49 – L53 (1986)

[3] M. Breger: Main-sequence pulsation in open clusters. Astrophys. J. **176**, 373 – 380 (1972)

[4] J. M. Burgers: *Flow Equations for Composite Gases*, Academic Preff, New York (1969)

[5] S. Chapman: Convection and diffusion in giant stars. MNRAS **77**, 540 (1917)

[6] S. Chapman, T. G. Cowling: *The Mathematical Theory of Non-Uniform Gases*, Cambridge University Press (1970)

[7] P. Charbonneau, G. Michaud: Two-dimentional particle transport in HgMn and FmAm stars. Astrophys. J. **327**, 809 – 816 (1988)

[8] J. Christensen-Dalsgaard, C. R. Proffitt, M. J. Thompson: Effects of diffusion on solar models and their oscillation frequences. Astrophys. J. **403**, L75 – L78 (1993)

[9] H. Cugier: Carbon abundance in the primaries of six Algol-type stars. Astron. Astrophys. **214**, 168 – 178 (1989)

[10] H. Cugier, W. Dziembowski, A. A. Pamyatnykh: Nonadiabatic observables in β Cephei star models. *in* Pulsation, Rotation and Mass Loss in Early-Type Stars IAU Symposium **162** (1993), eds. L. A. Balona and H.Huib (Kluver)

[11] H. Cugier, W. Dziembowski, A. A. Pamyatnykh: Nonadiabatic observables in β Cephei models. Astron. Astrophys. (in press)

[12] H. Cugier, J. Hardorp: Carbon abundance in β Persei and λ Tauri. Astron. Astrophys. **202**, 101 – 108 (1988)

[13] J. P. De Greve, H. Cugier: Evolution of the surface abundance of carbon in mass-exchanging binaries. Astron. Astrophys. **211**, 356 – 360 (1989)

[14] W. A. Dziembowski, A. A. Pamyatnykh, R. Sienkiewicz: Helium content in the solar convective envelope from helioseismology. MNRAS **249**, 602 – 605 (1991)

[15] A. S. Eddington: *The Internal Constitution of the Stars*, Cambridge University Press (1926)

[16] A. S. Eddington: MNRAS **90**, 54 (1929)

[17] F. Fricke, R. Kippenhahn: Evolution of rotating stars. Ann. Rev. Astron. Astrophys. **10**, 45 – 72 (1972)

[18] J. A. Guzik, A. N. Cox: Using solar *p*-modes to determine the convection zone depth and constrain diffusion-produced composition gradients. Astrophys. J. **411**, 394 – 401 (1993)

[19] C. A. Iglesias, F. J. Rogers, B. G. Wilson: Spin-orbit interaction effects on the Rosseland mean opacity. Astrophys. J. **397**, 717 – 728 (1992)

[20] R. Kippenhahn, G. Ruschenplatt, H. C. Thomas: The time scale of thermohaline mixing in stars. Astron. Astrophys. **91**, 175 – 180 (1980)

[21] A. G. Kosovichev, J. Christensen-Dalsgaard, W. Dappen, W. A. Dziembowski, D. O. Gough, M. J. Thompson: Sources of uncertainty in direct seismological measurements of the solar helium abundance. MNRAS **259**, 536 – 558 (1993)

[22] G. Michaud: Diffusion processes in peculiar A stars. Astrophys. J. **160**, 641 – 658 (1970)

[23] G. Michaud: Diffusion, meridional circulation and mass loss in main sequence and horizontal branch stars. *in* Evolution of Stars: The Photospheric Abundance Connection IAU Symp. No. **145** (1991), eds. G. Michaud and A. Tutukov Kluwer Academic Publishers, pp. 111 – 124

[24] G. Michaud, P. Charbonneau: The Lithium abundance in stars. Space Sci. Rev. **57**, 1 – 58 (1991)

[25] G. Michaud, Y. Charland: Mass loss in A and F stars: the λ Bootis stars. Astrophys. J. **311**, 326 – 334 (1986)

[26] D. Mihalas, W. Dappen, D. G. Hummer: The equation of state for stellar envelopes. II. Algorithm and selected results. Astrophys. J. **331**, 815 – 825 (1988)

[27] D. Muchmore: Diffusion in white dwarf stars. Astrophys. J. **278**, 769 – 783 (1984)

[28] P. D. Noerdlinger: Diffusion of helium in the Sun. Astron. Astrophys. **57**, 407 – 415 (1977)

[29] B. Paczynski: Evolutionary processes in close binary systems. Ann. Rev. Astron. Astrophys. **9**, 183 – 208 (1971)

[30] C. R. Proffitt: Effects of heavy-element settling on solar neutrino fluxes and interior structures. Astrophys. J. **425**, 849 – 855 (1994)

[31] C. R. Proffitt, G. Michaud: Diffusion and mixing of Lithium and Helium in population II dwarfs. Astrophys. J. **371**, 584 – 601 (1991)

[32] R. Rebolo: Lithium and Beryllium in main sequence stars. *in* Evolution of Stars: The Photospheric Abundance Connection IAU Symp. No. **145** (1991), eds. G. Michaud and A. Tutukov Kluwer Academic Publishers, pp. 85 – 97

[33] J. Richer, G. Michaud, C. R. Proffitt: Helium gravitational settling in the envelopes of evolving main-sequence A and F stars. Astrophys. J. Suppl. **82**, 329 – 350 (1991)

[33] M. Sarna: Chemical evolution of Algol-type stars: mass-exchanging binary systems in cases AB and early B. MNRAS **259**, 17 – 36 (1992)

[34] E. Schatzman: White Dwarfs, North Holland Publ. Comp. Amsterdam, (1958)

[35] E. Schatzman: Gravitational separation of the elements and turbulent transport. Astron. Astrophys. **3**, 331 – 346 (1969)

[36] M. J. Seaton, Y. Yan, D. Mihalas, A. K. Pradhan: Opacities for stellar envelopes. MNRAS **266**, 805 – 828 (1994)

[37] M. E. Stern: Tellus **12**, 172 (1960)

[38] P. A. Sweet: The importance of rotation in stellar evolution. MNRAS **110**, 548 – 558 (1950)

[39] J. Tomkin, D. L. Lambert, M. Lemke: The chemical composition of Algol systems - V. Confirmation of carbon deficiencies in the primaries of eight systems. MNRAS 265, 581 – 587 (1993)

[40] R. K. Ulrich: Thermohaline convection in stellar interiors. Astrophys. J. **172**, 165 – 177 (1972)

[41] S. Vauclair: Turbulence and the Lithium Abundance in Giants and Main-Sequence Disk and Halo Stars. *in* Evolution of Stars: The Photospheric Abundance Connection IAU Symp. No. **145** (1991), eds. G. Michaud and A. Tutukov Kluwer Academic Publishers, pp. 327 - 340

[42] S. Vauclair, G. Vauclair: Element segregation in stellar outher layers. Ann. Rev. Astron. Astrophys. **20**, 37 - 60 (1982)

[43] G. Vauclair, S. Vauclair, A. Pamyatnykh: Diffusion processes in the envelopes of main-sequence A stars: model variations due to helium depletion. Astron. Astrophys. **31**, 63 - 70 (1974)

[44] H. Von Zeipel: MNRAS **84**, 665 (1924)

[45] H. Vogt: Zum Strahlungsgleichgewicht der Sterne. Astron. Nachr. **223**, 229 - 232 (1925)

[46] W. D. Watson: Element diffusion and abundance anomalies in metallic A stars. Astrophys. J. Lett. **162**, L45 - L48 (1970)

Surface Diffusion and Lateral Interactions in Adsorbed Layers

A. G. Naumovets, M. V. Paliy and Yu. S. Vedula

Institute of Physics, Ukrainian National Academy of Sciences,
46 Prospekt Nauki, UA-252650, Kiev 22, Ukraine

Abstract: We present and discuss a collection of experimental data illustrating effects of lateral interactions on surface diffusion of adsorbed particles which proceeds in a concentration gradient.

1 Introduction

Atoms and molecules adsorbed on solid surfaces are known to produce a multitude of two-dimensional (2D) structures whose symmetry in the ordered state is determined by the symmetry of the substrate surface and by lateral interactions between the adsorbed particles. For a given adsorbate/substrate system one usually observes a number of (2D) phases corresponding to various adsorbate concentrations (coverages) and temperatures (for a review see e.g. [1]). Therefore, if surface diffusion proceeds in a concentration gradient, it should generally be accompanied by formation of 2D phases in the diffusion zone. In turn, this process should affect the kinetics of surface diffusion, which is evident from the relationship

$$\mathbf{J} = -L\nabla\mu\,, \tag{1}$$

where $\mathbf{J}$ is the diffusion flux, L the transport coefficient and $\nabla\mu$ the chemical potential gradient, all the values being concentration dependent. Lateral repulsion should promote the spreading of adsorbed particles (adparticles) whereas lateral attraction should impede it and, when sufficiently strong, may even cause an uphill diffusion. These considerations are quite general and were established previously in theoretical and experimental studies of volume diffusion (see e.g. [2]). Contrary to this, investigations of *multiphase surface diffusion* were started only comparatively recently, since it was necessary to elaborate the appropriate experimental techniques.

There are a few experimental possibilities to get insight into effects of lateral interactions on surface diffusion. One can follow the correlated walks of individual atoms [3, 4], the density fluctuations in macroscopically uniform adlayers

[5, 6] or the evolution of adsorbate concentration profiles purposely set up on a surface under study [7]. In this paper we shall focus on the latter approach, since the concentration profiles manifest the phase effects in surface diffusion in a rather graphical way and allow one to study them in a wide coverage range.

In principle, any technique which locally probes the adsorbate concentration with a sufficient accuracy and spatial resolution can be utilized to record the diffusion profiles. Data are available acquired by contact potential techniques, Auger electron spectroscopy, photoelectron emission, ellipsometry and other methods [7-12]. However, since the experiments are usually very laborious, only a few systems have been investigated in considerable detail. Below we give some representative examples of multiphase surface diffusion. We have chosen these examples among the results obtained for electropositive (alkali, alkaline-earth) adlayers on refractory metals. These systems are remarkable for several reason.

First, the lateral interaction in them are particularly strong [13], which results in pronounced phase effects. Second, these interactions are diverse even for the same system, e.g. a strong repulsion at small coverages may change for an attraction at medium coverages and again for the repulsion at coverages approaching a monolayer [1, 7, 13]. Finally, the electropositive adlayers cause strong work function changes, which gives a possibility to determine the coverage with a high accuracy ($\sim 10^{-2}$ of monolayer).

The experiments are carried out in ultra-high vacuum and with careful outgassing of materials used, because surface diffusion is extremely sensitive to impurities. The results presented in this paper were obtained with a scanning technique based on contact potential measurements [7]. Its spatial resolution is $\approx 15\mu$m. The work function distributions are converted into coverage distributions over the surface using calibrated curves work function versus coverage. The structural data were obtained by low-energy electron diffraction. For experimental details see [7] and [14]. Evaporating an adsorbate on the substrate through different masks, one can set up initial coverage profiles shaped as steps ("semi-planes"), stripes or trenches. The diffusion profiles are recorded by scanning the surface with the electron beam, the energy of electrons being below 1 eV.

2 Experimental results and discussion

Fig. 1 shows a typical coverage profile obtained in surface diffusion of barium on the (011) surface of molybdenum [15]. This surface is densely packed and has a pseudohexagonal structure. The initial Ba deposit had a step-like shape with the boundary at $x = 0$ and the coverage about 3 monolayers. (In fig. 1 the coverage is measured in close-packed Ba monolayers. For the Ba–Mo(011) system, the close-packed Ba monolayer has a hexagonal symmetry and is incommensurate with the substrate structure). The diffusion profile is seen to have a complex shape which is completly different from the symmetrical smooth shape predicted for diffusion of noninteracting particles and represented by a well-known combination of the error functions [2]. The coverage terraces and steps seen in the profile correlate with the formation of the specific Ba phases marked in the right part of fig. 1. The terraces corresopnd to the commensurate Ba lattices $c(6\times2)$ and $c(2\times2)$ where the concentrations of Ba adatoms amount respectively to 1/12 and 1/4 of the surface concentration of Mo atoms on the (011) plane. The longest terrace emerges in the vicinity of the $c(2\times2)$ phase.

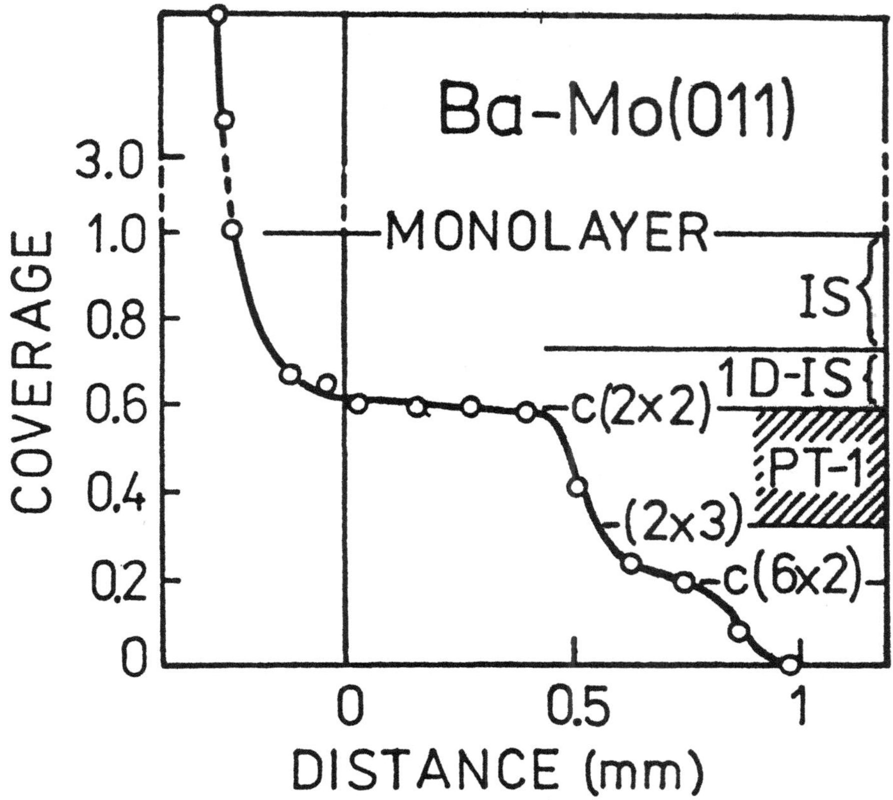

Fig. 1. A coverage profile for Ba diffusion out of a step-like initial deposit (notations are explained in the text). The substrate is Mo(011).

It is evident even intuitively that flat sections of the diffusion profile should mark the coverage intervals where the diffusivity is higher. Quantitatively, this follows from the Boltzmann-Matano equation which enables one to extract the diffusivity $D(\Theta_0)$ at a particular coverage Θ_0 from the concentration profile $\Theta(x)$ (see e.g. [2, p. 11]):

$$D(\Theta_0) = -\frac{\int_0^{\Theta_0} x \, d\Theta}{2t\left(\frac{\partial\Theta}{\partial x}\right)_{\Theta_0}} . \tag{2}$$

Here t is the diffusion time and x is measured form the Matano plane which divides the $\Theta(x)$ profile in two parts having equal areas.

The structure $c(2\times 2)$ is specific in the sense that it is the densest commensurate lattice. At higher Θ's, there occurs a transition to incommensurate structures (IS). The transition is initiated by one-dimensional (1D) compression of the $c(2\times 2)$ lattice which is effected by formation of linear incommensurate walls separating the commensurate domains [16]. Owing to mutual repulsion, the domain walls form their own superlattices. The coverage range corresponding to this region of the phase diagram is labeled 1D–IS in fig. 1. It has been suggested that the fast Ba diffusion in this Θ interval (revealed by the wide terrace in the $\Theta(x)$ profile) is due to high mobility of the incommensurate domain walls [7]. These walls can be treated analytically as topological solitons in the commensurate structure, so the suggested diffusion mechanizm was termed the soliton mechanism [17].

Actually the displacement of the domain walls is predicted to be effected by random walks of kinks which are generated in the walls by thermal excitation. The random walks of the kinks result in meandering and, in the final account, in the random walks of the walls. Of course, it should be kept in mind that these walks are influenced by the interaction of the domain walls (solitons) [1, 17]. Anyway it is clear that the diffusion under such conditions has a pronounced collective character. A model of the soliton diffusion mechanism is depicted in fig. 2. The adatoms from the initial deposit intrude into the commensurate phase and form there solitons which move to the edge of the spreading layer and shift it forward.

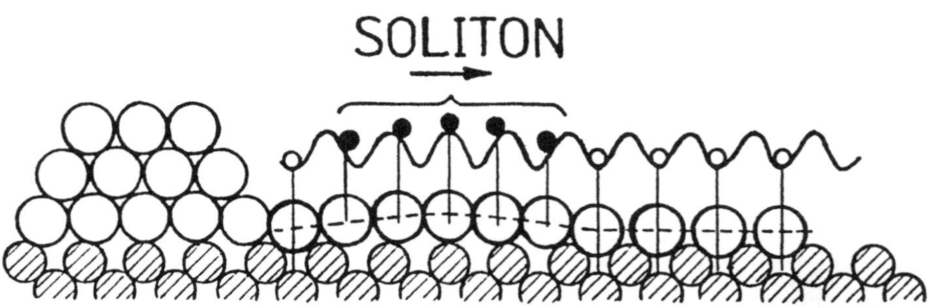

Fig. 2. A model of the soliton mechanism of surface diffusion. Positions of adatoms with respect to surface potential relief are shown schematically.

The coverage range marked PT-1 in fig. 1 corresponds to the region of first-order phase transition where the phases (2×3) and $c(2\times 2)$ coexist in equilibrium (the relative concentration of Ba atoms in the (2×3) phase with respect to Mo surface atoms equals 1/6). It is seen that this coverage interval is represented as a "step" in the $\Theta(x)$ profile. Similar results have been obtained in volume diffusion [2]. This behaviour can be readily understood in terms of Eqn. (1). If one considers a quasi-stationary diffusion along the x axis and writes $(\partial\mu/\partial x) = (\partial\mu/\partial n)(\partial n/\partial x)$, then in the first-order phase transition one should have $(\partial\mu/\partial n) = 0$ and $(\partial n/\partial x) \to \infty$.

These results illustrate that in the state of local equilibrium the diffusion zone represents actually a section of the phase diagram at the diffusion temperature, i.e. a kind of a "collective portrait" of all 2D phases appearing at various coverages.

Very graphical results consistent with this statement were obtained in computer simulations in which surface diffusion was displayed for some model systems with a realistic set of lateral interaction parameters [18].

In our opinion, the formation of 2D phases within the diffusion zone may be considered as a kind of reaction between the diffusing adatoms and vacant adsorption sites. The arising "products" can be characterized by a specific geometry of the bonds, the 2D structure, and the stoichiometry. Thus, such a system may belong to the class of the reaction-diffusion systems with initially separated components (see e.g. [19]).

For Ba on Mo(011) and other similar adlayers it was impossible to follow the diffusion of adatoms at $\Theta \approx 1$ and $\Theta > 1$, because the adatom mobility at such coverages occurs to be essentially lower than at $\Theta < 1$. Such behaviour is typical of the systems with strong repulsive interaction within the first monolayer which, in addition to the concentration gradient, favours the spreading.

Let us now consider some data for adlayers where this factor is less pronounced and the observation of the diffusion is possible at supermonolayer coverages, too. We have recently obtained such results for lithium on the molybdeum (112) face. The Mo(112) surface has a highly anisotropic structure: it consists of close-packed rows of Mo atoms separated by atomically deep channels (fig. 3). This imposes strong anisotropy both on the lateral interactions of adatoms [13] and on their diffusion. The models of Li structures based on low-energy electron diffraction experiments [20] are shown in fig. 3. The chain structures $p(1\times4)$ and $p(1\times2)$ are formed at $\Theta = 0.25$ and $\Theta = 0.5$, respectively (for Li adlayers the coverage is defined as the ratio of surface concentrations of Li and Mo atoms). The structure denoted 1D-IS is a one-dimensionally incommensurate structure: within the range $0.67 < \Theta < 1.0$ there occurs a uniaxial compression of the adlayer along the channels which finishes up with the formation of the $p(1\times1)$ structure. The second monolayer ($\Theta > 1$) fills the channels between the rows of Li adatoms in the $p(1\times1)$ structure.

The Li coverage profiles which evolve in the course of diffusion out of a step-like initial deposit approximately 5 monolayers thick are given in fig. 4. The diffusion which is followed along the channels is seen to result in spreading of a monolayer with a rather sharp edge. The concentration gradient at the edge remains virtually constant, and the displacement of the edge is proportional to $\sqrt{t}$. The shape of the profiles reveals the highest mobility in the second layer ($1 < \Theta < 2$). Thus the diffusion mechanism seems to correspond to the model shown in fig. 4. This mechanism was suggested for the cases when adatoms within the first monolayer are less mobile than in the second and subsequent monolayers [21]. Equation (2) was used to calculate the diffusivity from measured concentration profiles. Since D is higly coverage-dependent, we shaped initial step-like

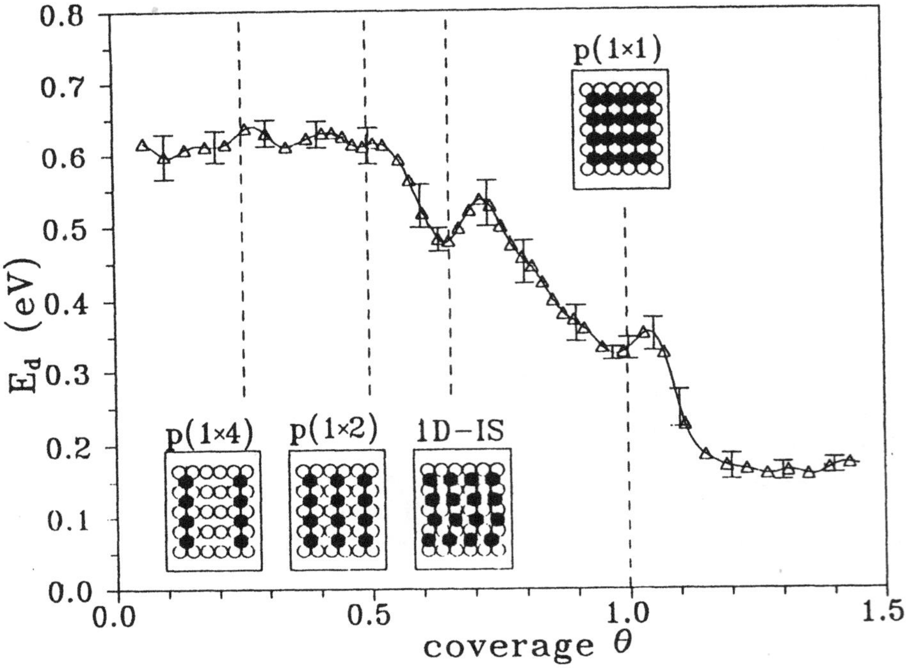

Fig. 3. Coverage dependence of the activation energy E_d for Li diffusion on Mo(112). Instets: Li structures on Mo(112) [20]. See text for notations.

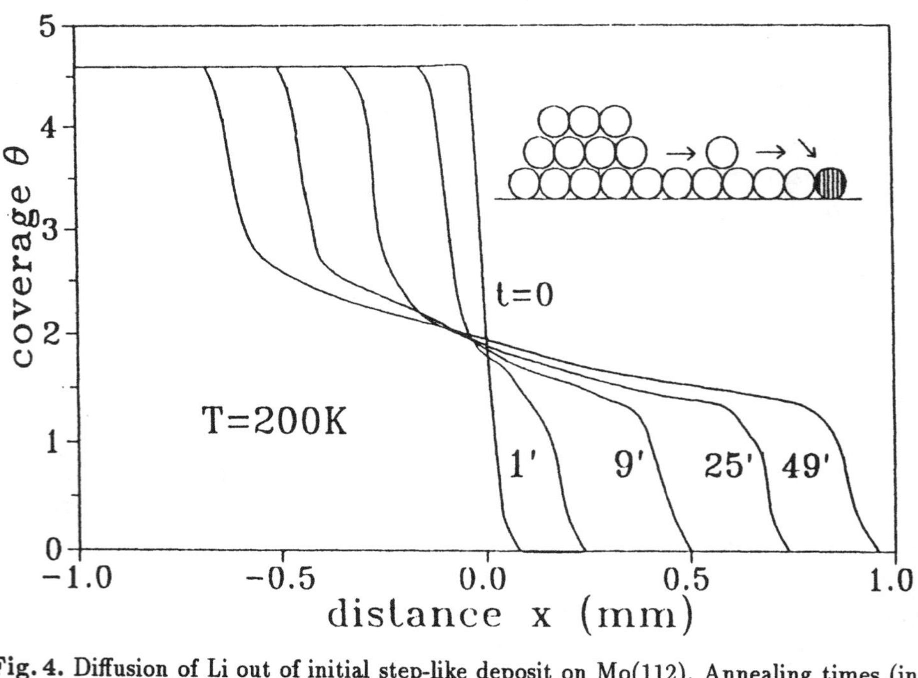

Fig. 4. Diffusion of Li out of initial step-like deposit on Mo(112). Annealing times (in minutes) are shown. Inset: assumed diffusion mechanism.

profiles with different initial coverages, and their evolution was recorded at different temperatures. Some of the data on the diffusivity obtained in this way are depicted in fig. 5. The full set of the data was utilized to plot Arrhenius lines (fig. 5) as well as to extract the activation energies (fig. 4) and the preexponen-

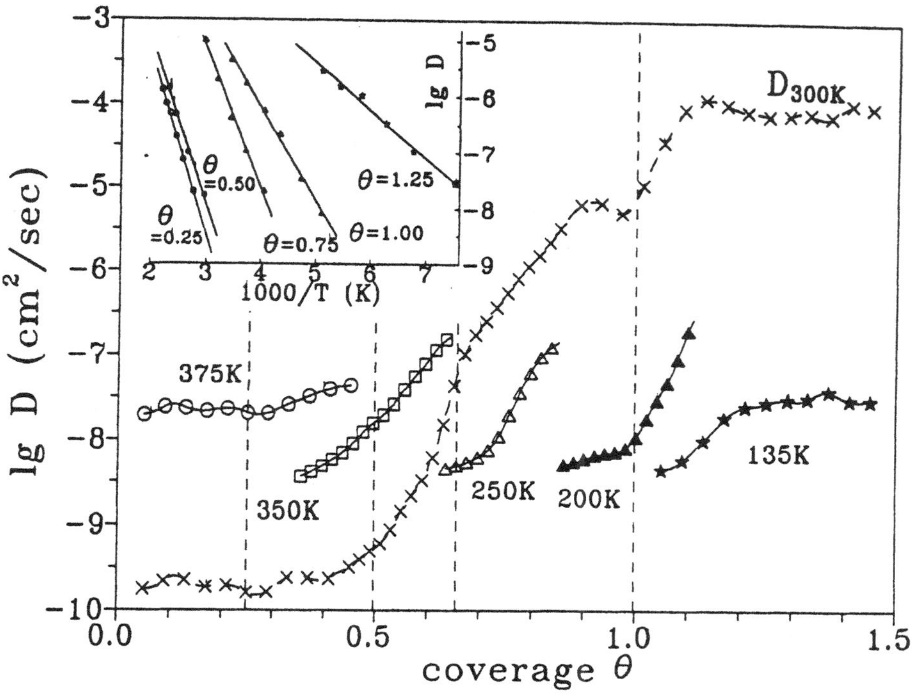

Fig. 5. Diffusivity versus coverage for Li on Mo(112) at different temperatures. Dashed curve is the calculated dependence $D(\Theta)$ at T = 300K. Inset: Arrhenius plots at selected coverages.

tial factors (not shown here). In turn, these values were used to calculate D as a function of Θ for the whole coverage range $0.05 < \Theta < 1.5$ at a constant temperature 300 K (fig. 5). It is seen that the growth of the diffusivity in this Θ interval spans more than 5 orders of magnitude and that the activation energy of diffusion decreases (with same nonmonotonities) by a factor of three. The strongest variations occur in the region of incommensurate structures ($0.5 < \Theta < 1$) and just above $\Theta = 1.0$. These changes mirror the strong repulsive interaction in the incommensurate structures and the sharp enhancement of the adatom mobility on passing to the second lithium monolayer.

Interesting peculiarities of Li diffusion in the vicinity of $\Theta = 1$ and in the second monolayer were observed on the Mo(011) surface (fig. 6). To reveal them more cleary, we shaped a step-like Li profile on the surface previoulsy covered with a *continuous* ("base") Li adlayer corresponding to a coverage of 0.8. Here

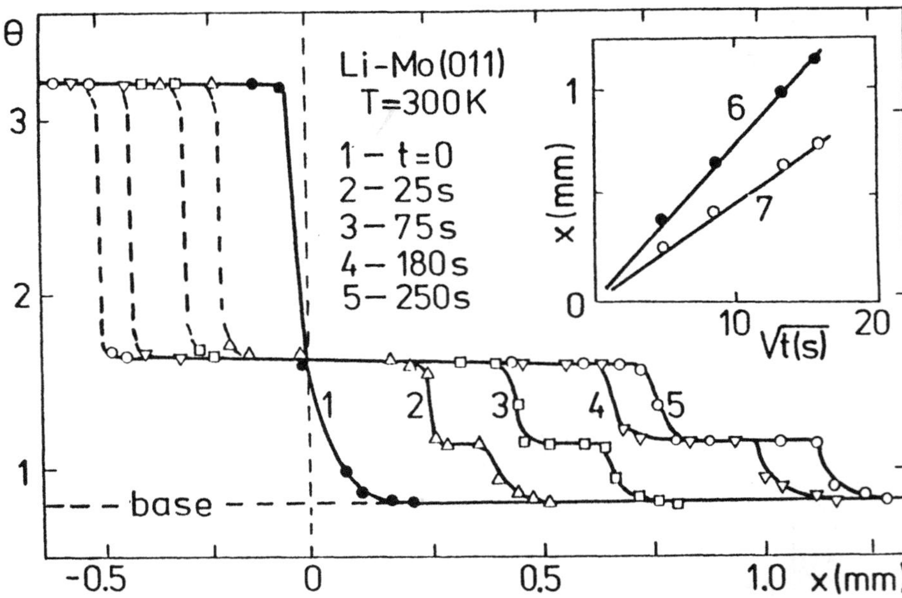

Fig. 6. Diffusion of Li out of an initial step-like deposit on the Mo (011) surface precovered with a base adlayer ($\Theta = 0.8$). (1) to (5): coverage profiles; inset: Displacement of the leading edge versus $\sqrt{t}$ for $\Theta = 1.07(6)$ and $\Theta = 1.6(7)$. Data were obtained with V. V. Poplavsky.

we observed the emergence of clear-cut terraces at $\Theta \approx 1.07$ and $\Theta \approx 1.6$. The former correspond approximately to the close-packed Li monolayer. The terraces at $\Theta = 1.6$ may be associated with the formation of a double-layer film. It was inferred that during the building-up of the second monolayer a part ($\approx 20\%$) of Li atoms of the first (somewhot compressed) monolayer may be expelled to the second one. The edges of both terraces move linearly with $\sqrt{t}$, but their spreading rates are somewhat different (fig. 6). These results exemplify a process of layering of the spreading film. Similar behaviour was found recently for spreading droplets of oily liquids on solid surfaces [10].

Consider now some diffusion data relating to the limit of very small coverages ($\Theta < 0.1$). They were obtained lately for Li on the Mo(112) surface. An example of these results is presented in fig. 7. We continuously evaporated lithium on a half of the sample while another half was shadowed with a mask. The process of Li diffusion (along the atomic channels) was followed in the shadowed region. The measured coverage profiles show two peculiarities: a terrace at an hollow behind the terrace (more experimental data can be found in [14]). These features are still not understood.

It might be supposed that the appearance of the low-coverage terrace reflects the filling of some surface defects (traps) by diffusing Li atoms. However, this assumption seems to be not compatibile with the formation of the coverage hollow. It is therefore speculated that the terrace could be due to the growth

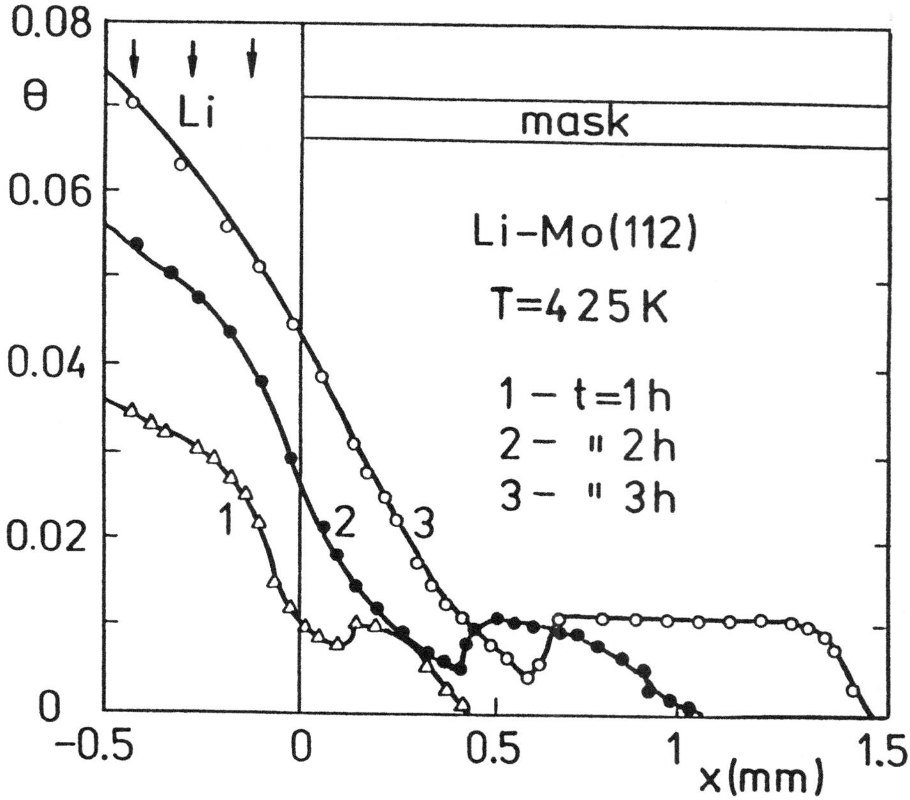

Fig. 7. Evolution of concentration profiles during continuous evaporation of Li on a part of Mo(112) crystal. The flux of Li atoms is $1.4 \cdot 10^{10} cm^{-2} s^{-1}$.

of a 2D lithium phase with large (up to $\sim$ 30Å in the coverage) interatomic distances. The surface like Mo(112) are predicted to provide an especially long-range substrate–mediated lateral interaction ($U \sim r^{-1}$) along the channels. This is due to the circumstance that the Fermi surface has some flat areas oriented normally to the channel axis [13]. Another effect which could ensure the long-range interaction and stabilize the rarefied phase is connected with mesoscopic self-organization of adlayers observed recently [22]. The driving force for such a self-organization can stem both from elastic strain [23] and electrostatic interaction [23, 25].

As to the concentration hollow (fig. 7), it must signal on some peculiarity in the chemical potential which is able to produce a region of uphill diffusion. Since this region is displaced along the surface while its shape is conserved, there must be some factor stabilizing it. Its origin (electrostatic? elastic strain?) is so far unclear. It cannot be ruled out that this concentration feature should be considered in terms of spatio-temporal dissipative structures [12, 26-28], since the processes in the diffusion zone seem to be strongly nonlinear and nonequilibrium.

The above suggestion are of course only hypotetical. It is definitely necessary to accumulate more experimental data and first of all to provide a better spatial resolution in the surface diffusion measurements.

3 Conclusion

The experimental results discussed in this brief review were selected to demonstrate that interaction of diffusing particles makes surface diffusion an intricate phenomenon. Due to difference in the diffusion rates inherent to various adsorbate phases there occurs a competition between them in the diffusion zone, the largest area on the surface being occupied by the phase with the highest spreading rate. Thus one may speak of a self-organization of the diffusion zone in which a dynamic phase portrait of the spreading layer is developed. The strongly nonlinear properties of adlayers can bring about some interesting effects.

It seems safe to predict that further experimental and theoretical studies of surface diffusion will substantially contribute both to basic understanding of diffusion processes and to practical applications in thin films technologies, catalysis and other fields.

Acknowledgements

We are grateful to Professor A. Pękalski for the invitation to prepare this paper. Support by the State Committee for Science and Technologies of Ukraine and from the Batory Foundation are also gratefully acknowledged.

References

[1] I. Lyuksyutov, A. G. Naumovets, and V. L. Pokrovsky, *Two-Dimmensional Crystals,* Academic Press, Boston, 1992.

[2] J. Philibert, *Atom movements. Diffusion and mass transport in solid,* Les Éditions de Physique, Les Ulis, 1991.

[3] T. T. Tsong, Surf. Sci. Rep. **8**, 127 (1988).

[4] G. Ehrlich, Surf. Sci. **246**, 1 (1991).

[5] R. Gomer, Rep. Progr. Phys. **53**, 917 (1990).

[6] J. Bęben, Ch. Kleint and R. Męclewski, Surf. Sci. Rep. **213**, 224 (1989).

[7] A. G. Naumovets and Yu. S. Vedula, Surf. Sci. Rep. **4**, 365 (1985).

[8] R. Butz and H. Wagner, Surf. Sci. **87**, 69 and 75 (1977).

[9] M. Browker and D. A. King, Surf. Sci. **78**, 208 (1978).

[10] J. De Conink, N. Fraysse, M. P. Valignat and A. M. Cazabat, Langmuir **9**, 1906 (1993).

[11] M. Ondřejček, W. Stenzel, H. Conrad, V. Cháb, Z. Chvoj, W. Engel and A. M. Bradshaw, Chem. Phys. Lett. **215**, 528 (1993).

[12] W. Święch, C. S. Rostomjec, R. Imbihl, J. W. Evans, B. Rausenberger, W. Engel, A. K. Schmid, A. M. Bradshaw and E. Zeitler, Surf. Sci. **307–309**, 138 (1994).

[13] O. M. Braun and V. K. Medvedev, Sov. Phys. Uspekhi **32**, 327 (1989).

[14] A. G. Naumovets, M. V. Poliy and Yu. S. Vedula, Phys. Rev. Lett. **71**, 105 (1993).
[15] Yu. S. Vedula, A. T. Loburets ans A. G. Naumovets, Sov. Phys. JETP **50**, 391 (1980).
[16] E. V. Klimenko, E. M. Litvinova, I. F. Lyuksyutov, A. G. Naumovets and I. N. Zasimovich, Surf. Sci. **271**, 244 (1992).
[17] I. F. Lyuksyutov and V. L. Pokrovsky, Sov. Phys JETP Lett. **33**, 326 (1981).
[18] A. Natori and H. Otsubo, Surf. Sci. **171**, 13 (1986); **184**, 289 (1987).
[19] H. Taitelbaum, Y. E. Koo, S. Havlin, R. Kopelman and G. H. Weiss, Phys. Rev. A **46**, 2151 (1992).
[20] M. S. Gupalo, V. K. Medvedev, B. M. Polyukh and T. P. Smereka, Sov. Phys. Sol. State **21**, 568 (1979).
[21] R. Gomer, R. Wortman, and R. Lundy, J. Chem. Phys. **26**, 1147 (1957).
[22] K. Kern, H. Niehus, A. Schatz, P. Zeppenfeld, J. George and G. Comsa, Phys. Rev. Lett. **67**, 855 (1991).
[23] V. I. Macharenko, Sov. Phys. JETP Letters, **33**, 381 (1981).
[24] D. Andelman, F. Brochard, P.-G. de Gennes et J.-F. Joanny, C. R. Acad. Sc. Paris **301**, 675 (1985).
[25] D. Vanderbilt, Surf. Sci. **268**, L300 (1992).
[26] G. Ertl, Surf. Sci. **287 / 288**, 1 (1993).
[27] J. H. Block, M. E. Ehsasi and V. V. Grodetskii, Progr. Surf. Sci. **42**, 143 (1993).
[28] R.Imbihl, Progr. Surf. Sci. **44**, 185 (1993).

Simple Quantum Dynamics in Presence of Topological Disorder

Andrzej Lusakowski[1] *and Lukasz A. Turski*[2]

[1] Institute of Physics, Polish Academy of Sciences
Al. Lotników 32/46, 02-668 Warszawa, Poland
[2] Center for Theoretical Physics, Polish Academy of Sciences
Al. Lotników 32/46, 02–668 Warszawa, Poland

Abstract: The theory of particles moving in topologically random disordered environment is just in its infancy. Following our previous work on classical random walk in the presence of topological disorder due to randomly distributed line defects, we analyze the problem of a quantum, spinless, charged particle moving on a topologically disordered lattice. Specifically we discuss the motion on a d=2 square lattice penetrated by an array of randomly distributed Bohm–Aharonov magnetic fluxes. Next, we discuss the situation in which particle moves over a d=3 lattice containing random array of screw-dislocation dipoles. We show that the particle dynamics becomes highly anisotropic and that the underlying dynamics can be approximated by that of a quantum particle with anisotropic effective mass lighter in the direction towards which the dislocation dipoles are aligned.

1 Introduction

One of the fascinating problems in the mathematical theory of random processes is the so called Bochner problem [1] in which a marble is rolled, without the slip, down the plane along the random walk trajectory, and the questions are posed as to the statistical properties of the trace of that trajectory on marble surface. Bochner problem focuses our attention on necessity of developing *statistical differential geometry*, and the theory of *stochastic processes on differentiable manifolds* in particular. Potential applications of these theories are diverse and range from image processing problems to general relativity and cosmology. In this work we shall report on our attempts to analyze quantum theory for a single, charged and spinless, particle moving in various topologically disordered environments. We repeat our analysis of random Bohm-Aharonov problem and outline our recent analysis of the dislocation dipole lattices.

It is often said that the Bohm-Aharonov effect has no classical analogy, and that topological properties of a manifold on which the particle moves are of no real importance in classical physics. We believe that this is not entirely true and that classical dynamics in the presence of topological defects exhibits some features akin to the Bohm–Aharonov effect. Indeed, elsewhere [2] we have have shown that the simple random walk [4] executed by a particle in a crystal containing random array of straight, parallel to each other screw disclinations shows essentially different properties than that on a empty lattice.

Since the diffusion process involves large traveling distances of a particle it is then convenient to use the continuum theory of defects initiated by Kröner[5, 6]. This theory emerges from a lattice description in the limit of vanishing lattice spacing where however local angular directions are retained. The natural mathematical framework for such a theory is the differential geometry. In Ref.[2] we have shown that the proper description of random walk on the differential manifold containing topological defects, like dislocations and disclinations is via the Fokker–Planck equation for the probability distribution $P(\mathbf{x}, t)$. Using systematically the apparatus of the path integral approach, we were able to show how the diffusion changes *qualitatively* in the presence of line defects.

When the defect ensemble is assumed to be Gaussian with zero mean, and a translational-invariant second moment which includes the possibility of screening of the topological charges [2], we obtain the root mean square particle displacement behavior as:

$$\overline{\langle x^2 \rangle}(t) \sim t^{1-\gamma/(8\pi m^2)} \quad , \tag{1}$$

where m is the screening length, and γ is the strength of the defects correlation function second moment.

When the topological defects are not screened then above, subdiffusive behavior changes dramatically, and we obtain:

$$\overline{\langle x^2 \rangle}(t) \sim (\ln t)^2 \quad . \tag{2}$$

The above result, sometimes referred to as Sinay diffusion has been well understood for a one dimensional models [7]. We believe that our result is the first more than d=1 example of the Sinay diffusion. Referring the reader to the Ref.[2] for details of the classical theory we turn now to description of quantum propagation.

2 Quantum Particle Moving in Random Flux Field

In this section we shall analyze the motion of a quantum particle influenced by a particular type of the topological disorder namely the random magnetic fluxes system. The model discussed below [8], originated as an obvious generalization of a conventional the Bohm–Aharonov effect [9]. The question is how this effect will look like in the presence of many flux line and when they become randomly distributed.

The problem of a quantum particle moving in the random potential field has been subject of extensive experimental and theoretical investigation [10].

In contrast, very little is known about the behavior of the quantum particle in presence of a random magnetic field. Recently Pryor and Zee have analyzed numerically motion of a spinless quantum particle in the random magnetic flux arrangement [11]. As we shall see in what follows, their results do agree quite well with our analytic theory.

Imagine two – dimensional square lattice (see *Fig. 1*) and assume that each plaquette is penetrated by a magnetic flux $\phi(x + a/2, y + a/2)$.

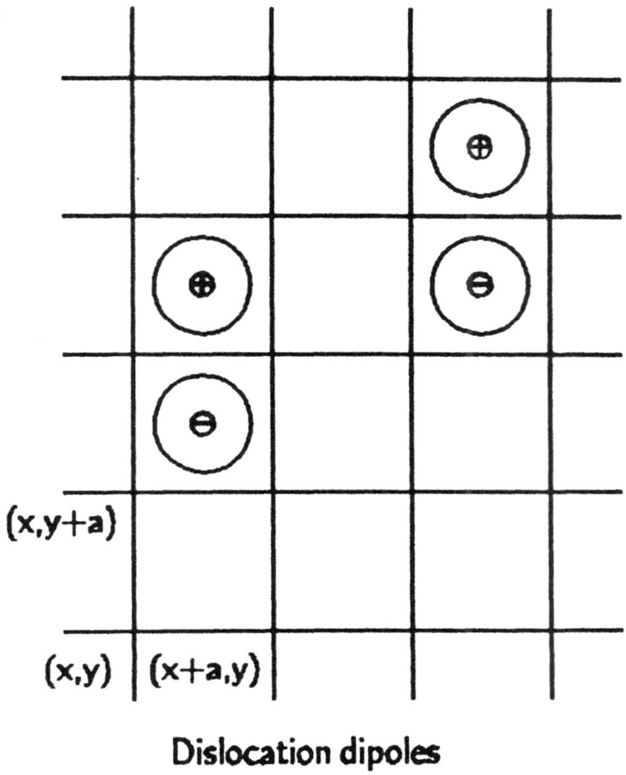

Fig. 1. Magnetic Flux Lines

The fluxes in different plaquettes are assumed to be independent and homogeneously distributed over the interval $(0, 2\pi)$. (The values of fluxes are measured in units of elementary flux h/e). Furthermore, assume that a single, spinless, quantum particle move over this lattice, and that its motion is governed by the tight binding Hamiltonian, which we write in the form

$$\mathcal{H}_{x,y;x'y'} = -K_x \left(\delta_{x,x'+a} e^{iaA_x(x-a/2,y)} + \delta_{x,x'-a} e^{-iaA_x(x+a/2,y)}\right) \delta_{y,y'} - K_y \left(\delta_{y,y'+a} e^{iaA_y(x,y-a/2)} + \delta_{y,y'-a} e^{-iaA_y(x,y+a/2)}\right) \delta_{x,x'} \,, \qquad (3)$$

where K_x and K_y are the coupling constants along x and y directions of the lattice, respectively. A_x and A_y are the components of the magnetic field vector potential,which, as in the lattice gauge theory, are defined on the lattice bonds.

The main difference, however, between the present problem and that of particle moving in random potential rests on later *locality*. Indeed, the particle needs to move just one lattice constant to experience changes induced by random scalar potential. In our case of random magnetic fluxes the particle must traverse a closed loop around the region containing the flux to notice its presence at all. This non-locality, known from the theory of the Aharonov-Bohm effect [9], results in technical difficulties in the analysis to follow.

If the magnetic fluxes were distributed periodically our model will be similar to that analyzed theoretically by Rammer and Shelankov [12] and experimentally by Bending *et al.* [13]. We believe that experimental arrangements of fluxes analyzed in our paper can be achieved using one of the new high T_c materials in which fluxes might form random arrangements in contrast to usual superconductors in which fluxes form periodic lattices.

In this work we shall be interested in the averaged density of states $\langle \varrho(\omega) \rangle$, where $\langle \cdots \rangle$ denotes the average over the flux distribution. $\varrho(\omega)$ is given by the difference of the retarded and advanced propagators

$$\varrho(\omega) = i \sum_{\mathbf{r}} \left(G^R(\mathbf{r}, \mathbf{r}'; \omega) - G^A(\mathbf{r}, \mathbf{r}'; \omega) \right) \quad , \tag{4}$$

where

$$G^{R/A}(\mathbf{r}, \mathbf{r}'; \omega) = \sum_i \frac{\phi_i(\mathbf{r}) \phi_i^*(\mathbf{r}')}{\omega - \varepsilon_i \pm i\eta} , \tag{5}$$

and, $\phi_i(\mathbf{r})$ and ϵ_i are normalized eigenfunctions and eigenenergies of the one particle Hamiltonian H.

In order to calculate the above quantity we adopted here generating functional approach developed for the random potential problem by Bausch and Leschke [14]. To this end let us define two "vectors"

$$\Phi(\mathbf{r}, \omega) = \begin{pmatrix} \psi(\mathbf{r}, \omega) \\ \hat{\psi}(\mathbf{r}, \omega) \end{pmatrix} \tag{6}$$

$$\tilde{\Phi}(\mathbf{r}, \omega) = \left(\hat{\psi}^*(\mathbf{r}, \omega), -\psi^*(\mathbf{r}, \omega) \right) \tag{7}$$

where ψ and $\hat{\psi}$ are complex fields, and the star denotes complex conjugation. The generating functional $Z\{\hat{\ell}\}$ depends on the matrix source field $\hat{\ell}$ and has the following form:

$$Z\{\hat{\ell}\} = \int D\Phi D\tilde{\Phi} exp(J + S) \quad , \tag{8}$$

where

$$J = \frac{1}{2} \int \frac{d\omega}{2\pi} \sum_{\mathbf{r}} \Bigg[\tilde{\Phi}(\mathbf{r},\omega) \begin{pmatrix} \omega + i\eta & 0 \\ 0 & \omega - i\eta \end{pmatrix} \Phi(\mathbf{r},\omega)$$
$$+ K_x e^{iaA_x(x+a/2,y)} \tilde{\Phi}(x,y;\omega) \cdot \Phi(x+a,y;\omega)$$
$$+ K_x e^{-iaA_x(x+a/2,y)} \tilde{\Phi}(x+a,y;\omega) \cdot \Phi(x,y;\omega)$$
$$+ K_y e^{iaA_y(x,y+a/2)} \tilde{\Phi}(x,y;\omega) \cdot \Phi(x,y+a;\omega)$$
$$+ K_y e^{-iaA_y(x,y+a/2)} \tilde{\Phi}(x,y+a;\omega) \cdot \Phi(x,y;\omega) \Bigg] \ . \qquad (9)$$

The functional S, the so-called source term, is given by:

$$S = \frac{1}{2} \int \frac{d\omega}{2\pi} \frac{d\omega'}{2\pi} \sum_{\mathbf{r}} \sum_{\alpha,\beta} \tilde{\Phi}_\alpha(\mathbf{r},\omega)\, l_{\alpha\beta}(\mathbf{r};\omega,\omega')\, \Phi_\beta(\mathbf{r},\omega') \ . \qquad (10)$$

The virtue of the generating functional approach is that by differentiating $Z\{\hat{\ell}\}$ with respect to source field $l_{\alpha\beta}$ we can obtain the relevant quantity. For example the functional derivative of $Z\{\hat{\ell}\}$ with respect to $\ell_{11}(\mathbf{r};\omega,\omega')$ at $\hat{\ell}=0$ equals $-2\pi\delta(\omega-\omega')G^R(\mathbf{r},\omega)$.

Now, since $Z\{\hat{\ell}=\mathbf{0}\}=\mathbf{1}$ we may perform the averaging over the (quenched) disorder *before* attempting to calculate either of the propagators. In process of evaluation of the mean value of $Z\{\hat{\ell}\}$ one encounters the problem of averaging expressions like

$$\exp\left(\sum_{\mathbf{r}} [U(\mathbf{r})\chi_1(\mathbf{r}) + U^*(\mathbf{r})\chi_2(\mathbf{r}) + V(\mathbf{r})\chi_3(\mathbf{r}) + V^*(\mathbf{r})\chi_4(\mathbf{r})] \right) \ , \qquad (11)$$

with $U(\mathbf{r}) = \exp\{iaA_y(x,y+a/2)\}$ and $V(\mathbf{r}) = \exp\{iaA_x(x+a/2,y)\}$, and $\chi_i(\mathbf{r})$, i $=1\ldots4$ being arbitrary functions of $\mathbf{r}$.

For an arbitrary type of magnetic disorder, the calculation of those averages become formidable mathematical task. In the case of magnetic disorder we have chosen, the average is conveniently done in the Landau gauge :

$$A_x = 0$$
$$A_y(x,y+a/2) = \sum_{x_1=-\infty}^{x-a} \phi(x_1+a/2,y+a/2) \ , \qquad (12)$$

what implies that U-factors, for different values of y coordinate, are statistically independent. Rather tedious analysis presented in Ref.[8] permits us to express the effective functional $J_1 = \ln\langle \exp J \rangle$ as

$$J_1 = \frac{1}{2} \int \frac{d\omega}{2\pi} \sum_{\mathbf{r}} \Bigg[\tilde{\Phi}(\mathbf{r},\omega) \begin{pmatrix} \omega + i\eta & 0 \\ 0 & \omega - i\eta \end{pmatrix} \Phi(\mathbf{r},\omega) +$$

$$K_x\left(\tilde{\Phi}(x,y;\omega)\Phi(x+a,y;\omega)+\tilde{\Phi}(x+a,y;\omega)\Phi(x,y;\omega)\right)\Bigg]+$$
$$\sum_{\mathbf{r}}\ln I_0\left(2K_y\sqrt{TrR(x,y)\,R(x,y+a)}\right)\quad, \tag{13}$$

where we have introduced the matrix $\hat{R}$

$$R_{\alpha\beta}(\mathbf{r};\omega;\omega')=\frac{1}{2}\Phi_\alpha(\mathbf{r};\omega)\tilde{\Phi}_\beta(\mathbf{r};\omega')\quad. \tag{14}$$

Here and in what follows Tr denotes usual matrix trace and the integration over ω's.

Notice that the term with $\ln I_0$ produces infinite number of vertices. This should be compared with the random potential case, where after averaging we get only one vertex of the fourth order in the fields ψ and $\hat{\psi}$. This is the direct manifestation of the non-locality of the present problem.

The averaged value of the Z is then

$$\langle Z\{\hat{\ell}\}\rangle=\int D\Phi D\tilde{\Phi}\exp\{J_1+S\} \tag{15}$$

where the source term is written as $S=\sum_{\mathbf{r}}Tr\hat{\ell}(\mathbf{r})\hat{R}(\mathbf{r})$.

The mean filed analysis of the averaged path integral Eq. (15) has been presented in Ref.[8]. Here we present our main results.

Since our physical system is spatially homogeneous we are looking for the $R_{\alpha\beta}$ function in the form $R^0_{\alpha\beta}(\mathbf{r},\omega,\omega')=2\pi\delta(\omega-\omega')\delta_{\alpha\beta}R_\alpha(\omega)$ This assumption permits us to evaluate exactly the imaginary part of $R_\alpha(\omega)$ which in turns prowide us with the averaged density of states $\varrho(\omega)$. In Fig. 2 we display $\varrho(\omega)$ for three different values of α.

Now, the main result is the narrowing of the band. In the case of the free particle the allowed energies extend between $-2K_x-2K_y$ and $2K_x+2K_y$. In the presence of magnetic disorder the band is shrinking ($\omega_0<2K_x+2K_y$). This is in contrast with the potential fluctuation case where we observe the tails of the averaged density of states in the energy range forbidden for the free particle.

For the symmetric lattice, $K_y=K_x$ we may compare our result with those of Pryor and Zee [11]. In Ref.[11] the motion of a quantum particles was analyzed on a finite lattice in the presence of the magnetic flux disorder analogous to that discussed in our paper. The calculated quantity was $p(E)$, defined as the probability density of finding the state with its energy between E and $E+dE$. The main results in Ref.[11] was the narrowing of the band. States with eigenenergies $3.4K_y<|E|<4K_y$ were found to be extremely improbable. This result coincides with our band shrinking (lowest curve in Fig. 1) obtained within the mean field approximation. Pryor and Zee [11] considered also a different type of magnetic disorder than those discussed so far. They assumed that each plaquette of the lattice may be penetrated by the magnetic flux which takes two values - 0 or π only, and with equal probability. Our analysis can be directly applied

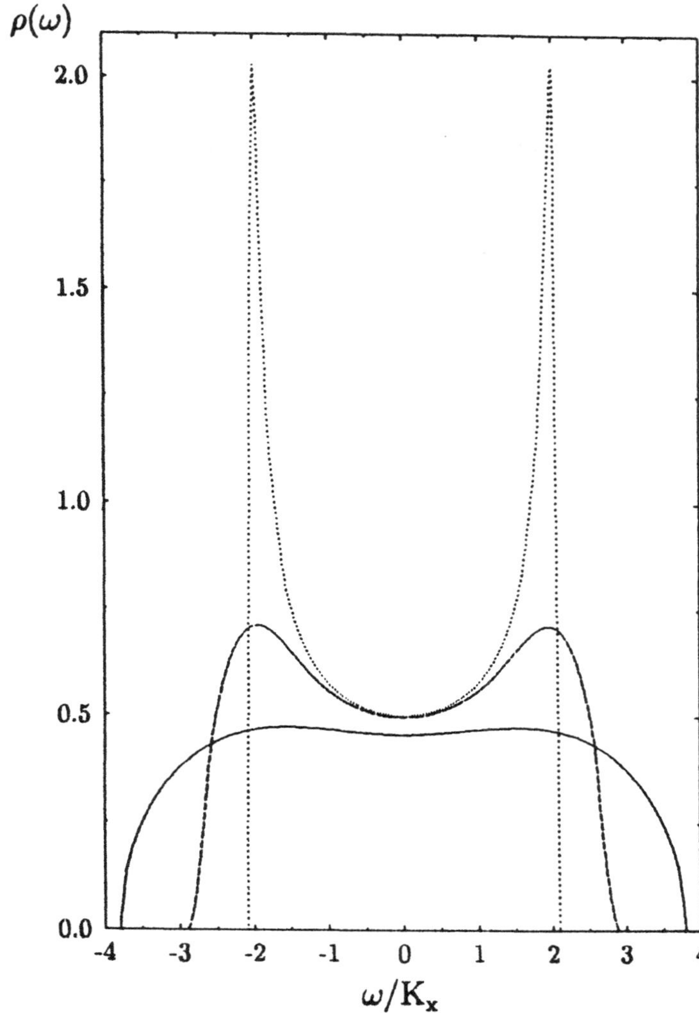

Fig. 2. Density of States plotted for various values of the ratio $\alpha = K_x/K_y$. Upper line $\alpha = 0.1$, heavy line $\alpha = 0.5$. Bottom line $\alpha = 1$.

also to that case by averaging the generating functional $Z\{\hat{\ell}\}$ over such magnetic disorder. In order to do so we have to modifiy our saddle point method, as outlined in Ref.[8], but finally we obtain the same density of states as previously. It is tempting to interpret that results as a mean field indicator of some sort of statistical universality.

3 Dislocation Dipoles

Real crystalline materials are far from being ideal. Among possible crystalline media defects, those carrying topological charges, like dislocations and disclinations are of particular interest. [5, 6, 15, 16, 17] Indeed, screw dislocations and disclinations might serve as kind of elastic field analogues of the Bohm–Aharonov fluxes. In the previous Section we have discussed the motion of a quantum particle hopping on a two-dimensional lattice which cells contained randomly distributed magnetic field fluxes. In this Section we would like to approach a model which is more general for it indeed deals with fully three dimensional case, and instead of random magnetic fluxes uses more realistic model of elastic screw dislocation array.

The problem of a quantum particle interaction with a topological defects was studied before [18], where the influence of electorn scattering off the dislocations on the electric and thermal conductivity was analyzed. This analysis was based on the Boltzmann equation, thus all the important quantum interference efects were lost. As our work shows these effects are of predominant importance.

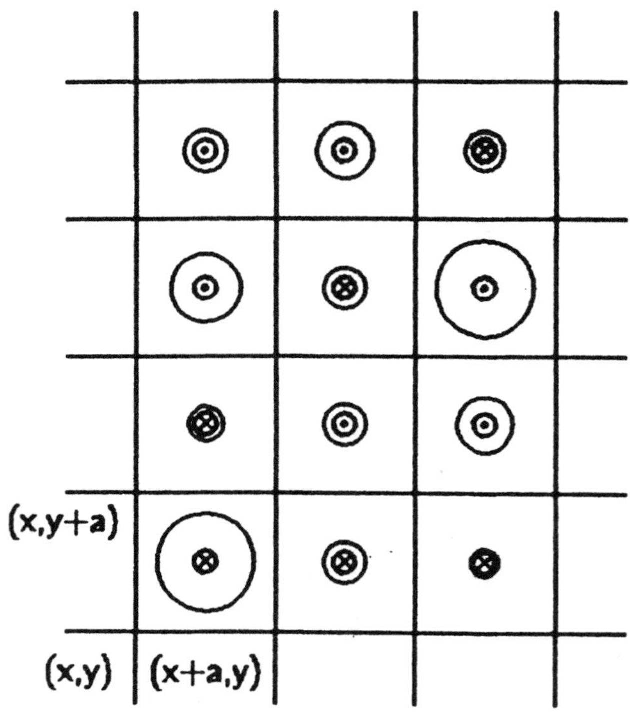

Two - dimensional lattice with

fluxes penetrating its cells.

Fig. 3. Screw Dislocation Dipoles Array

To begin with consider now a two-dimensional slice of a simple cubic lattice as shown in Fig. 3. Consider now two screw dislocations parallel to the z axis of the lattice, piercing the slice at the location $(x-a/2, y+a/2)$ and $(x+a/2, y+a/2)$ where a is the lattice spacing. The Burgers vectors of these dislocations having the same length b are pointing in opposite directions. We shall refer to such an arrangements of a *two* dislocation lines as *dislocation dipole*, or as d-dipole, located on the bond joining lattice positions $(x,y) \rightarrow (x, y+a)$. Analogously we can talk about d-dipoles located on the bond joining $(x,y) \rightarrow (x+a,y)$. If the bond $(x,y) \rightarrow (x,y+a)$ is occupied then this bond connects lattice sites with different z-components, namely $(x,y,z) \rightarrow (x, y+a, z+b)$. It is therefore natural to introduce the Burgers function $\mathcal{A}(x,y)$ such that, if $\mathcal{A}(x, y+a/2) = 0$ the bond joins lattice sites $(x,y,z) \rightarrow (x,y+a,z)$ for each z. If $\mathcal{A}(x,y+a/2) = \pm b$ then the bond joins lattice sites $(x,y,z) \rightarrow (x, y+a, z \pm b)$.

When the lattice is populated with the dislocation dipoles, we introduce the randomness by assumption that, for each point (x,y)

$$\begin{aligned} \mathcal{A}(x,y+a/2) &= 0 \text{ with probability } 1-2p \\ \mathcal{A}(x,y+a/2) &= b \text{ with probability } p \\ \mathcal{A}(x,y+a/2) &= -b \text{ with probability } p \end{aligned} \tag{16}$$

and similarly for bonds parallel to x-axis.

The motion of a quantum, spinless particle moving on the lattice populated by such a d-dipoles is described in the tight–binding approximation, and write its (discrete) Hamiltonian as

$$\begin{aligned} \mathcal{H}(\mathbf{r},\mathbf{r}') = &- \gamma \left(\delta_{x,x'+a}\delta_{z,z'+\mathcal{A}(x-a/2,y)} + \delta_{x,x'-a}\delta_{z,z'-\mathcal{A}(x+a/2,y)}\right)\delta_{y,y'} \\ &- \gamma \left(\delta_{y,y'+a}\delta_{z,z'+\mathcal{A}(x,y-a/2)} + \delta_{y,y'-a}\delta_{z,z'-\mathcal{A}(x,y+a/2)}\right)\delta_{x,x'} \\ &- \gamma \left(\delta_{z,z'+a} + \delta_{z,z'-a}\right)\delta_{x,x'}\delta_{y,y'} \end{aligned} \tag{17}$$

Now, the main quantity we are interested in, is the single particle Green function $\mathcal{G}(\mathbf{r},\mathbf{r}',t)$. This function satisfies the Schrödinger equation , which we found convenient to write immediately in its time–Fourier transform form

$$\sum_{\mathbf{r}'} \left[(\omega + i\eta)\delta_{\mathbf{r},\mathbf{r}'} - \mathcal{H}(\mathbf{r},\mathbf{r}')\right] \mathcal{G}(\omega,\mathbf{r}',\mathbf{r}'') = \delta_{\mathbf{r},\mathbf{r}''} \tag{18}$$

Following its construction the d-dipoles array possess full lattice translational symmetry along the z-axis. We write then the Fourier transform of relevant functions as:

$$\mathcal{H}(\mathbf{r},\mathbf{r}') = a\int_{-\pi/a}^{\pi/a} \frac{dk}{2\pi} \exp(ik(z-z'))\mathcal{H}_k(\rho,\rho') \ , \tag{19}$$

where ρ denotes a two dimensional space vector. Similar expression holds for the Green function $\mathcal{G}$.

Now, the Fourier transform of the Hamiltonian Eq(17) can be split naturally into its bare, $\mathcal{H}_k^0(\rho,\rho')$ and perturbed $\mathcal{H}_k^1(\rho,\rho')$ parts. We have

$$\begin{aligned} \mathcal{H}_k^0(\rho,\rho') = &- \gamma \left(\delta_{x,x'+a} + \delta_{x,x'-a}\right)\delta_{y,y'} \\ &- \gamma \left(\delta_{y,y'+a} + \delta_{y,y'-a}\right)\delta_{x,x'} - 2\gamma\cos(ka) \ , \end{aligned} \tag{20}$$

and

$$\mathcal{H}_k^1(\rho,\rho') =$$
$$= -\gamma\left[\delta_{x,x'+a}\left(e^{-ikA(x-a/2,y)}-1\right)+\delta_{x,x'-a}\left(e^{ikA(x+a/2,y)}-1\right)\right]\delta_{y,y'}$$
$$-\gamma\left[\delta_{y,y'+a}\left(e^{-ikA(x,y-a/2)}-1\right)+\delta_{y,y'-a}\left(e^{ikA(x,y+a/2)}-1\right)\right]\delta_{x,x'} \quad (21)$$

One should compare this Hamiltonian with that in Eq.(3).

Having done so, we can follow the standard perturbation theory and write the Dyson equation for the Green function $\mathcal{G}(\omega,\mathbf{k},\rho,\rho')$. The perturbation series can then be averaged over the d-dipoles ensemble. In the final expression for the mass operator we shall retain terms which are linear in the probability p. Doing this we encounter situation somwhat analogous to already discussed in Section 2. Indeed the quantity analogous to $U(\mathbf{r})$ and $(V(\mathbf{r})$ from Eq.(11) are now :

$$\begin{aligned} U_1(x,y+a/2) &\equiv \exp(-ikA(x,y+a/2))-1\\ V_1(x+a/2,y) &\equiv \exp(+ikA(x+a/2,y))-1\\ U_{-1}(x,y+a/2) &= U_1^*\\ V_{-1}(x+a/2,y) &= V_1^* \ . \end{aligned} \quad (22)$$

The averaging procedure, for the perturbation series, details of which will be published elsewhere [19], is straightforward although quite tedious. Following all the steps we obtain the explicit expression for the Fourier transform of the mass operator $\hat{\Sigma}$

$$\Sigma(\omega,k,k_x,k_y) =$$
$$= 8\gamma p\sin^2\left(\frac{kb}{2}\right)\frac{[1-2\gamma G_0(a)][\cos(k_xa)+\cos(k_ya)]+4\gamma G_0(0)}{1+4\sin^2(kb/2)[\gamma^2G_0^2(a)-\gamma G_0(a)-\gamma^2G_0^2(0)]}\ , \quad (23)$$

where $G_0(a)$ is the the Green function for unperturbed Hamiltonian equals to $G_0(\omega k;x,y;x,y+a)$ related to the value $G_0(0)$ of the same function via the relation $1-2\gamma G_0(a) = 1/2+(1/2)(\omega+2\gamma\cos(ka))G_0(0)$.

This complicated expressions simplifies considerably in the long wavelength limit ($ka \ll 1$) and close to the zone bottom, i.e. $\omega \approx -6\gamma$. We obtain then

$$\Sigma(\omega,k,k_x,k_y) \approx 2p\gamma a^2\left(\frac{b}{a}\right)^2 k^2\ . \quad (24)$$

Since in the long wavelength limit $\gamma a^2 = 1/2m$ where m is the particle mass, we obtain from Eq.(24) that the particle mass becomes anisotropic assuming bare value in both directions transverse to the dislocations, and along the d-dipole direction equals to

$$\frac{1}{2m_z^*} = \frac{1}{2m}\left(1+2p\left(\frac{b}{a}\right)^2\right)\ . \quad (25)$$

Notice, that Eq.(25) predicts that particle is *lighter* along the dislocation lines direction, the result which can be easily understood on physical grounds, and which clearly reflects the topological property of the screw dislocation. One of the remarkable features of our result is that, within our approximation, *i.e.* in the long wavelength limit and with accuracy up to linear terms in p the mass operator Eq.(24) is purely real. That implies that the our quasiparticle (the electron moving in the d-dipole field) has infinite life time.

The above outlined theory of quantum motion in the random d-dipole field can be extended for dislocation dipoles with the variable dipole moment *i.e* for d-dipoles with the length $L = (2\ell + 1)a$, $\ell = 0, 1, 2, \cdots$. Working again in the long wavelength limit, $kb \ll 1$ we can now derive also the expression for the mass operator in the large ℓ limit when still $\ell q_0 a \ll 1$. Here q_0 is the wave vector in the direction transverse to the dislocation lines. Again, we observe reduction of the particle mass along the preferred direction, which is slowly varying function of the d-dipole length ℓ.

$$\frac{1}{2m_z^*} = \frac{1}{2m}\left(1 + \frac{4p}{\pi}\left(\frac{b}{a}\right)^2 \ln(2\ell + 1)\right) . \tag{26}$$

4 Conclusions

In the previous Section we have outlined the theory of random dislocation dipoles induced modification in the quantum propagation of a particle. We have shown that in the continuum limit, *i.e* with accuracy up to $\mathbf{k}^2$, the presence of the topological disorder results only in the change of the effective particle mass. The imaginary part of the mass operator $\hat{\Sigma}$ is zero. The change in the effective mass results in small modification of the density of states. As compare to the d=3 free particle case the only change is in the slope of the density of states, which becomes smaller in d-dipole case, reflecting lower value of the particle effective mass.

Note that the analysis in Section 3 was not self consistent and we expect to report on the self consistent analysis, similar to that presented for magnetic flux case in Section 2 shortly.

The two models presented above represents just the "tip of an iceberg" of problems in which topological defects may totally alter the dynamical properties of the system. The analysis presented here is far from being complete and this talk was intended rather as introduction to the class of problems rather than a presentation of advanced theory. Both quantum models discussed in the paper are of considerable applied interest. In order to appreciate the physical significance of the predicted band narrowing we should analyze two points Green functions for our particle. Only then we would be able to compare the case of random fluxes to that discussed in [12]. Whether the departure form $|\mathbf{B}|$ (periodic fluxes) or the usual $\mathbf{B}^2$ (homogeneous field) [20] behavior of the magneto–resistivity will be observed also in that case remains an open question. Some work along this line is now in progress.

Acknowledgments

Financial help of the German DFG-SFB Nr.237 extended to LAT in various stages of this paper preparation is kindly acknowledge. This work was also supported in part by the Polish State Science Council (KBN) grant 2P 30203804.

References

[1] N. Ikeda and S. Watanabe, *Stochastic Differential Equations and Diffusion Processes* North–Holland, Amsterdam (1981). Chapt.V, Sec.4.

[2] R. Bausch, R. Schmitz, and L.A. Turski, submitted for publication.

[4] N.G. van Kampen, *Stochastic Processes in Physics and Chemistry,* North–Holland, Amsterdam, (1981).

[5] E. Kröner, *Kontinuumstheorie der Versetzungen und Eigenspannungen*, Springer Verlag, Berlin (1958). For recent review of Kröner approach c.f. E. Kröner, in *Physics of Defects, Les Houches Session XXXV* edited by R. Ballian, North-Holland, Amsterdam (1981).

[6] M. Żórawski, *Theorie Mathématique des Dislocations*, Dunod, Paris (1967).

[7] T. Schneider, A. Politi and R. Badii, Phys. Rev. **A 34**, 2505 (1986).

[8] A. Lusakowski and L.A. Turski, Phys. Rev. **B 48**, 3835, 1993.

[9] For comprehensive review cf. M. Peshkin, Phys. Reports. **80**, 375 (1981). Elementary exposition of the Bohm-Aharonov effect is given in I. Białynicki-Birula, M. Cieplak and J. Kamiński, *Theory of Quanta* Oxford University press, Oxford (1992).

[10] *c.f* F. Wegner Phys. Rep.**67**, p. 15, (1980); L. Schaefer, F. Wegner Z. Phys. **B 38**, 113, (1980); G. Bergmann Phys. Rep. **107**, p. 1, (1984); *Localization, Interaction and Transport Phenomena* edited by B. Kramer, G. Bergmann, Y. Bruynserade, Springer Series in Solid State Sciences 61 (Springer Verlag, Berlin 1985); Lee P. A., Ramakrishnan T. V., Rev. Mod. Phys. **57**, 287, (1985); S. Chakravarty, A. Schmid Phys. Rep. **140**, 193 (1986).

[11] Craig Pryor, A. Zee, Phys. Rev. **B 46**, 3116 (1992).

[12] J. Rammer, A. L. Shelankov Phys. Rev. **B 36**, 3155 (1987).

[13] S. J. Bending, K. v. Klitzing, K. Ploog, Phys. Rev. Lett. **65**, 1060, (1990).

[14] R. Bausch, H. Leschke Zeit. f. Phys.**B 53**, p. 249, (1983).

[15] A. Kosevitch, *Foundations of the Crystalline Lattice Mechanics*, Science, Moscow (1972) (in russian).

[16] L.A. Turski, Bull. Polon. Acad. Sci. Ser.IV (Mechanics) **14**, 281 (1966) and in *Methods of Geometry in Physics and Technology*, edited by P. Kucharczyk, WNT, Warsaw (1966).

[17] H. Kleinert, *Gauge Fields in Condensed Matter*, World Scientific, Singapore (1989).

[18] K. Kawamura, Z. für Physik **B 29**, 101 (1978), ibid **B 30**, 1 (1978).

[19] A. Lusakowski, L.A. Turski, to be published.

[20] B.L. Altshuler, A.G. Aronov, D.E. Khmelnitski, and A.I. Larkin, in *Quantum Theory of Solids* edited by I.M. Lifshitz, (MIR Publishers, Moscow 1982).

Initially-separated reaction-diffusion systems: theory and experiment

Haim Taitelbaum

Department of Physics, Bar-Ilan University, Ramat-Gan 52900, ISRAEL

1 Introduction

The $A + B \rightarrow C$ reaction-diffusion process with initially separated components is more readily amenable to experimental studies than similar non-convective systems with initially uniformly mixed reactants, a condition which is difficult to achieve in a real chemical system. The initial separation of reactants leads to formation of a dynamic reaction front, which is a spatially localized region with non-zero reactant production, that can be visualized and measured. The presence of such a reaction interface is a characteristic of many processes in nature [1-12], and has recently attracted much research interest [13-34].

When the reaction constant is vanishingly small, the species will mix before reacting. This leads to an initial behavior, very rich in detail, prior to a crossover to the asymptotic behavior. The crossovers for a variety of characteristics of the reaction front have been found to depend on the system parameters, such as the microscopic reaction constant, (which can therefore be determined from macroscopic experiments), as well as the diffusion coefficients and the initial densities of the two species.

In this paper we review our recent theoretical and experimental studies on initially-separated reaction-diffusion systems, and describe the novel kinetic properties of the reaction front.

The following set of mean-field-type reaction-diffusion equations for the local concentrations ρ_a, ρ_b has been assumed to describe the $A + B \rightarrow C$ reaction-diffusion system [13] :

$$\frac{\partial \rho_a}{\partial t} = D_a \nabla^2 \rho_a - k \rho_a \rho_b \tag{1a}$$

$$\frac{\partial \rho_b}{\partial t} = D_b \nabla^2 \rho_b - k \rho_a \rho_b \tag{1b}$$

where D_a, D_b are the diffusion constants, and k is the microscopic reaction constant. The equations are subject to the initial separation condition along the x axis,

$$\rho_a(x,0) = a_0 H(x), \qquad \rho_b(x,0) = b_0[1 - H(x)], \tag{1c}$$

where a_0, b_0 are the initial densities and $H(x)$ is Heaviside step function. This means that initially the A's are uniformly distributed on the right side ($x > 0$), and the B's - on the left ($x < 0$).

Several dynamic quantities characterize the kinetic behavior of this system: The global reaction rate, $R(t)$, the location of the reaction front center, $x_f(t)$, the front width, $w(t)$, and the production rate of C at the center, $R(x_f, t)$. The local production rate of C, defined as the reaction term in Eq.(1),

$$R(x,t) = k\rho_a(x,t)\rho_b(x,t), \tag{2}$$

is the basis for defining all other quantities. For example, the global reaction rate is the spatial integration over $R(x,t)$, the center of the reaction front is defined as the position where $R(x,t)$ has a maximum, and the width is defined through its second moment [13].

2 Methods and Results

Two main analytical methods have been used to study the model system. The first is a *scaling theory* by Gálfi and Rácz [13], which is valid for the asymptotic time limit. They assume that in the asymptotic time domain, $R(x,t)$ has the scaling form

$$R(x,t) \sim t^{-\beta} F\left(\frac{x - x_f}{w}\right), \tag{3}$$

where F is a scaling function. Their study yields $\beta = 2/3$, with $x_f \sim t^{1/2}$ and $w = t^{\alpha} \sim t^{1/6}$, results which have been supported by numerical calculations [14-17].

The second analytical approach is a *perturbation theory* by Taitelbaum *et al* [17-18], which is valid for relatively short times, when very little reaction takes place before the species become effectively mixed. This initial behavior has been found, theoretically and experimentally, to be very rich in detail. The perturbation expansion is based on a description of the initial behavior using a small dimensionless reaction constant, given by

$$\epsilon = \frac{k}{\sqrt{a_0 b_0 D_a D_b}}. \tag{4}$$

On the assumption $\epsilon \ll 1$, one can apply perturbation theory to solve Eq.(1), by expanding the concentrations in the power series

$$\rho_a(x,t) = a_0 \sum_{j=0}^{\infty} \alpha_j \epsilon^j, \qquad \rho_b(x,t) = b_0 \sum_{j=0}^{\infty} \beta_j \epsilon^j, \tag{5}$$

where the dimensionless concentrations α_0 and β_0 satisfy ordinary diffusion equation under the initial separation condition in terms of dimensionless Heaviside step functions.

The combination of the results from these two theoretical methods, leads to a series of crossovers from the initial to the asymptotic behavior of the kinetic properties of the reaction front. These crossovers depend on the microscopic reaction constant, as well as on the ratios of diffusion coefficients and initial densities of the two species. For example, the global reaction rate $R(t)$ changes dramatically from an initial $t^{1/2}$ increase to a final $t^{-1/2}$ decrease, at time proportional to k^{-1} [17]. In practice, these crossovers take place on a real time scale of ***hours***, thus providing a useful means to extract microscopic parameters from macroscopic experiments.

We have confirmed the crossover predictions by solving Eq.(1) numerically [17], using a procedure which is based on a ***split-step*** algorithm that uncouples the diffusion and the reaction at each time unit [35]. The diffusion part has been solved by the exact enumeration method [36], which is equivalent to discretizing the evolution equation, whereas the reaction effect has been calculated according to the reaction term in Eq.(1). This algorithm is more efficient and less time-consuming than a standard numerical solution of the corresponding differential equations.

In Table 1 we summarize the crossover from the short-time limit to the asymptotic behavior for characteristics of the reaction front in the initially-separated system.

Table 1. A summary of the crossover behavior of the reaction front properties in the initially-separated $A + B \to C$ reaction-diffusion system.

quantity	notation	short-time	long-time
global rate	$R(t)$	$t^{1/2}$	$t^{-1/2}$
center of front	$x_f(t)$	non-universal	$t^{1/2}$
width of front	$w(t)$	$t^{1/2}$	$t^{1/6}$
local rate at x_f	$R(x_f, t)$	const.	$t^{-2/3}$

In the context of the mean-field results, it should be noted that Larralde *et al* [24] have found an asymptotic quantitative expression for $R(x,t)$ for the case of symmetrical system parameters. The mean-field description has been argued by Cornell *et al* [20] to be valid above an upper critical dimension $d_{up} = 2$. Results for systems below this dimension, in particular in one dimension, have been obtained by several groups, and are reviewed by Havlin *et al* [27,28].

3 Experiments

The experimental studies of this system have been performed at the University of Michigan by the group of Prof. R. Kopelman. The experiments are based on optical absorbance measurements of the reactants and the product, and have been performed in a 52 centimeter long glass reactor made of a rectangular capillary with 4 * 2 mm^2 cross section (Fig. 1a). The reactants were poured into each end arm of the reactor slowly by using syringes. They met at the center, forming a vertical boundary at $t = 0$. Light sources in appropriate wavelengths have been used for the optical excitation (with low-pass filters for noise reduction), and a photo-multiplier-tube (PMT) as a detector. A slit unit, composed of two slits with adjustable widths has allowed light to pass to and from the reactor. The slit widths were 0.05 mm on the light source side and 0.09 mm on the detector side to ensure a 0.1 mm mechanical resolution. The light source, solenoid with the filters, slit unit and the detector were all fixed on a stepping motor, which moved the system along the reaction front domain (Fig. 1b), the length of which was less than 5 cm after 24 hours. The glass reactor itself was separately fixed over the slit unit. A typical snapshot of one scan, which yields the absorbance profile of the product, is shown in Fig. 2. Further details about the experimental technique are described in [15].

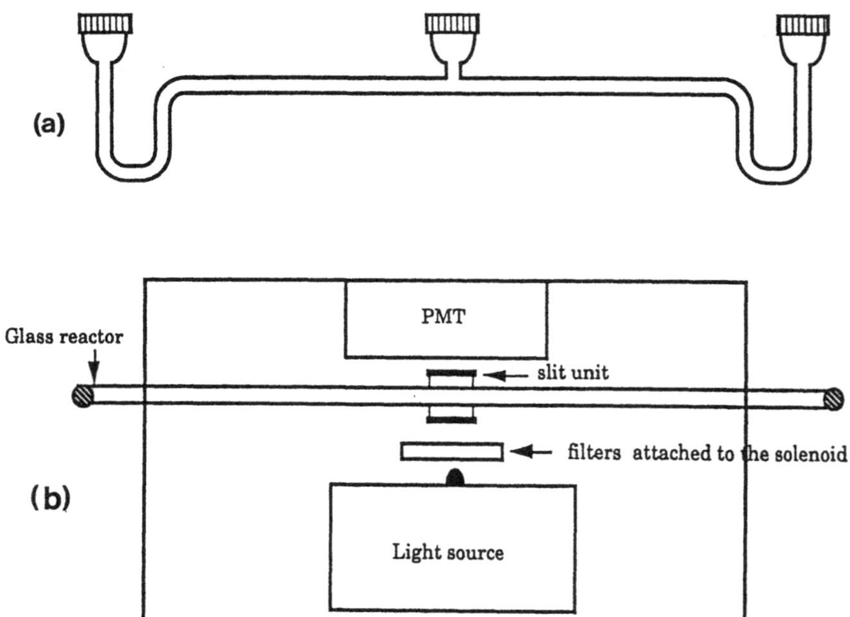

Fig. 1. (a) The glass reactor. (b) Top view of the experimental set-up for the absorbance measurements. From Koo and Kopelman [15].

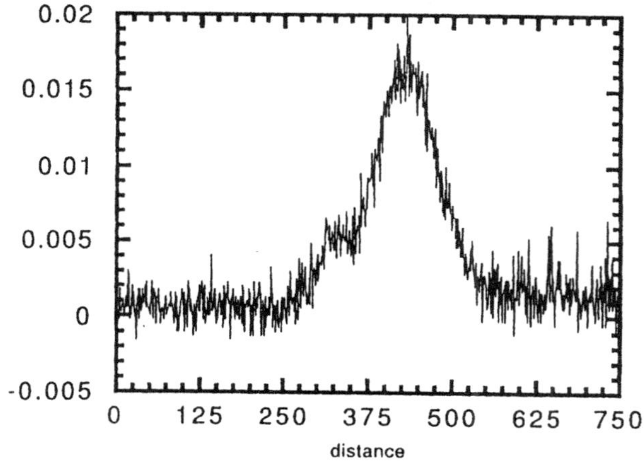

Fig. 2. A typical snapshot of an absorbance profile of a product.

In order to test experimentally the *asymptotic* predictions, Koo and Kopelman have chosen [15] the inorganic bimolecular reaction Cu^{2+}+ "tetra" $\rightarrow 1:1$ complex (in a gel solution), where "tetra" is the molecule disodium ethyl bis(5-tetrazolylazo)acetate trihydrate. This reaction is instantaneous, which means that k is large, and the reaction, being diffusion-limited, passes immediately to the asymptotic regime. In addition, one can easily monitor simultaneously both the reactants and the product through optical absorbance measurements, due to their very different colors. The use of a gel as a solvent resulted in a vertical boundary between the species at $t = 0$ and also prevented convection, still allowing efficient diffusion [15]. With these materials, Koo and Kopelman found the asymptotic results for the rate $R(t) \sim t^{-0.54}$, for the width $w \sim t^{0.18}$, and for the reaction center $x_f \sim t^{0.48}$, all exponents within 10% of those predicted by the theory $(-1/2, 1/6$, and $1/2$, respectively).

However, in order to test the *crossover* predictions, one has to use reactants which react slowly. For this purpose, Kopelman and co-workers [18,33] chose the reaction Cr^{3+}+ X.O. $\rightarrow 1:1$ complex in gel, where X.O. is Xylenol Orange, a molecule whose structure is shown in Fig. 3. In this slow reaction the color changes from orange (of the X.O.) to purple (of the product). The experimental results have confirmed the crossover predictions.

In Fig. 4 we show the behavior of the global production rate $R(t)$ which crosses over from an initial increase proportional to $t^{1/2}$ to a final decrease proportional to $t^{-1/2}$, as has been predicted above. The crossover point enables estimation of the microscopic reaction constant.

In Fig. 5 we show one of the most striking phenomenon exhibited by the initially-separated system. This is the non-monotonic motion of the reaction front center. The front changes its direction of motion along the separation axis, after reaching an extremum position. Indeed, such a behavior is predicted by the perturbation analysis. In what follows we review the interesting results for the non-universal behavior of the center of the reaction front in the short-time limit, which has been mentioned briefly in Table 1 above.

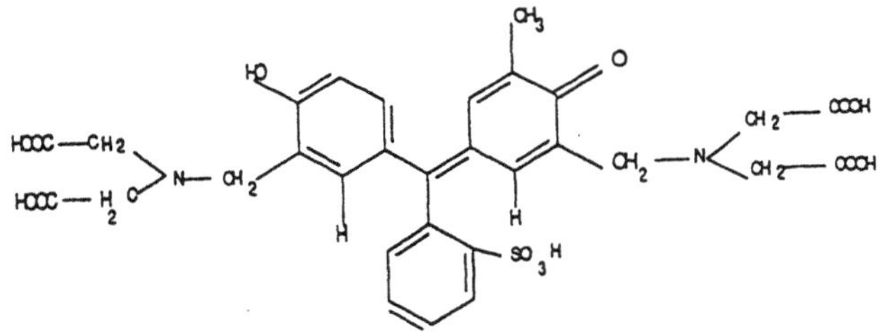

Fig. 3. The structure of a Xylenol Orange molecule.

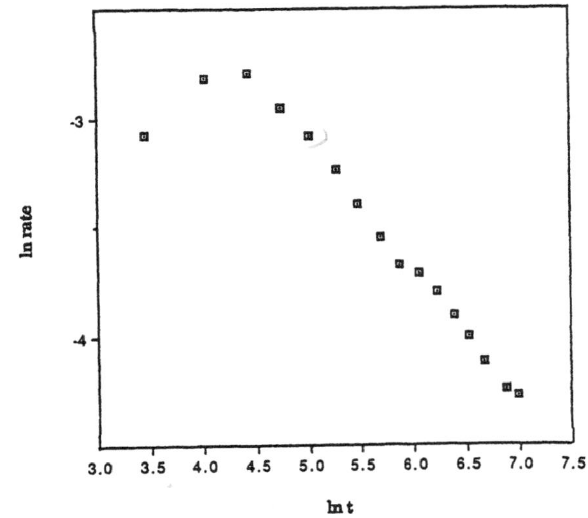

Fig. 4. Experimental results of the global reaction rate $R(t)$ as a function of time for the reaction Cr^{3+}+ X.O. $\rightarrow$ 1 : 1 complex.

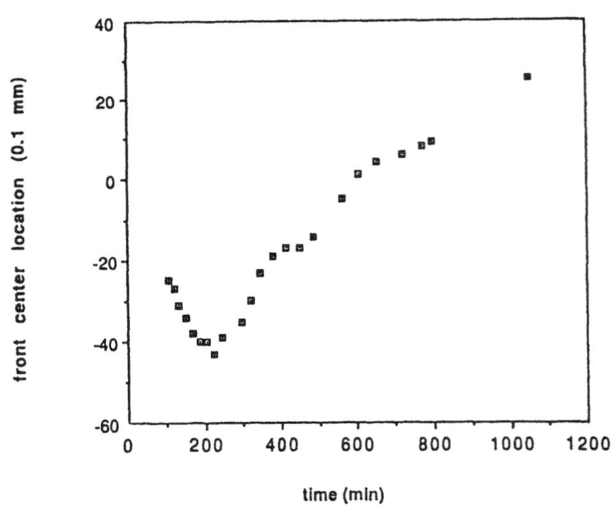

Fig. 5. Experimental results of the reaction center location x_f (in units of 0.1 mm), as a function of time (min), for the Cr^{3+}+ X.O. $\rightarrow$ 1 : 1 complex system. From [18].

4 Kinetics of the Reaction Front Center

The center of the reaction front, x_f, is defined as the position where the local rate $R(x,t)$ is a maximum. The general expression for x_f is found after maximization of $R(x,t)$ and expansion around the initial $x=0$, and is given by

$$x_f(t) \simeq \frac{\frac{1}{\sqrt{\pi}}\left(\frac{1}{\sqrt{D}} - \sqrt{D}\right) t^{-1/2} + \epsilon M(D,r) t^{1/2}}{\frac{1}{\pi} t^{-1} + \epsilon N(D,r)} \tag{6}$$

where

$$D = \sqrt{\frac{D_a}{D_b}}, \qquad r = \sqrt{\frac{a_0}{b_0}}, \tag{7}$$

and M and N are time-independent constants, with a non-trivial dependence on D and r. The behavior of x_f has been found to be extremely sensitive to these parameters, as is summarized in Table 2. Apart from the symmetric trivial case ($x_f = 0$ for all times), there can be as many as four different universality classes for describing the kinetics of the front in the little reaction regime, and as many as three distinguishable regimes in time.

Table 2. A summary of the various time-dependences of x_f, the center of the reaction front, as a function of the system parameters D and r.

$\{D, r\}$	short times	intermediate times	long times
$\{D=1,\ r=1\}$	0	0	0
$\{D=1,\ r\neq 1\}$	0	$t^{3/2}$	$t^{1/2}$
$\{D\neq 1,\ r=1\}$	$t^{1/2}$	$t^{1/2}$	$t^{1/2}$
$\{2+\sqrt{3} > D > 1,\ r<1\}$ or $\{1 > D > 2-\sqrt{3},\ r>1\}$	$t^{1/2}$	$t^{3/2}$ - non-monotonic	$t^{1/2}$
$\{D>1,\ r>1\}$ or $\{D<1,\ r<1\}$	$t^{1/2}$	$t^{3/2}$	$t^{1/2}$

As can be seen from the table, for short times $x_f \sim t^{1/2}$, provided that $D \neq 1$. When $D = 1$, (and $r \neq 1$), the first term in the numerator vanishes, and the center of the front remains fixed in position at the earliest times. As time increases, the second term in the numerator becomes significant, so that $x_f \sim \epsilon t^{3/2}$, finally changing to the asymptotic $t^{1/2}$ behavior. Thus, for $D = 1$ and $r \neq 1$, there are three distinct regions for the kinetic behavior of x_f. If also $r = 1$, than M vanishes as it should, and $x_f = 0$ for all t, due to symmetry.

The most interesting feature of x_f, is, of course, the possibility of the non-monotonic motion along the separation axis. The mathematical expression of this feature is that the function $x_f(t)$ has some extremum point as a function of time. The time t^* for which $x_f(t)$ has such an extremum point, depends in a complicated manner on D and r, through the rather cumbersome expressions for M and N. However, it turned out that t^* will be positive, and thus with a physical meaning, only if the parameters D and r satisfy the double condition $\{D > 1$ and $r < 1\}$, or, equivalently, $\{D < 1$ and $r > 1\}$. In fact, if D is too large ($D > 2+\sqrt{3} \approx 3.73$) or too small ($D < 2 - \sqrt{3} \approx 0.27$), t^* will be negative, and x_f will not have a physical extremum point, independent of the value of r. But if we restrict ourselves to values of D not too far from 1, than we can write $\{D > 1 , r < 1\}$ (or the opposite), as the condition which ensures the change of the direction of motion of the front.

This result can be physically understood. Suppose, e.g., that $D_a > D_b$ and $a_0 < b_0$. At very early times diffusion effects are dominant, and the direction of motion is determined by the penetration of the A-species to the left, B-side of the system. Later on, the reaction comes into play, and the species with higher concentration, B, will govern the direction of motion, which will be towards the right, A-side. Note that this result involves both two lowest orders in ϵ, so one expects the transition of the front from one direction to the other to occur at a rate proportional to $t^{3/2}$. Indeed, in the experiment whose results are plotted in Fig. 5, the parameters used were 5 x $10^{-5} M$ Xylenol Orange (diffusion constant 3.5 x 10^{-10} m^2/sec), and 7.5 x 10^{-4} M Cr^{3+} (diffusion constant 2.7 x 10^{-10} m^2/sec), so that $D = 1.14$ and $r = 0.26$ in accord with the requirements for an extremum position. In addition, one can clearly see that the change of direction occurs at a rate much higher than the asymptotic $t^{1/2}$. This formal short-time limit behavior takes place at time $t = 223$ min, which is definitely quite long. This is because of the very small reaction constant k in this chosen system, so that the corresponding small parameter ϵ defined in Eq.(4) is extremely small, and the short-time limit is extended over an easily measurable time interval.

For $D > 1$ and $r > 1$ (or vice versa), equation (6) predicts that x_f is monotonic in time. However, for small values of k, there are three regions of different kinetic behavior: at the very earliest times $x_f \sim t^{1/2}$, at later times $t^{3/2}$, (corresponding to the order of the perturbation expansion), and finally one obtains the asymptotic proportionality to $t^{1/2}$.

5 Two reaction fronts

Recently, while repeating the last experiments (with X.O.), some different and challenging results have also been obtained. The first is the global rate which exhibits more than one crossover as a function of time (as is seen in Fig. 6), and the second, which is also very surprising, is the existence of more than one localized front, as is exemplified by the two peaks in the results for the local production rate $R(x,t)$ as a function of x, as is shown in Fig. 7.

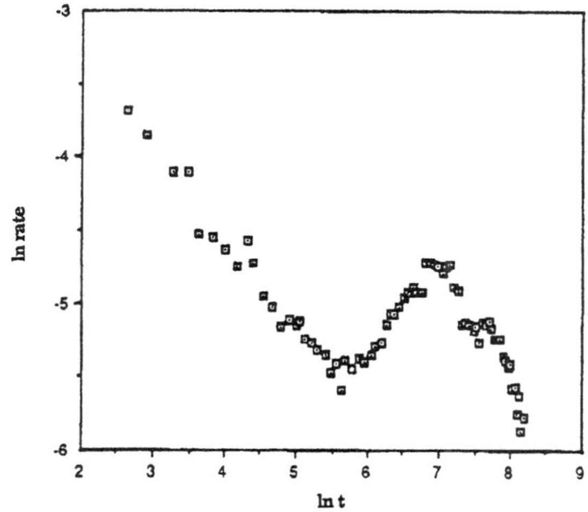

Fig. 6. Experimental results of the global reaction rate $R(t)$ as a function of time, with more than one crossover.

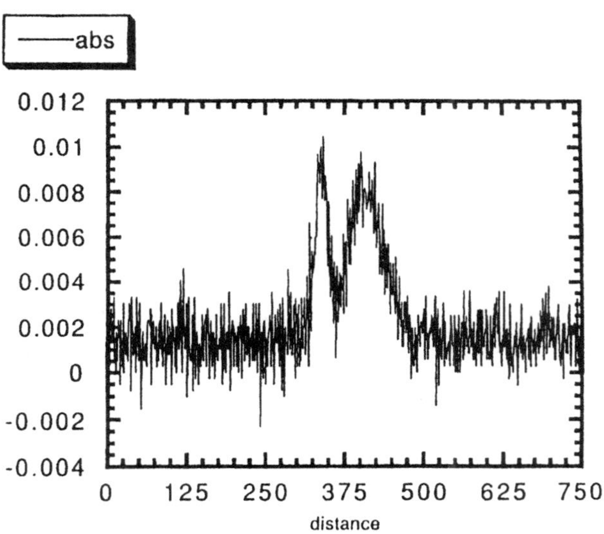

Fig. 7. Experimental results for the spatial profile of the local production rate $R(x,t)$, which exhibits two reaction fronts.

Recently, we have suggested a generalized theory to explain these results. The theory is based on the idea that there may be more than a single species on one of the sides of the initially-separated system. The model is based on the existence of two *similar* species A_1 and A_2 on the right side of the system, and a single species B on the left side. By similar species we mean that most of their properties with respect to the reaction-diffusion system are the same. For example, both react with B, and both have the same diffusion coefficient. However, each of the A's is assumed to react with a *different* reaction constant with the B, so that the reaction scheme is

$$A_1 + B \xrightarrow{k_1} C_1 \tag{8a}$$

$$A_2 + B \xrightarrow{k_2} C_2. \tag{8b}$$

The products C_1 and C_2 are assumed to be either identical or at least with very similar wavelength, so that they are indistinguishable by the detector in the experiment. Thus, the local reaction rate of this system will be a superposition of the form

$$R(x,t) = k_1\rho_{a_1}(x,t)\rho_b(x,t) + k_2\rho_{a_2}(x,t)\rho_b(x,t). \tag{9}$$

We have studied this system by means of the numerical split-step algorithm described above. The detailed results will be published elsewhere. However, the main point is that our results are very similar to those of the experiments, provided that k_1 and k_2 differ by several orders of magnitude, and the faster reacting A_i species is only a very small fraction of the A's. For example, when k_1 is 10^3 or 10^4 times larger than k_2, and A_1 ia abouts 2% of the entire A bulk, then the numerical results resemble the experiments. This is since the main reaction is indeed the slow one (with A_2), but there exist a very small fraction of slightly different version of A which reacts much faster with B, and introduces, e.g., an early decrease to the behavior of the global production rate, $R(t)$, prior to the crossover induced by the slow reaction of A_2. Such a system also explains the two peaks in the behavior of the local rate $R(x,t)$, which results from a superposition of two reaction fronts, the fast and the slow, as will be described in detail elsewhere.

However, referring to the experiments, the question is *who are* these A_1 and A_2 in the specific experimental system. Naturally, one would think about various sources of impurities or defects which result in a non-ideal experimental system. However, the puzzle is that A_1 and A_2 should be somehow related, since they yield the same, or at least a very similar product upon reacting with B.

In order to answer this question, we notice that one of the parameters that plays a crucial rule in such reactions is the pH value of the solution. The reaction under study (Cr^{3+}+ X.O. $\rightarrow$ 1 : 1 complex) is known to be highly dependent on the pH value [37]. Back in 1960, in a paper written in German by Řehák and Körbl [38], the authors perform a physico-chemical study of Xylenol Orange.

They show that X.O. can appear in *ten* different ionic transformations as a function of the pH value of the aqueous solution. The percentage of each chelate as a function of pH is shown in Fig. 8. Indeed, for pH=4.5, which was the average pH value in most of the experiments, one can see that there exist simultaneously two forms of X.O., 3% of H_4XO^{2-}, and 97% of H_3XO^{3-}, which is exactly the expected picture for the relative portions of the A_i's in the two-front system. Note that the two transformations differ only by a single proton, thus one expects that the diffusion coefficients are the same, the product of the bimolecular reaction is the same, and the only difference, which can be explained on chemical grounds, is the very different reaction constant which represents the reaction efficiency.

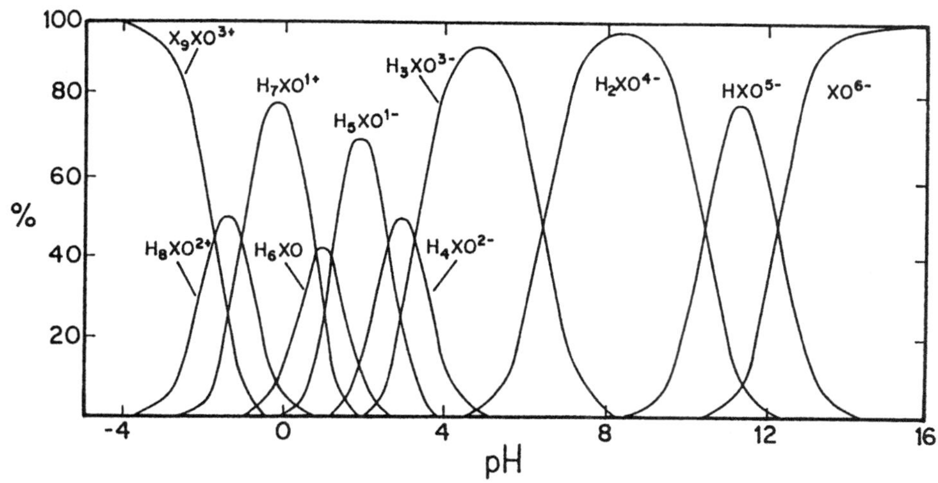

Fig. 8. Percentage of each ionic transformation of Xylenol Orange as a function of the pH of the solution. After Řehák and Körbl [38].

6 Summary

In summary, we have presented various characteristics of the reaction-diffusion system $A + B \to C$ with initially separated components, which exhibits a series of crossovers from an initial to an asymptotic time behavior. We have focused on the kinetics of the reaction front center and showed that its motion can be classified into four different universality classes, depending on the system parameters. In particular, we showed theoretically and experimentally that the front can exhibit a non-monotonic motion, which occurs when one of the species has larger diffusion coefficient but smaller initial density. In addition, we have discussed the case of a two-front system, which results from a simultaneous presence of two similar species on one of the sides of the system, an example of which is the case of two ionic transformations of the same molecule.

Acknowledgements. I would like to thank B. Vilensky, S. Rabinovich and S. Havlin from Bar-Ilan University, J.E. Kiefer, B. Trus and G.H. Weiss from the National Institutes of Health, and Y-E.L. Koo, A. Lin, A. Yen and R. Kopelman from the University of Michigan for fruitful collaboration and useful discussions. I thank R. Kopelman for permission to use unpublished figures.

References

[1] D. Avnir and M. Kagan, Nature, **307**, 717 (1984) and references cited therein.
[2] G.T. Dee, Phys. Rev. Lett., **57**, 275 (1986).
[3] B. Heidel, C.M. Knobler, R. Hilfer and R. Bruinsma, Phys. Rev. Lett., **60**, 2492 (1988).
[4] R.E. Liesegang, Naturwiss. Wochenschr. **11**, 353 (1896).
[5] K.F. Mueller, Science, **225**, 1021 (1984).
[6] D.A. Smith, J. Chem. Phys,. **81**, 3102 (1984).
[7] G. Venzl, J. Chem. Phys., **85**, 2006 (1986).
[8] J. Willemsen and R. Burridge, Phys. Rev. A, **43**, 735 (1991).
[9] H.C. Kang and W.H. Weinberg, Phys. Rev. E, **47**, 1604 (1993).
[10] Proceedings of the NIH Meeting on <u>Models of Non-Classical Reaction Rates</u>, J. Stat. Phys., **65**, No. 5/6, Dec. 1991.
[11] I.M. Sokolov and A. Blumen, Int. J. Mod. Phys. B, **5**, 3127 (1991).
[12] F. Leyvraz and S. Redner, Phys. Rev. A, **46**, 3132 (1992).
[13] L. Gálfi and Z. Rácz, Phys. Rev. A, **38**, 3151 (1988).
[14] Y.E. Koo, L. Li and R. Kopelman, Mol. Cryst. Liq. Cryst., **183**, 187 (1990).
[15] Y-E.L. Koo and R. Kopelman, J. Stat. Phys., **65**, 893 (1991).
[16] Z. Jiang and C. Ebner, Phys. Rev. A, **42**, 7483 (1990).
[17] H. Taitelbaum, S. Havlin, J.E. Kiefer, B. Trus and G.H. Weiss, J. Stat. Phys., **65**, 873 (1991).
[18] H. Taitelbaum, Y-E.L. Koo, S. Havlin, R. Kopelman and G.H. Weiss, Phys. Rev. A, **46**, 2151 (1992).
[19] B. Chopard and M. Droz, Europhys. Lett., **15**, 459 (1991).
[20] S. Cornell, M. Droz and B. Chopard, Phys. Rev. A, **44**, 4826 (1991).
[21] S. Cornell, M. Droz and B. Chopard, Physica A, **188**, 322 (1992).
[22] M. Araujo, S. Havlin, H. Larralde and H.E. Stanley, Phys. Rev. Lett., **68**, 1791 (1992).
[23] E. Ben-Naim and S. Redner, J. Phys. A, **25**, L575 (1992).
[24] H. Larralde, M. Araujo, S. Havlin and H.E. Stanley, Phys. Rev. A, **46**, 855 (1992).
[25] H. Larralde, M. Araujo, S. Havlin and H.E. Stanley, Phys. Rev. A, **46**, R6121 (1992).
[26] M. Araujo, H. Larralde, S. Havlin and H.E. Stanley, Physica A, **191**, 168 (1992).
[27] S. Havlin, M. Araujo, H. Larralde, H.E. Stanley and P. Trunfio, Physica A, , 143 (1992).
[28] S. Havlin, M. Araujo, H. Larralde, A. Shehter, H.E. Stanley, Fractals, **1**, 405 (1993).
[29] B. Chopard, M. Droz, T. Karapiperis and Z. Rácz, Phys. Rev. E, **47**, R40 (1993).
[30] S. Cornell and M. Droz, Phys. Rev. Lett., **70**, 3824 (1993).

[31] M. Araujo, H. Larralde, S. Havlin and H.E. Stanley, Phys. Rev. Lett., **71**, 3592 (1993).
[32] H. Taitelbaum, Physica A, **200**, 155 (1993).
[33] Y-E. Koo, R. Kopelman, A. Yen and A. Lin, MRS Symp. Proc., **290**, 273 (1993).
[34] B. Vilensky, S. Havlin, H. Taitelbaum and G.H. Weiss, J. Phys. Chem., **98**, 7325 (1994).
[35] E.S. Oran and J.P. Boris, "Numerical Simulation of Reactive Flows", Elsevier, New-York, 1987.
[36] S. Havlin and D. Ben-Avraham, Adv. Phys., **36**, 695 (1987).
[37] F.D. Snell, "Photometric and Fluorometric Methods of Analysis" - Part 1, Wiley, New-York, 1978.
[38] B. Řehák and J. Körbl, Collection Czechoslov. Chem. Commun., **25**, 797 (1960).

Anomalous Reaction-Diffusion Systems

S. Havlin[1,2], *M. Araujo*[1], *H. Larralde*[1], *A. Shehter*[2], *H. E. Stanley*[1] *and P. Trunfio*[1]

[1] Center for Polymer Studies and Department of Physics, Boston University, Boston, MA 02215, USA
[2] Department of Physics Bar-Ilan University Ramat-Gan, Israel

Abstract: We review recent developments in the study of the diffusion reaction systems of the type $A + B \to C$ in which the reactants are initially separated. We consider the case where the A and B particles are initially placed uniformly in Euclidean space at $x > 0$ and $x < 0$ respectively. We find that whereas for $d \geq 2$ the mean field exponent characterizes the width of the reaction zone, fluctuations are relevant in the one-dimensional system. We also present analytical and numerical results for the reaction rate on fractals and percolation systems.

1 Introduction

The dynamics of diffusion controlled reactions of the type $A + B \to C$ has been studied extensively since the pioneering work of Smoluchowski [1, 2]. Most studies have focused on homogeneous systems, i.e., when both reactants are initially uniformly mixed in a d-dimensional space, and interesting theoretical results have been obtained. When the concentrations of the A and B reactants are initially equal, i.e., $c_A(0) = c_B(0) = c(0)$, the concentration of both species is found to decay with time as, $c(t) \sim t^{-d/4}$ for Euclidean $d \leq 4$-dimensional systems [3-10] and as $c(t) \sim t^{-d_s/4}$ for fractals [5,6] with fracton dimension $d_s \leq 2$. Also, self-segregated regions of A and B in low dimensions ($d \leq 3$) [4] and in fractals [9] have been found. Quantities such as the distributions of domain sizes of segregated regions and interparticle distances between species of the same type and different types have been calculated [11–13]. These systems were also studied theoretically and numerically under steady state conditions and several interesting predictions have been obtained [14–17]. However, the above numerical and theoretical predictions have not been observed in experiments, in part because of difficulties to implement the initially uniformly-mixed distribution of reactants.

In recent years it was realized that diffusion reaction systems in which the reactants are initially separated [18], can be studied experimentally [19,20] and that the dynamics of such a system have many surprising features [20–27]. These systems are characterized by the presence of a dynamical interface or a front that separates the reactants. Such a reaction front appears in many biological, chemical and physical processes [28–34].

Gàlfi and Ràcz [18] studied the diffusion-controlled reaction system with initially separated reactants. They studied the kinetics of the reaction diffusion process using a set of mean-field (MF) type equations,

$$\frac{\partial c_A}{\partial t} = D_A \frac{\partial^2 c_A}{\partial x^2} - k c_A c_B \tag{1a}$$

$$\frac{\partial c_B}{\partial t} = D_B \frac{\partial^2 c_B}{\partial x^2} - k c_A c_B . \tag{1b}$$

Here $c_A \equiv c_A(x,t)$ and $c_B \equiv c_B(x,t)$ are the concentrations of A and B particles at position x at time t respectively, $D_{A,B}$ are the diffusion constants and k is the reaction constant. The rate of production of the C-particles at site x and time t, which we call the reaction-front profile, is given by $R(x,t) \equiv k c_A c_B$. The initial conditions are that the A species are uniformly distributed on the right-hand side of $x = 0$ and the B species are uniformly distributed on the left-hand side.

Using scaling arguments, Gàlfi and Ràcz [18] find that the width w of the reaction front $R(x,t)$ scales with time as, $w \sim t^{\alpha}$ with $\alpha = 1/6$ and the reaction rate at the center of the front, which is called the reaction height, scales as $h \sim t^{-\beta}$ with $\beta = 2/3$.

Experiments [19] and simulations [19,21–24] for $d \geq 2$ systems in which both reactants diffuse, support the above predicted values for α and β. Indeed, Cornell et al [23] argue that the upper critical dimension is $d = 2$ and the MF approach should therefore be valid for $d \geq 2$. Moreover, numerical simulations of 1D systems show that the width exponent appears to be $\alpha \simeq 0.3$ and the height exponent $\beta \simeq 0.8$ [23,24]. Recently [25] it was argued that α varies between 1/4 and 3/8, depending on the moment at which the width is calculated. For a more detailed discussion, see Sec. 3. The origin of the difference between the exponents of 1D systems and those of higher dimensional systems is due to fluctuations in the location of the front which are important in low dimensions and are neglected in the MF approach.

Taitelbaum et al [20,22] studied analytically Eqs. (1) and presented experiments for the limit of small reaction constant or short time. The main results are that several measurable quantities undergo interesting crossovers. For example, the global reaction rate changes from $t^{1/2}$ in the short time limit to $t^{-1/2}$ at the asymptotic time regime. The center of the front can change its direction of motion as found in experiments [20]. Ben-Naim and Redner [26] studied the solution of (1) under steady-state conditions.

2 The Form of the Reaction Front, $R(x,t)$, in the Mean-Field Approach

In a recent work [27] we consider the symmetric case in which both diffusion constants and initial concentrations are equal, i.e., $D_A = D_B \equiv D$ and

$c_A(x,0) = c_B(x,0) = c_0$. If we define $F(x,t) \equiv c_A(x,t) - c_B(x,t)$, then from Eq. (1) follows,

$$\frac{\partial F}{\partial t} = D\frac{\partial^2 F}{\partial x^2}, \tag{2}$$

subject to the conditions that initially the A particles are uniformly distributed to the right of the origin while the B particles are uniformly distributed to the left of the origin. Equation (2) has the solution $F(x,t) = c_0 \operatorname{erf}(x/\sqrt{4Dt})$.

We rewrite the concentrations of A and B particles as (see Fig. 1),

$$c_A(x,t) = G_1(x,t) + \delta c_1(x,t), \qquad c_B(x,t) = G_2(x,t) + \delta c_2(x,t) \tag{3}$$

where

$$G_1(x,t) = \begin{cases} F(x,t) & [x > 0] \\ 0 & [x < 0], \end{cases} \tag{4}$$

and $G_2(x,t) = G_1(-x,t)$. It is easy to see that under the above conditions, $\delta c_1(x,t) = \delta c_2(x,t) \equiv \delta c(x,t)$. Substituting Eqs. (3) into Eqs. (1) yields

$$\frac{\partial(\delta c)}{\partial t} = D\frac{\partial^2(\delta c)}{\partial x^2} - k\left(c_0 \operatorname{erf}\left(\frac{x}{\sqrt{4Dt}}\right) + \delta c\right)\delta c \tag{5}$$

The asymptotic solution for this equation that vanishes as $x \to \infty$ is,

$$\delta c(x,t) \sim t^{-1/3}\left(\frac{x}{t^{1/6}}\right)^{-1/4} \exp\left[-\frac{2}{3}\left(\frac{\lambda x}{t^{1/6}}\right)^{3/2}\right], \qquad t^{1/6} \ll x \ll (4Dt)^{1/2}, \tag{6}$$

where $\lambda = (ka/D)^{1/3}$, $a \equiv c_0/(\pi D)^{1/2}$. As may be confirmed by direct substitution, this expression is a solution of Eq. (5) up to terms of order $(\delta c)/t$, which can be neglected for large t.

Using Eq. (6) we can write an expression for the reaction-front profile $R(x,t)$ defined in (1) as,

$$R(x,t) \simeq \frac{kax}{t^{1/2}}(\delta c) \sim t^{-2/3}\left(\frac{x}{t^{1/6}}\right)^{3/4} \exp\left[-\frac{2}{3}\left(\frac{\lambda x}{t^{1/6}}\right)^{3/2}\right]. \tag{7}$$

It is seen that the width of the reaction front grows as $t^{1/6}$, whereas the height can be identified with the prefactor $t^{-2/3}$ in Eq. (7), consistent with the exponents found by Gàlfi and Ràcz [18]. Equation (7) provides a more quantitative solution of Eqs. (1) than the previous scaling arguments [18], as well as information on the dependence of the form of the reaction front on the parameters c_0, k, and D, for the symmetric case.

For the case in which one reactant is static no analytical solution (of Eq. (1)) exists for the form of the reaction front profile. However, numerical solutions of Eq. (1) with $D_B = 0$ shown in Fig. 2, suggest that $R(x',t) \sim t^{-\beta} g(x'/t^{\alpha}) \cdot \cdot \exp(-|x'|/t^{\alpha})$, where $x' \equiv x - \gamma t^{1/2}$. The excellent scaling in Fig. 2b suggests that the width does not increase with time, i.e., $w \sim t^{\alpha}$ with $\alpha = 0$ and $h \sim t^{-\beta}$ with $\beta = 1/2$, consistent with the scaling arguments in ref. [21].

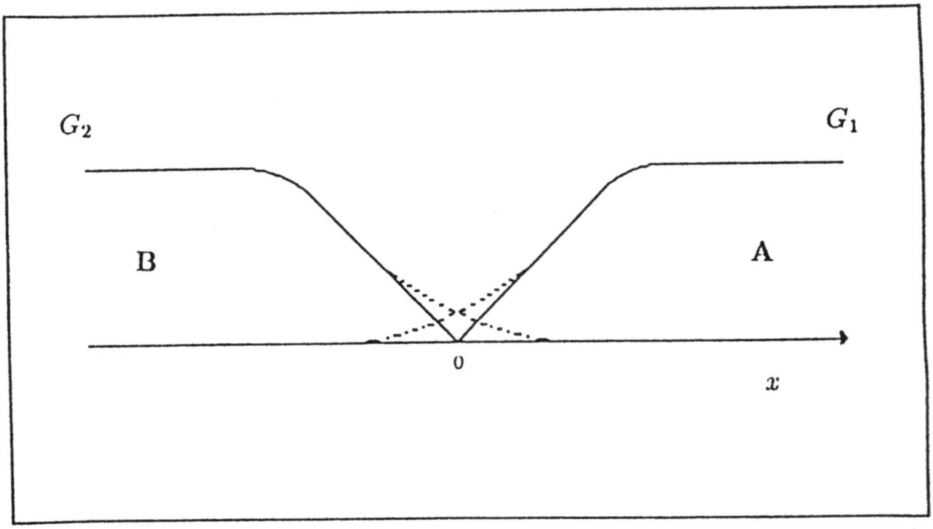

Fig. 1. Schematic picture of the reactant concentration profiles near the origin. The solid lines represent the $G_{1,2}(x,t)$ part of the profile, the dashed lines represent the complete form $G_{1,2}(x,t) \pm \delta c(x,t)$. Note that the profile of species A is given solely by $\delta c(x,t)$ on the left of the origin. (From Ref. [27].)

3 The Front, $R(x,t)$, in $d = 1$

The reaction front profile in $d = 1$ systems, $R(x,t)$, when both reactants are diffusing with the same diffusion constant, $D_A = D_B \neq 0$, has been calculated numerically [24]. The data shown in Fig. 3 suggests that

$$c_C(x,t) \equiv \int_o^t R(x,t')dt' \sim \exp(-a|x|/t^\alpha), \tag{8}$$

with $\alpha = 0.33 \pm 0.05$.

For the case $D_A \neq 0, \quad D_B = 0$, analytical and numerical studies [37] yield for the reaction front profile

$$R(x,t) = \frac{1}{4t^{3/4}}\left(\frac{2\gamma^2}{\mu\pi}\right)^{1/2} \exp\left(-\frac{(x-\gamma t^{1/2})^2}{2\mu t^{1/2}}\right)\left(1+\frac{x-\gamma t^{1/2}}{2\gamma t^{1/2}}\right), \tag{9}$$

where γ and μ are constants. From eq. (9) follows that $\alpha = 1/4$ and $\beta = 3/4$. It is interesting to note that the time integral of $R(x,t)$, which is the total production of the C-particles at x up to time t, is given by

$$c_C(x,t) = \int_0^t R(x,\tau)d\tau = \frac{1}{2}\text{erfc}\left(\frac{x-\gamma t^{1/2}}{\sqrt{2\mu t^{1/2}}}\right). \tag{10}$$

To summarize the case of $A + B \to C$ with initially separated reactants, we list in table 1 the four sets of exponents discussed above.

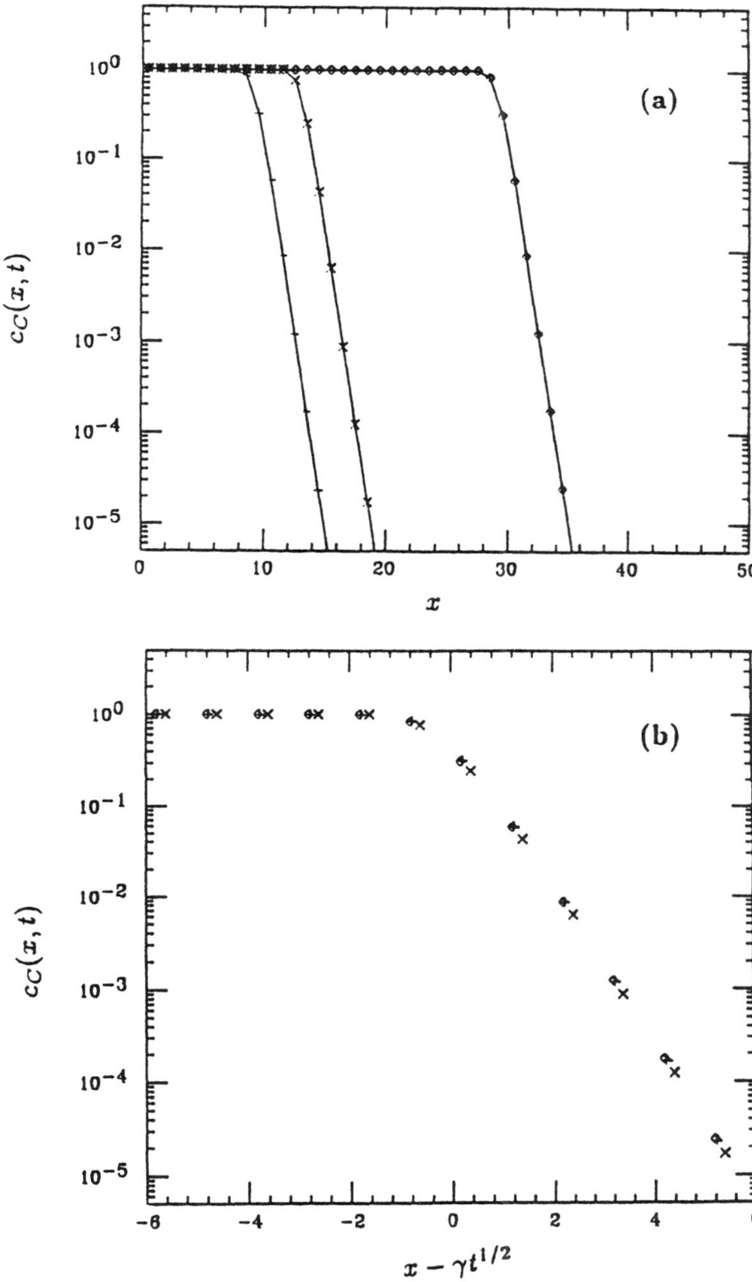

Fig. 2. Numerical solution of Eq. (1) for the case $D_B = 0, D_A \neq 0$: (a) plot of $c(x,t)$ as a function of x for $t = 500$, 1000 and 5000; (b) the good scaling plot of $c(x,t)$ as a function of $(x - \langle x \rangle)$ indicates that $\alpha = 0$ and $\beta = 1/2$.

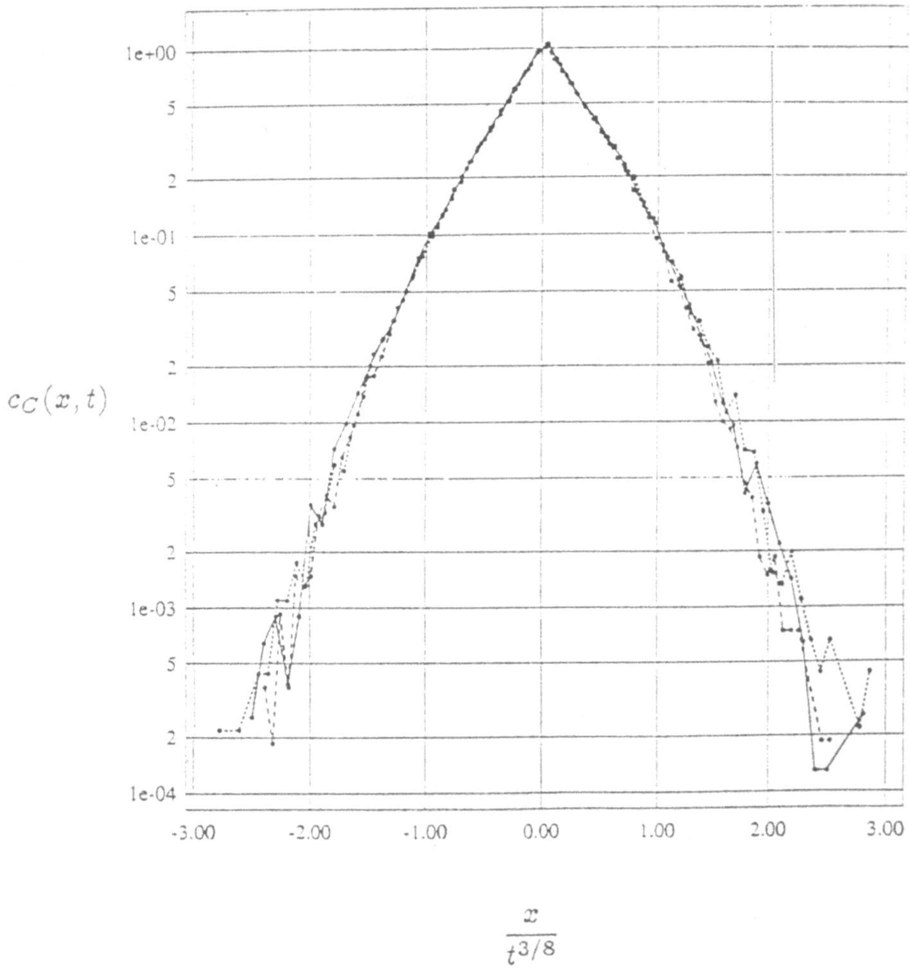

Fig. 3. Plot of $c_C(x,t)$ defined in Eq. (8) for $d = 1$ system with system size $L = 10000$ and $t = 1000, 5000$ and 10000.

Table 1. The values of the exponents α and β ($w \sim t^{\alpha}$, $h \sim t^{-\beta}$).

	$d = 1$	MF
Both moving	$\alpha \cong 0.3$ $\beta \cong 0.8$	$\alpha = 1/6$ $\beta = 2/3$
One static	$\alpha = 1/4$ $\beta = 3/4$	$\alpha = 0$ $\beta = 1/2$

4 The Reaction Rate in $d = 2$ Percolation

The case of A+B→C with initially-separated reactants on fractal systems was studied on the $d = 2$ infinite percolation cluster at criticality [22], for a demonstration of the system see Fig. 5.

It is expected that the total number of reactants up to time t,

$$\int_{-\infty}^{\infty} c_C(x,t)dx,$$

scales as the mean displacement of a random walker on a fractal, i.e., $\langle r^2 \rangle^{1/2} \sim t^{1/dw}$, where d_w is the anomalous diffusion exponent [38]. From this follows that the reaction rate

$$R(t) \equiv \int_{-\infty}^{\infty} R(x,t)dx \sim t^{-\gamma}, \qquad \gamma = 1 - 1/d_w. \tag{15}$$

One has to distinguish between reaction-diffusion on the infinite cluster and in the percolation system, containing also small clusters [39]. The reaction rate on the infinite cluster is smaller and decrease slower compared with the system containing clusters of all sizes. This can be understood as follows. At any finite time we can divide all clusters into two groups according to their sizes: *active* - clusters of mass $s > s^*$, in which at time $t < t^* \sim s^{*d_w/d_f}$ particles are not aware of the finitness of their cluster (this group contains the infinite cluster), and *inactive* clusters of mass $s < s^*$ on which at least one of the reactants has been vanished and the reaction rate is zero. According to this picture, in the full percolation system at any time there are active clusters of finite size, that can contribute to the reaction rate. Therefore, the rate of reaction in the percolation system is always higher, than on the infinite cluster. Also, at any time there are some finite clusters that become inactive causing additional (comparing with the infinite cluster network) decrease of the rate of reaction in the percolation system. Since the system is self-similar one expect a change in the reaction rate exponent.

To quantify the above considerations we can look on each cluster of mass s and linear size $r \sim s^{1/d_f}$ as a reservoir of particles divided by the *front line* into A- and B-parts. We introduce an *active front* of a cluster as the sites belonging both to the cluster and to the front line. The length ℓ_s of the *active front* of a single cluster of size s is expected to be

$$\ell_s \sim r^{d_f - 1} \sim s^{\frac{d_f - 1}{d_f}} \tag{16}$$

Next we assume that rate of reaction on a cluster of mass s per unit length of active front is

$$R_0(t) \sim \begin{cases} t^{-\gamma} & t < t^* \\ 0 & t > t^* \end{cases}, \tag{17}$$

where $t^* = s^{d_w/d_f}$. Therefore, the total contribution of active clusters of size s to the reaction rate is

$$R_s(t) \sim \varphi_s s^{\frac{d_f - 1}{d_f}} t^{-\gamma}, \tag{18}$$

Fig. 4. Reaction-diffusion on the 2 — dimensional infinite percolation cluster at criticality at $t = 2000$. White squares represent sites of the infinite claster. Blue and red circles represent the A and B particles. Initially all sites of cluster in the right half plane and left half plane were occupied by A and B particles respectively.

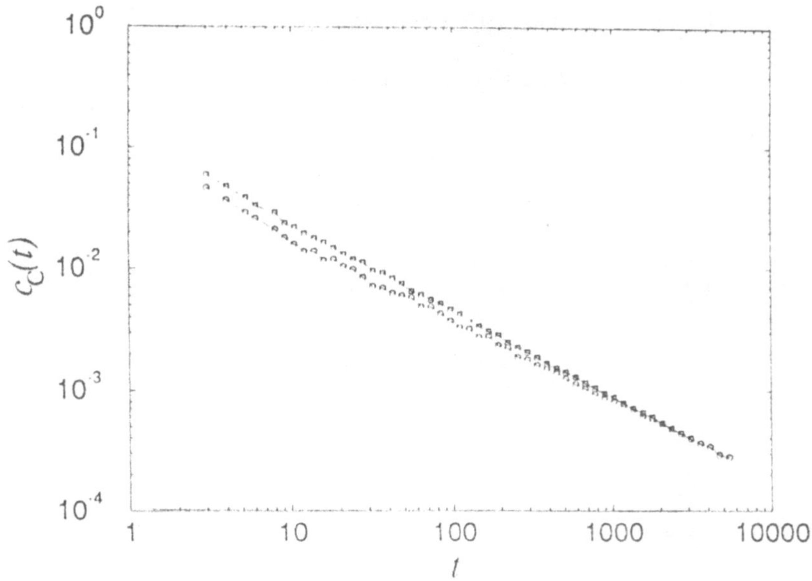

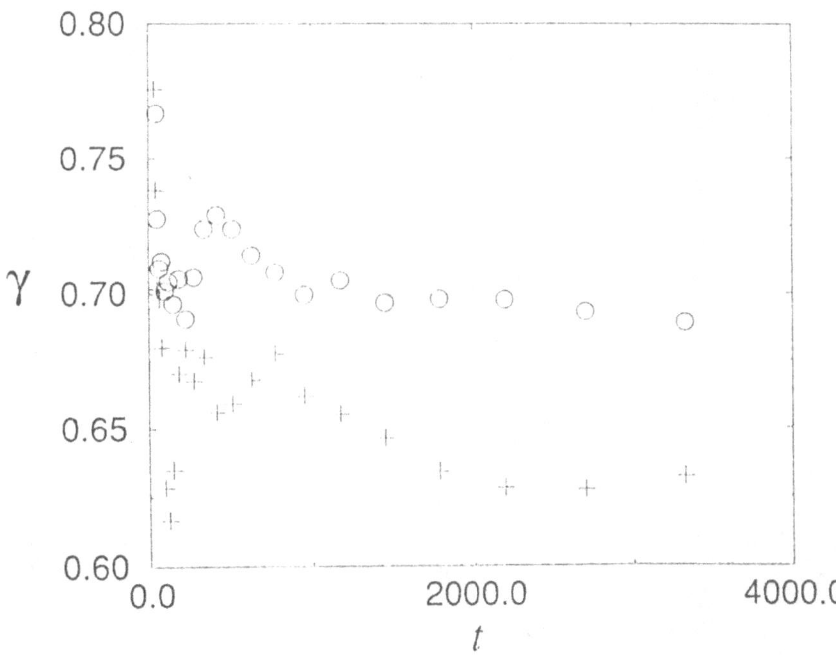

Fig. 5. (a) Plot of the rate $c_C(t)$ for the percolation system ($\diamond$) and for the infinite percolation cluster (o). (b) Plot of successive slopes of the data in (a). The exponent γ of reaction rate on the infinite claster (+) and on the percolation system (o).

where φ_s is number of clusters of size s that intersect the front line. One can estimate φ_s as follows. In a percolation system of size $L \times L$ there are n_s clusters of mass s. Only a small part of them intersects the front line, those in a strip of width $w \sim s^{1/d_f}$ around the front line. Their fraction is w/L. Therefore, $\varphi_s \sim \frac{s^{1/d_f}}{L} n_s \sim s^{1/d_f} n_s$. Substituting this in (18) we get

$$R_s(t) \sim s^{1/d_f} n_s s^{\frac{d_f-1}{d_f}} t^{-\gamma} = t^{-\gamma} s n_s. \tag{19}$$

Thus, the reaction rate in the percolation system is

$$R(t) = \sum_{s=s^*}^{\infty} R_s(t) = t^{-\gamma}(s^*)^{2-\tau} = t^{-\gamma} t^{-\delta} = t^{-\gamma'}$$

$$\delta = \gamma' - \gamma = \frac{d_f}{d_w}(\tau - 2). \tag{20}$$

These results are in good agreement with our simulations, see Fig. 6.

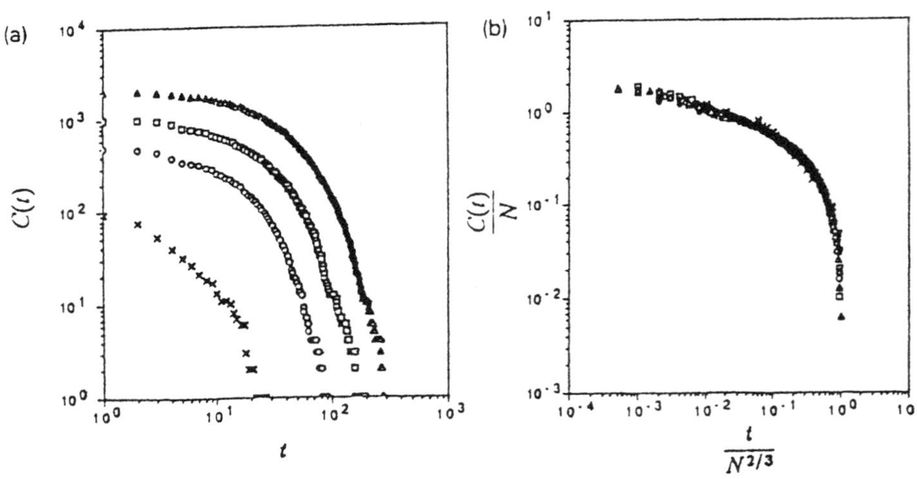

Fig. 6. Numerical simulations of $A + B \to C$ where N diffusing particles of type A are initially at the center of a lattice and particles of type B are static and located at each site of the $d = 3$ lattice: (a) plot of $C(t)$ for $N = 100$ (×), 500 (•), 1000 (□), and 2000 (△) particles; (b) plot of $C(t)$ in the scaling form Eq. (22a). Note that the results are of a *single* Monte Carlo run and not averaged—showing that fluctuations are negligible in this process.

We also study the finite size effects on the reaction-diffusion system. For a percolation system of size $L \times L$ we expect that for the infinite cluster,

$$R(t) = L^{d_f-1} t^{-\gamma}, \tag{21a}$$

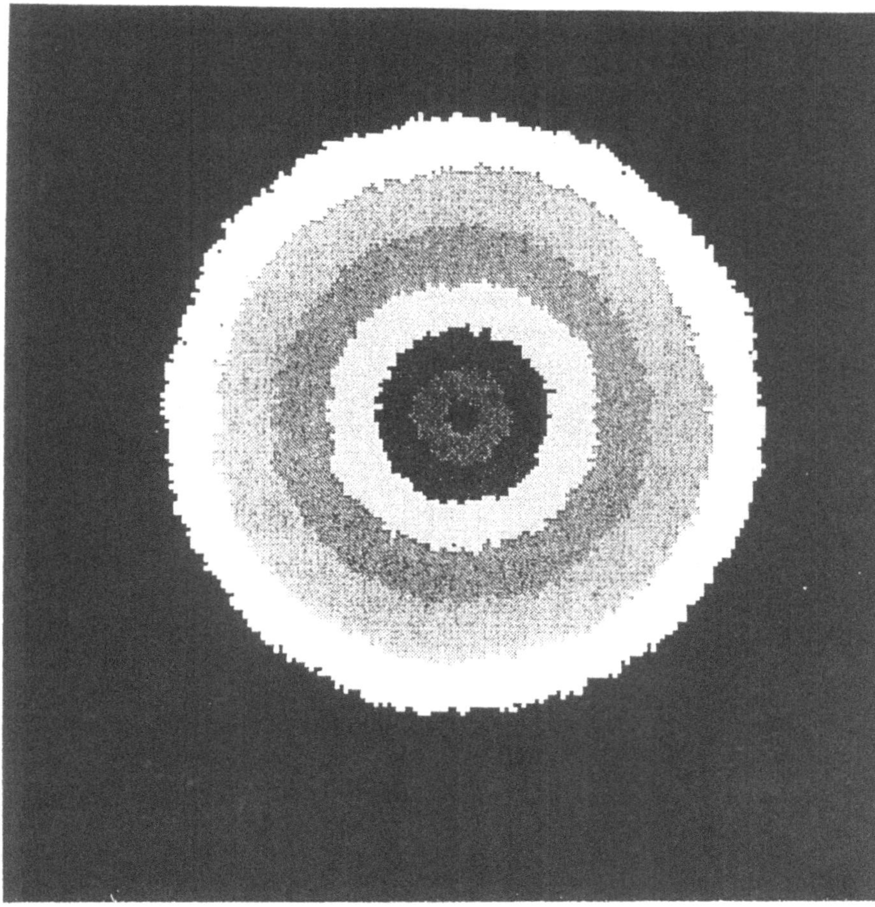

Fig. 7. Numerical simulations of $A + B_{(\text{static})} \to C_{(\text{inert})}$ where A particles are injected at rate $\lambda = 5$ at the center of the lattice and particles of type B are static and are located at each site of a $d = 2$ lattice. Plot of the reactant area after $t = 20$, 160, 540, 1280, 2500, 4320 and 6860.

while for the percolation system

$$R(t) = Lt^{-\gamma'}. \tag{21b}$$

We expect that at time $t^* \sim L^{d_w}$ these two rates become equal, since no "small" active cluster exist in the system above t^*. Indeed, equating the last two expressions of (21) reproduces Eq. (20). The prefactor L in (21b) assures that the reaction rate for percolation system is larger for $t < t^*$ than the reaction rate on the infinite cluster. Indeed, the ratio of the reaction rates at $t = 1$ is L^{2-d_f}.

5 The Reaction Rate $A+B_{static} \to C_{inert}$: Localized Source of A

Another system in which the reactants are initially separated and which is amenable to experiment, is the reaction $A+B_{static} \to C_{inert}$ with a localized source of A species. There exist many systems in nature in which a reactant A is "injected" into a d-dimensional substrate B where upon it reacts to form an inert product C. Recently such an experiment has been performed [40] by injecting iodine at a point of a large silver plate and measuring quantities of the reaction $I_{2\,gas} + 2Ag_{solid} \to 2AgI_{solid}$

First we consider N particles of type A that are initially at the origin of a lattice. The B particles are static and distributed uniformly on the lattice sites. Using an approximate quasistatic [41] analytical approach for trapping in a moving boundary we derived expressions for $C(t)$, the time-dependent growth size of the C-region and for $S(t)$ the number of surviving A particles at time t. For extremely short time $t < t_\times \sim \ln N$ we find $C(t) \sim t^d$. For $t > t_\times$ we find [42]

$$C(t) \sim Nf\left(\frac{t}{N^{2/d}}\right) \qquad \text{and} \qquad S(t) = N - C(t). \tag{22a}$$

The scaling function $f(u)$ is the solution to the differential equation

$$\frac{df}{d\tau} \sim k_d f^{-2/d}[1-f], \tag{22b}$$

and k_d is a constant, depending only on dimension. Figure 4 shows simulation data supporting (22a).

Now consider the case in which λ particles of type A are injected per unit time at the origin of the lattice. For this case we find [40]

$$C(t) \sim \begin{cases} \sqrt{8Dt\ln(\lambda^2 t/2D)} & d=1 \\ \pi\alpha t & d=2 \\ \lambda t & d=3, \end{cases} \tag{23}$$

and

$$S(t) \sim \begin{cases} \lambda t & d=1 \\ (\lambda-\pi\alpha)t & d=2 \\ C_3(\lambda)t^{2/3} & d=3. \end{cases} \tag{24}$$

In (23) and (24), α is the solution of $\alpha\pi = \lambda\exp(-\alpha/4D)$ and $C_3(\lambda) = (\lambda/4D)\cdot$ $\cdot(3\lambda/4\pi)^{2/3}$. Moreover, we find that for both one- and three-dimensional systems $C(t)$ satisfies the scaling relation

$$C(t) \sim \lambda^{d/(d-2)} g\left(\frac{t}{\lambda^{2/(d-2)}}\right). \tag{25}$$

Equations (23)–(25) have been supported by numerical simulations [40].

Equations (22) can be generalized for fractals;

$$C(t) \sim Nf\left(\frac{t}{N^{2/d_s}}\right), \tag{26a}$$

where $f(u)$ is the solution of the differential equation,

$$\frac{df}{du} \sim k_{d_s} f^{-2/d_s}[1-f]. \tag{26b}$$

Here d_s is the fracton dimension defined by $d_s = 2d_f/d_w$, in which d_f is the fractal dimension and d_w the diffusion exponent [43]. For the case of constant injection rate on a fractal we do not have an analytical derivation. However, we recently calculated [38] the number of distinct sites visited on a fractal by N random walkers starting from the origin, $S_N(t) \sim (\ln N)^{d_f/\delta} t^{d_s/2}$ with $\delta = d_w/(d_w-1)$. This result can be shown to be valid also for the number of distinct site visited by random walkers injected at the origin with a constant rate λ when replacing $N = \lambda t$. Thus we obtain that $(\ln \lambda t)^{d_f/\delta} t^{d_s/2}$ is an upper bound for $C(t)$.

Acknowledgements

We wish to thank F. Leyvraz, S. Redner, H. Taitelbaum and G. H. Weiss for useful discussions.

References

[1] M. v. Smoluchowski, Z. Phys. Chem. **92**, 129 (1917).
[2] See, e.g., S. A. Rice, *Diffusion-Limited Reactions* (Elsevier, Amsterdam, 1985).
[3] A. A Ovchinnikov and Y. B. Zeldovich, Chem. Phys. **28**, 215 (1978).
[4] D. Toussiant and F. Wilzeck, J. Chem. Phys. **78**, 2642 (1983).
[5] P. Meakin and H. E. Stanley, J. Phys. A **17**, L173 (1984).
[6] K. Kang and S. Redner, Phys. Rev. Lett. **52**, 955 (1984); Phys. Rev. A **32**, 435 (1985).
[7] K. Lee and E. J. Weinberg, Nucl. Phys. B**246**, 354 (1984).
[8] G. Zumofen, A. Blumen and J. Klafter, J. Chem. Phys. **82**, 3198 (1985).
[19] R. Kopelman, Science, **241**, 1620 (1988).
[10] M. Bramson and J. L. Lebowitz, Phys. Rev. Lett. **61**, 2397 (1988); J. Stat. Phys. **65**, 941 (1991).
[11] P. Argyrakis and R. Kopelman, Phys. Rev. A **41**, 2121 (1990).
[12] G. H. Weiss, R. Kopelman and S. Havlin, Phys. Rev. A **38**, 466 (1989).
[13] F. Leyvraz and S. Redner, Phys. Rev. Lett. **66**, 2168 (1991); S. Redner and F. Leyvraz, J. Stat. Phys. **65**, 1043 (1991).
[14] L. W. Anacker and R. Kopelman, Phys. Rev. Lett. **58**, 289 (1987); J. Chem. Phys. **91**, 5555 (1987).
[15] K. Lindenberg, B. J. West and R. Kopelman, Phys. Rev. Lett. **60**, 1777 (1988).
[16] D. ben-Avraham and C. R. Doering, Phys. Rev. A**37**, 5007 (1988).
[17] E. Clément, L. M. Sander and R. Kopelman, Phys. Rev. A **39**, 6455 (1989).
[18] L. Gálfi and Z. Rácz, Phys. Rev. A **38**, 3151 (1988).
[19] Y.E. Koo, L. Li, and R. Kopelman, Mol. Cryst. Liq. Cryst. **183**, 187 (1990); Y.E. Koo and R. Kopelman, J. Stat. Phys. **65**, 893 (1991).
[20] H. Taitelbaum, Y.E Koo, S. Havlin, R. Kopelman and G.H. Weiss, Phys. Rev. A **46** 2151 (1992)

[21] Z. Jiang and C. Ebner, Phys. Rev. A **42**, 7483 (1990).

[22] H. Taitelbaum, S. Havlin, J. Kiefer, B. L. Trus, and G.H. Weiss, J. Stat. Phys. **65**, 873 (1991).

[23] S. Cornell, M. Droz, and B. Chopard, Phys. Rev. A **44**, 4826 (1991).

[24] M. Araujo, S. Havlin, H. Larralde, and H. E. Stanley, Phys. Rev. Lett. **68**, 1791 (1992).

[25] M. Araujo, H. Larralde, S. Havlin, and H. E. Stanley, Phys. Rev. Lett. **71**, 3592 (1993).

[26] E. Ben-Naim and S. Redner, J. Phys. A **25**, L575 (1992).

[27] H. Larralde, M. Araujo, S. Havlin and H. E. Stanley, Phys. Rev. A **46**, 855 (1992).

[28] D. Avnir and M. Kagan, Nature, **307**, 717 (1984)

[29] G. T. Dee, Phys. Rev. Lett. **57**, 275 (1986).

[30] B. Heidel, C. M. Knobler, R. Hilfer and R. Bruinsma, Phys. Rev. Lett. **60**, 2492 (1986).

[31] R. E. Liesegang, Naturwiss, Wochensch, **11**, 353 (1896).

[32] T. Witten and L. M. Sander, Phys. Rev. Lett. **47**, 1400 (1981).

[33] R. A. Ball, Ausr. Gemmol. **12**, 89 (1984).

[34] K. F. Mueller, Science, **255**, 1021 (1984).

[35] H. Larralde, M. Araujo, S. Havlin and H. E. Stanley, Phys. Rev. A **46**, R6121 (1992).

[36] A. Bunde and S.Havlin eds.*Fractals and Disordered Systems* (Springer, Heidelberg 1991).

[37] S. Havlin and D. Ben-Avraham, Adv. in Phys. **36**, 695 (1987).

[38] H. Larralde, Y. Lereah, P. Trunfio, J. Dror, S. Havlin, R. Rosenbaum and H. E. Stanley, Phys. Rev. Lett. **70**, 1461 (1993).

[39] J. Crank, *Free and Moving Boundary Problems* (Clarendon Press, Oxford, 1984).

[40] H. Larralde, P. Trunfio, S. Havlin and H. E. Stanley, to be published.

[41] S. Alexander and R. Orbach, J. Physique Lett. **43**, L625 (1987).

Some aspects of pattern formation in reaction-diffusion systems

Michel Droz

Département de Physique Théorique
Université de Genève,
CH-1211 Genève 4, Switzerland

1 Introduction

In the recent years, reaction-diffusion problems have stimuled a large body of work in many different directions. On the experimental side, new techniques have been developped to investigating Turing patterns. The availability of computers with massive parallel architecture makes it possible to perform extensive numerical simulations based on microscopic (or mesoscopic) models. Thus the role of the fluctuations, not included in a description based on mean-field like equations, is now quite well understood. Moreover, on the theoretical side, several advances, based on renormalisation group analysis of the field theories associated to these reaction-diffusion processes, have been achieved.

The purpose of this contribution is to briefly review some aspects of these recents developpements.

In a first section, we revisit the problem of anomalous kinetics in simple reaction diffusion models as $A + A \to 0$ or $A + B \to 0$ in view of the recents field theoretical developments. In the next section we consider the problem of reaction–diffusion fronts and particularly the case of a stationnary front obtained by imposing particles flux at the boundaries. A simple scaling theory leads to exact predictions concerning the long time proporties of the front. Then, the problem of Liesegang pattern formation is discussed. A cellular automata microscopic model reproducing the generic properties observed experimentally is presented. Finaly, in the last section, the problem of the role of the microscopic fluctuations in the formation of Turing patterns is briefly considered.

2 Anomalous kinetics

Chemical reactions in which several species react are usually described in terms of macroscopic rate equations. These rate equations give the time evolution for the local averaged concentrations of the different species. They assume that the reaction is completely described by the local average densities, i.e. the reaction introduces no correlations between the reacting species. This is reminiscent of a mean-field approximation in statistical physics, and therefore neglects an important aspect of the problem, namely the microscopic fluctuations.

For homogeneous diffusion-reaction systems, it is well known that the fluctuations play an important rôle. The simplest example is may be provided by the annihilation reaction $A + A \to 0$ [1]. The solution of the rate equations predicts that the concentration of A will decrease in time as $a \simeq t^{-1}$ while a calculation taking into account the microscopic fluctuations in the particle density a gives $a \simeq t^{\alpha}$, with $\alpha = \mathrm{Min}(1, \frac{d}{2})$.

The situation is even more surprising in the case $A + B \to 0$. For an initial state for which the number of A and B particles are the same, the solution of the rate equations predicts again that the concentration of A (or B) will decrease in time as $a \simeq t^{-1}$. A calculation taking into account the microscopic fluctuations gives for the particle densities n $n \simeq t^{\alpha}$, with $\alpha = \mathrm{Min}(1, \frac{d}{4})$ [2]. Thus the mean-field prediction is only valid above an upper critical dimension $d_u = 4$. Below this upper critical dimension, there is a spatial segregation of the two constituants. The system splits in patches reach in A or B. Thus, the componants can only react at the borders of the domains, hence the anomalous kinetics. The reason for which one has segregation is very subtle. As first noticed by Cornell [3], this is mainly an effect of the initial conditions. Providing that the particles in the initial state are randomly distributed (Poissonian state), it suffices to integrate the mean-field equations to obtainthe anomalous behavior described above. Moreover, if there are some correlations in the initial state, different behaviors can be expected.

An large number of papers have been devoted to the problem of anomalous kinetics and we do not want to review them here. However,we would like to make the following remark. Although non mean-field, the critical exponents involved in these kinetic problems are "simple", i.e. basicaly simple rational numbers. Is this accidental or is there some more fundamental reasons for that? The answer to such question can only be given in the framework of a global theoretical approach. One is interested in the long time behavior of systems which dynamics is "critical", i.e. characterised by power laws. Hence it seems natural to look for a renormalisation group approach. Two problems arise; first how to construct a field theory for such reaction–diffusion and second how to defined and implement a renormalisation group scheme.

The first question has been revisited recently by Droz and McKane [4]. To describe these nonequilibrium situations, one has to start at the master equation level to keep track of all the fluctuations. However, to explain the scaling behaviour which takes place in the long time regime, it is natural to seek a coarse-grained theory. It is, however, crucial that this coarse-grained theory keeps track of all the important fluctuations contained in the system.

Two different approaches, apparently disconnected one from the other, have been proposed in the literature: the Fock space formalism [5] and the Poisson representation [6]. As shown explicitly in [4], although starting from different point of view, the two above approaches are equivalent. Namely, there is a one to one correspondence between the Liouvillians and the field theories of the two methods. Once the Liouvillan obtained from the master equation, one can derive a path-integral representation for it and coarse-graining gives a field-theoretic description of the process under consideration. We shall not go into the details of these two formalisms here but simply quote the form of the Lagrangian of the field theory for the $A + A \to 0$ model [7]:

$$L = \int d^d r \left[i\tilde{\phi}(\dot{\phi} - D\nabla^2\phi + 2Dk\phi^2) - Dk\tilde{\phi}^2\phi^2 \right] \tag{1}$$

The field $\phi(\mathbf{r}, t)$ corresponds approximately to the local particle density, while the auxiliary field $\tilde{\phi}(\mathbf{r}, t)$ has no particular physical meaning.

A renormalisation group analysis can then be performed following the lines of the dynamical renormalisation group approach used in dynamical critical phenomena. The first work along this line has been made by Droz and Sasvari[7] for $A + A \to 0$ with external sources. Generalisation to the case $nA \to 0$ has been investigated by Lee[8]. Extensions to $A + B \to 0$ have been studied by Cardi and Lee[9].

Let us return to the simple case $A + A \to 0$. The renormalization group analysis procedure is the following. First, one eliminates the Fourier components of the fields belonging to the shell $\Lambda/b < k < \Lambda$. Then, one rescales the lengths and time according to

$$k' = bk, \qquad t' = b^{-z}t. \tag{2}$$

Fields and vertices are rescaled as required by their dimensions.

It turns out that no two-legs diagram is generated. The bare propagator is not renormalized, hence the fields do not have an anomalous dimension (the theory is super-renormalisable). This is basically the reason why the critical exponents discribing the anomalous kinetics are "simple".

Moreover, the new couplings absent in the initial Lagrangian and generated during the renormalization procedure, are irrelevant. In $d > 2$, the stable fixed point corresponds to the limiting case in which the reaction is very slow in comparison with the diffusion. Thus the diffusion mixes the particles efficiently enough so that a homogeneous reaction takes place for which the mean-field theory is valid.

In $d < 2$, the behavior of the system is governed by the nontrivial fixed points. We have, in fact, a line of fixed points parametrized by the initial values of the couplings. On this fixed line, diffusion and reaction proceed at a comparable rate, resulting in an interplay of the two processes, hence the breakdown of the mean-field theory. The critical exponent are found exactly in all dimension.

For the the $A + B \to 0$, the situation is more subtle. As shown by Cardi and Lee, the corresponding field theory has also an upper critical dimension $d_u = 2$. However, the term in the Lagrangian describing the (poissonian) initial state is very important in this case. New terms are generated during the renormalisation procedure which are relevant for a dimension $d \leq 4$. These new terms reflect the apparition of segregation in the system.

Thus the renormalisation group approach of the corresponding field theories gives an unified description of the problem.

3 Reaction-diffusion fronts

A more complicated class of problem is that for which the initial condition is inhomogeneous. The inhomogeneities of the problem often arise from the presence of sources or sinks of particles. This may lead to the formation of reaction-diffusion fronts.

3.1 The time dependant problem

A simple model displaying a reaction front has been analysed at the level of the rate equations by Gálfi and Rácz[10]. They consider two species A and B diffusing on a two-dimensional substrate and reacting to form a new species C. Provided that the two diffusing reagents A and B are initially separated in space, they will react in a confined region ("reaction front"). An appropriate choice of geometry reduces the problem to one dimension in this approximation. In particular, it is assumed that, at time $t = 0$, the system is uniformly filled with A for $x \leq 0$, and uniformly filled with B for $x > 0$. The study of the properties of this front provides relevant information about the production of C particles. The analytic treatment of Gálfi and Rácz assumes that the reaction rate is expressed in terms of the product of mean particle densities, which we shall henceforth refer to as the "mean field" approximation. Their main result is that the distribution of the procuction of C particles, i.e. the reaction front, has a scaling form in the large time limit. In particular, the width of the front behaves as $W(t) \simeq t^{1/6}$.

The rôle of fluctuations in this system has been investigated within several approaches.

1. A. Cellular automata simulations[11, 12]
2. B. Scaling theories[10, 12]
3. C. Mapping on quantun Hamiltonian systems[13]

The main results of these approaches is that the fluctuations play an important rôle in two dimension $d \leq 2$. However, a precise determination of the exponents by numerical simulations can be a very difficult task. Indeed, very slow transients can be present and one is never sure if the true assymptotic regime has been reached. Accordingly, it is useful to consider a different, but related model in which the front is stationnary.

3.2 The steady state problem

One considers the more general reaction-diffusion system $mA + nB \to C$. One study[14, 15] the scaling behavior of the front formed in the steady state reached by imposing antiparallel current densities (henceforth just "currents") $J_A = m|J|$ and $J_B = -n|J|$ of A- and B-particles at $x = -\infty$ and $x = +\infty$ respectively. This situation is much easier to investigate, since the front is no longer time-dependent and there are only three relevant parameters (J, diffusion constant D, reaction constant k). However, the results for the scaling exponents and critical dimension may be directly applied to the time-dependent case, where the front is formed quasistatically by currents $\propto t^{-\frac{1}{2}}$. One finds[15] the surprising result that dimensional analysis, coupled with consistency arguments, are sufficient to show that the critical spatial dimension is $d_c \equiv 2/(m+n-1)$. Above d_c, the scaling limit $J \longrightarrow 0$ is equivalent to the mean-field limit $k \longrightarrow 0$. For $d < d_c$ scaling is equivalent to $k \longrightarrow \infty$, and one finds an exact prediction that corresponds to $\alpha = \frac{1}{4}$ in the time-dependent case for $d = 1$, $(m,n) = (1,1)$. The theoretical predictions are well supported by numerical simulations for several values of n and m. This approach shows clearly that there is only one length into the problem. However, one has to keep in mind that the above results are derived under the as sumption that the front is formed quasistatically.

4 Liesegang structures

4.1 Phenomenological description

A more complicated class of problems concerns Liesegang patterns formation. These patterns are produced by precipitation in the wake of a moving reaction front.

Typical experiments exhibiting such a pattern formation consist of a test tube containing a gel in which a chemical species B (for example $AgNO_3$) is uniformly distributed with concentration b_0. Another species A, with concentration a_0 (for example HCl) is allowed to diffuse into the tube from its open extremity and chemically react with B. As this reaction goes on, formation of consecutive bands of precipitate (AgCl in our example) is observed in the tube, provided that the concentration a_0 is large enough compare to b_0 so that the reaction propagates in the tube. A striking feature of this process is that, after a transient time, these bands appear at some positions x_i and times t_i that obey simple laws. More precisely, it is first observed that the center position x_n of the nth band is related to the time t_n of its formation through the so-called time law $x_n \sim \sqrt{t_n}$. Second, the ratio $p_n \equiv x_n/x_{n-1}$ of the positions of two consecutive bands approaches a constant value p for large enough n. This last property is known as the Jablczynski law[16] or the spacing law. Finally, the width w_n of the the nth band is an increasing function of n. The presence of bands is related to the geometry of the experiment, i.e the use of a test tube with axial symmetry and most of the experiments have been performed in this case. However, for more complicated geometries, different shapes may be obtained. A well known example is provided by the rings formed in agate rocks[16],[17],[18]. In the reaction–diffusion process described above, the reaction front position $x_f(t)$ obeys the relation $x_f(t) \sim \sqrt{t}$, with an amplitude depending on the difference of the concentrations a and b. This behavior is mainly a consequence of the diffusive character of the motion of the particles. As the Liesegang patterns are formed in the wake of a moving reaction front, the time law appears to be a simple consequence of the diffusive dynamic. However, spacing and width laws are not a direct consequence of a simple reaction-diffusion dynamics. To produce patterns which are not infinitely slim, it is crucial to introduce an extra nucleation-aggregation mechanism. The formation of Liesegang patterns have been investigated by many researchers, both from an experimental and a theoretical point of view. The models proposed so far belong to three categories[19]: sol coagulation models, competitive particle growth models and supersaturation models. None of these models are able to account for all experimental observations. For example, particular situations, called inverse banding [20] where the distance between successive rings decreases as time increases are not directly explained by these models. New ingredients, expressing the capacity of the gel to dissolve the precipitates should be introduced. However, one believes, following Zeldovitch[21], Dee,[22] and Le Van et al.[23] that the supersaturation mechanism based on Ostwald ideas plays a crucial role in the band formation.

In the most recent scenario proposed by Dee[22], the two species A and B react to produce a new species C which also diffuses in the gel. When the local concentration of C reaches some threshold value, nucleation occurs. The droplets of nucleated particles (let us call them D) formed at the reaction front deplete their surroundings of the reaction product C. As a result, the level of supersaturation drops dramatically and the nucleation process stops. After some time, the reaction front has moved away and the concentration of product at the moving front reaches a value large enough allowing the nucleation to occur again. As a result, separated bands will appear. The above two scenario has been investigated at the mean-field level[22]. However, in view of the importance of the fluctuations in such systems, one would like to consider such models at a microscopic level. This has been done recently[17], [18] in term of cellular automata models.

4.2 A cellular automata model

The model is defined on a two-dimensional square lattice. For the axial system (test tube), the initial conditions are the following: at time $t = 0$, the left part of the system ($x \leq 0$) is randomly occupied by A particles, with a density a_0 and the right part ($x > 0$) is filled with B particles with a density b_0.

The particles are restricted to move along the main directions of the lattice, according to a discrete time clock. The particles which meet at a same site interact or transform according to the following rules of the cellular automaton. Four basic mechanisms are introduced in our model:

1. (i) diffusion
2. (ii) reaction
3. (iii) spontaneous precipitation
4. (iv) aggregation

Diffusion corresponds to a simultaneous random walk of all particles on the lattice[24]. We will use four bits at each site to represent the particles of each diffusing species. Each of these four bits describe the absence or the presence of a particle of the given species traveling in one of the four possible directions of the lattice (up, right, down or left). Our dynamics is such that it never happens that two or more particles of a given species enter simultaneously the same lattice site in the same direction. This fact (known as an exclusion principle) guarantees that four bits are enough, in two-dimensions, to describe each species at any time.

In absence of diffusion and reaction, the particles would simply move in straight lines, hopping at each time step to a neighboring site. Diffusion is produced as follows: at each time step, the four bits representing the configuration of particles entering at each site undergo a random permutation. In practice, we simply consider a random rotation (of 0, $\pi/2$, π or $-\pi/2$) of the lattice directions. Then, the particles move to a next site of the lattice, according to their new velocity direction. In this way, the number of particle is conserved during the updating and the exclusion principle is always obeyed without having to deal with conflicting particle motions. The probabilities of each of the random rotations allow us to adjust the diffusion coefficient of each species.

The production of C particles takes place according to the reaction $A+B \to C$. At the level of the cellular automata rule, this process is modeled as follows: when two particles of species A and B collide at a given site, they disappear with some given probability k and produce a C particle. This reaction is only possible if there is still room for this new C particle (the exclusion principle restricts their number to four at each site). If no reaction takes place, the two initial particles ignore each other and continue their own motion. It has been shown [25] that the above microdynamics reproduce the usual reaction-diffusion equations when a mean-field approximation is introduced.

Nucleation and aggregation phenomena are implemented in our model according to general principles of supersaturation theory.

First, the C particles, once created, will diffuse until their local density, computed as the number of particle in a small neighborhood, reaches a threshold value K_{ps}. Then, they spontaneously precipitate and become D particles at rest (nucleation). We have considered 3×3 neighborhoods centered around each site. Larger neighborhoods could possibly be envisaged, but they would have the tendency to average too much local density fluctuations.

Second, the C particles located in the vicinity of precipitate D particles will aggregate provided that their density is larger than an aggregation threshold $K_p < K_{sp}$. A C particle sitting on the top of a D always becomes a D. The parameters K_p and K_{ps} are the two main control parameters of the model. The introduction of these critical values refers to the qualitative models of solidification theory, relating supersaturation and growth behavior[19]. From a microscopic point of view, it is more common to describe the aggregation process in terms of a noise reduction algorithm : C particles aggregate on a D cluster only after several encounters. In order to separate the different time scales in our model (diffusion versus aggregation), it is more convenient to use the technique described above in terms of the threshold K_p.

Exemples of bands are given on Figure 1. The three generic laws are well satisfied [17], [18].

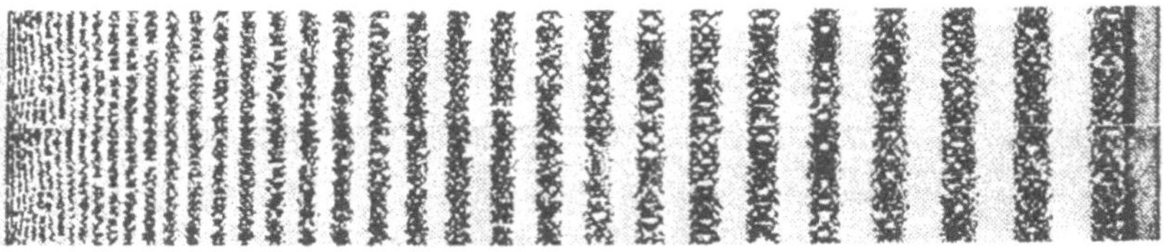

Fig. 1. Formation of Liesegang bands in a test tube, as obtained from the microscopic model described above.

It is experimentally well known that Liesegang patterns are only found if the parameters of the experiment are thoroughly adjusted. Outside of the region where Liesegang patterns are formed, we have observed from our simulation that other types of patterns are obtained. These various patterns can be classified in a qualitative phase diagram, as discussed in [18].

Moroever, another interesting case is the formation of rings or spirals [18] obtained when the reactant A is injected in the central region of a two-dimensional gel initially filled with B particles (cylindrical symmetry). An example is given on Figure 2. Here, a local defect produced by a density fluctuation develops and a spiral of precipitate appears instead of rings.

Thus, such cellular automata model for the formation of Liesegang patterns are able to reproduce many aspects observed experimentally. A significant advantage of this approach as compared to other ones previously used in the literature is the fact that fluctuations are included. This can lead to solutions with broken symmetry (spirals) that are experimentally observed and that cannot be obtained in a theory based on deterministic equations.

5 Turing patterns

Under certain conditions, reaction-diffusion systems can produce heterogeneous steady state spatial patterns which evolve by diffusion driven instabilities. These patterns, called Turing patterns [26] are characterized by an intrinsic wavelength independent of the size of the system.The domain of parameters space for which such patterns exist is called the Turing space.

This is a facinating subject that we shall not develop here in details. We simply would like to make of fiew remarks concerning the fluctuations.

These systems are generally described by macroscopic equations for the evolution of the local average densities of the chemicals. This mean-field approximation neglects an important element of the problem, namely the microscopic fluctuations. Here again, probabilistic cellular automata models taking the fluctuations into account in a natural way have been introduced [27, 28, 29]. These

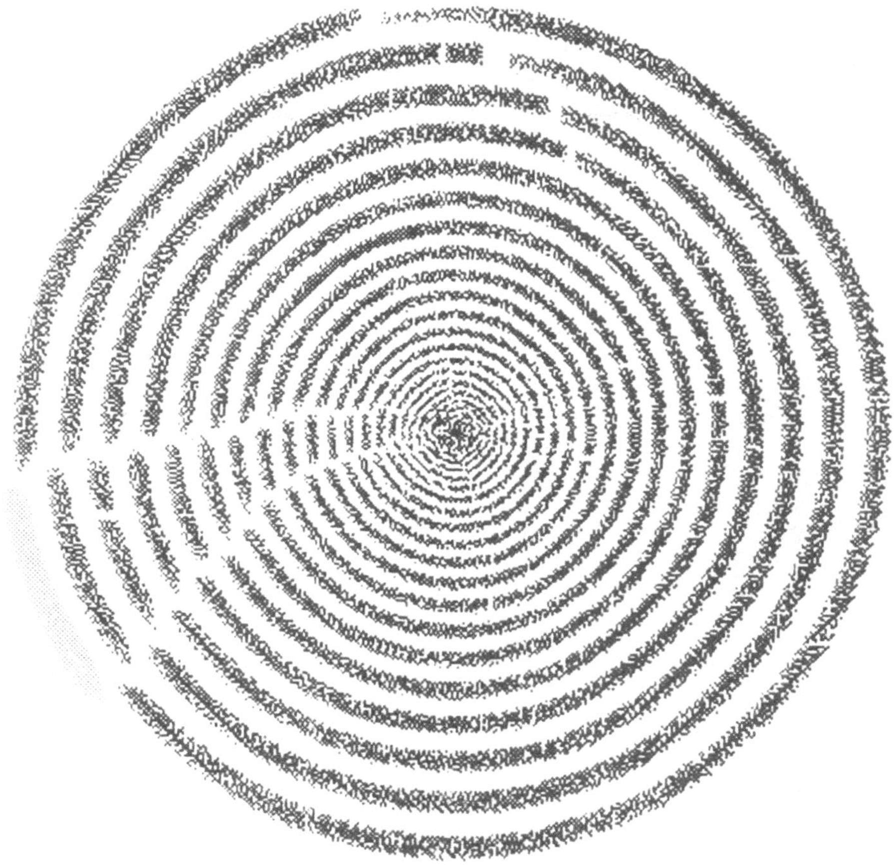

Fig. 2. Formation of Liesegang rings spiral in two dimensions.

models capture the essential features of the problem but are simple enough to permit large simulations.

The fluctuations have two main effects [27]:

1. i). The Turing space is enlarged.
2. ii). The evolution of the modes is drastically affected in short and intermediate time regime.

Some modes which grow exponentially with time in the deterministic theory do not have this exponential growth in the CA model. This can be understood in terms of couplings among the modes induced by the fluctuations.

6 Conclusions

We have seen in the different examples treated above that reaction-diffusion syt ems offer a vast choice of problems which can be approached by using many different theoretical tools. As often in nonequilibrium statistical mechanics, even to solution of the simplest models is very chalenging both from the experimental and theoretical point of view.

References

[1] D. Toussain and F. Wilczek, J. Chem. Phys **78**, 2642, (1983)
[2] M. Bramson and J. Lebowitz, J. Stat. Phys. **62**, 297 (1991)
[3] S. Cornell, Unpublished.
[4] M. Droz and A. McKane, J. Phys. (1994), to appear.
[5] M. Doi, J. Phys. **A9**, 1465 (1976)
[6] C. W. Gardiner and S. Chaturvedi, J. Stat. Phys. **17** 429 (1977); J. Stat. Phys. **18** 501 (1978).
[7] M. Droz and L. Sasvari, Phys. Rev. **E48** R2343 (1993).
[8] B. Lee, J. Phys. (1994), to appear.
[9] J. Cardi and B. Lee, (1994), to appear.
[10] L. Gálfi and Z. Rácz, Phys. Rev. **A38**, 3151 (1988)
[11] B. Chopard and M. Droz, Euro. Phys. Lett. **15**, 459 (1991).
[12] S. Cornell, M. Droz, and B. Chopard, Phys. Rev. **A44**, 4826 (1991)
[13] F. C. Alcaraz, M. Droz, M. Henkel, and V. Rittenberg, Annals of Physics **230**, 250 (1993)
[14] E. Ben-Naim and S. Redner, J. Phys. A **25**, L575 (1992)
[15] S. Cornell and M. Droz, Phys. Rev. Lett. **70**, 3824 (1993)
[16] K. Jablczynski, Bull. Soc. Chim. France **33**, 1592 (1923)
[17] B. Chopard, P. Luthi and M. Droz, Phys. Rev. Lett. **72**, 1384 (1994)
[18] B. Chopard, P. Luthi and M. Droz, Jour. Stat. Phys. (1994), to appear.
[19] H. K. Henisch, "*Periodic precipitation*", Pergamon Press, (1991)
[20] B. Mathur and S. Ghosh, Kolloid-Zeitschrift **159**, 143 (1958)
[21] YA. B. Zeldovitch, G. I. Barrenblatt and R. L. Salganik, Sov. Phys. Dokl. **6**, 869 (1962)
[22] G. T. Dee, Phys. Rev. Lett. **57**, 275 (1986)
[23] M. E. Le Van and John Ross, J. Phys. Chem. **91**, 6300 (1987)
[24] B. Chopard and M. Droz, J. Stat. Phys. **64**, 859 (1991)
[25] B. Chopard, S. Cornell, M. Droz and L. Frachebourg in the proce edings of the "Workshop on Cellular Automata for Astrophysical Phenomena", (1993), edt. World Scientific.
[26] A. Turing, Phil. Trans. Roy. Soc. Lond, **B237** (1952) 37
[27] M. Droz and L. Frachebourg, Helv. Phys. Acta **66**, 97 (1993).
[28] D. Dab, J.-P. Boon and Y.-X. Li, Phys. Rev. Lett. **66**, 2535 (1991)
[29] R. Kapral, A. Lawniczak and P. Masiar, Phys. Rev. Lett. **66**, 2539 (1991)

Interface Bursting and Interface Depinning

J.-F. Gouyet

Laboratoire de Physique de la Matière Condensée
Ecole Polytechnique, 91128 Palaiseau, France

Abstract We recall the main features of the fluctuations of self similar interfaces which can be observed in diffusion, invasion of a porous medium by a fluid, thin magnetic films of even corrosion patterns. The relations of this behavior with other physical problems like fragmentation of clusters, cluster dynamics of a lattice gas of particles, or some features of dynamical percolation is also presented. Finally its connections with interface depinning and growth of rough surfaces is discussed.

1 Introduction

There has been a considerable renewal of interest, these last years, in the analysis of interface motion in presence of disorder. Fluctuations of fronts can be observed in many physical situations like for instance fluid invasion in porous media [1], growth in systems with quenched disorder (magnetic domains [2], thin films corrosion [3], charge density waves [4]), diffusion fronts [5], some earthquakes models [6], spreading of various epidemics or forest fires models [7], or sandpiles models [8].

The physical observation of fluctuating interfaces can be made by looking to the evolution of the geometry with time (this is the most frequent case that we shall discuss below), or of the injected flux in the case of fluid invasion [9], or looking to the time fluctuations of an electrical potential or of the conductivity in the case of mercury injection in rock [1, 10], to the time correlation of acoustic emission in the case of earthquakes [6], or to magnetooptics observations in the case of ultra-thin magnetic film [2], etc.

Actually, all these phenomena can in general be grouped under a common behavior called self-organized-criticality (S.O.C) [11].

We shall essentially describe here the behavior of self-similar interfaces, for which the fluctuations are related to sudden connection (or invasion) of clusters, called bursts. The fluctuations of these self-similar interfaces, can be called "interface bursting", to distinguish them from the fluctuations of self-affine interfaces, which are associated to "interface depinning". We shall also say some words on its relation with growing rough interfaces for which the randomness leads to such local pinning.

Interface bursting can in general be described by ordinary percolation [12], while various models are probably necessary to describe interface depinning (as for instance directed percolation [13, 14]). The first studies involving interface bursting were made on diffusion fronts and concerned their possible relation with $1/f^{\alpha}$ noise generation, in particular in the case of interdiffusion in electrical contacts [5, 15, 16]. The case of fluctuations of invasion fronts was then considered [17-19]. Similar ideas, involving interface bursting of percolating structures have been also suggested to interpret acoustic emission in seismic phenomena [20].

Interface depinning is found in imbibition phenomena, dislocations in crystals, charge density waves, growth of magnetic domains, etc. It creates stationary self-affine structures, in which the kinetics consist in a succession of depinning from quenched impurities. Robbins and his collaborators showed [19] how the two approaches of drainage and imbibition could be related in the case of two dimensional fluid invasion. An other interesting model has been introduced by Tang and Leschhorn [13] and by Buldyrev et al. [14]. They propose directed percolation as a generic mechanism at least in two dimensions for interface pinning in media with quenched disorder: the depinning only occurs when the slope is large enough so that, when there exists a directed path of pinning sites, the growth never reaches high enough slope to depin the interface which then comes to a halt. We shall discuss later some relations between such interface depinning and interface bursting.

The motion of these interfaces is very generally driven by a force which can be of a thermodynamic origin as in diffusion fronts (gradient of concentration of particles) or an external force as in fluid invasion (pressure applied to the injected fluid), in earthquakes (tectonic plates motion), in magnetic films (applied magnetic field), etc.

In addition it has appeared to be very convenient to consider systems in which a gradient of concentration is present [12, 14, 21] because this automatically leads to an interface located in the critical region, and because all the interface remains bounded.

As it is related to connections and disconnections of percolation clusters, considerations on interface bursting immediately makes a link with other domains of the physics of clusters: From the general behavior of connection-disconnection processes it is easy to derive the kinetics of dilution (or growth) rates [22] and the fragmentation properties [23] of percolation clusters. Also, the same kind of process may be used to examine the dynamic percolation problem [24] in the extreme case of a very fast diffusing tracer (or another species of particles) in a lattice gas of slow diffusing particles.

This paper is essentially a survey of several former publications, but new unpublished insights on noise and dynamical percolation will be discussed. In section 2 we shall present the main results concerning interface bursting, consid-

ering both fluctuations of surfaces and fluctuations of mass. The consequence of these fluctuations on noise generation will be examined in section 3, and relation with self-organized criticality will be pointed out. In section 4, a Smoluchowski equation will be written which describes the cluster-cluster behavior in diffusion problems or the percolation with cluster size dependent dilution rates, while section 5 will be devoted to the consequences on the fragmentation of percolation clusters. The dynamic percolation problem of a very fast tracer in a slowly evolving lattice gas will be considered in section 6. Finally in section 7, we shall discuss the relations between interface bursting and interface depinning

2 Generalities on the Dynamics of Interface Bursting

As indicated in the introduction we consider here essentially the fluctuation of surfaces of physical phenomena which can be mapped onto a percolation model. The interface has then a self-similar geometry which, depending on the particular situation and on the space dimension, corresponds to the hull (external perimeter), to the so called "Grossman-Aharony" perimeter (in $d = 2$) [25], or even to all the perimeters, of the infinite percolation cluster.

During time evolution, bursts appear which are simply due to the connection or the disconnection of finite percolation clusters (diffusion case) or to the invasion of a finite percolation cluster by a fluid (drainage) or to the sudden reversal of magnetic moments of a finite percolation cluster, etc.

In all these problems it has appeared that it is very convenient to have a gradient of concentration (diffusion) or of threshold of bursting sites (we call bursting site the site with largest radius in a drainage problem) as it gives, among other advantages, an upper limit to the size of the burst, and imposes the interface to be in the critical region. It is not the place here to enter into the details of the model which have been published in various places [12, 18, 26]. The gradient technique has been used in many other numerical studies where it proved to be extremely useful [21]. We shall only gives here the main results, taking as basic examples diffusion fronts (DF) and invasion fronts (IF).

In quasi-static situation, the interface is fractal with dimension D_{f}; this interface is for instance the external perimeter of the "infinite" percolation cluster connected to the source of particles (DF) or of injected fluid (IF), and due to the presence of a gradient its width remains bounded and equal to σ_{ft}. It is located in the critical region around the concentration $p = p_{\mathrm{c}}$. The clusters have themselves a fractal dimension D.

When time evolves the structure of the interface changes either by connection or disconnection of a neighboring cluster (DF), due to the jump of a particle, or by invasion (breakthrough via the easiest throat) of a neighboring cluster (IF). In both case the main theoretical framework is the same.

The ingredients of the problem [15, 16, 18] are:

a) The average number of clusters of size s, close enough to the front to be connected or disconnected in one jump (DF case), or the average number of clusters of size s close enough to the front to be invaded in one breakthrough

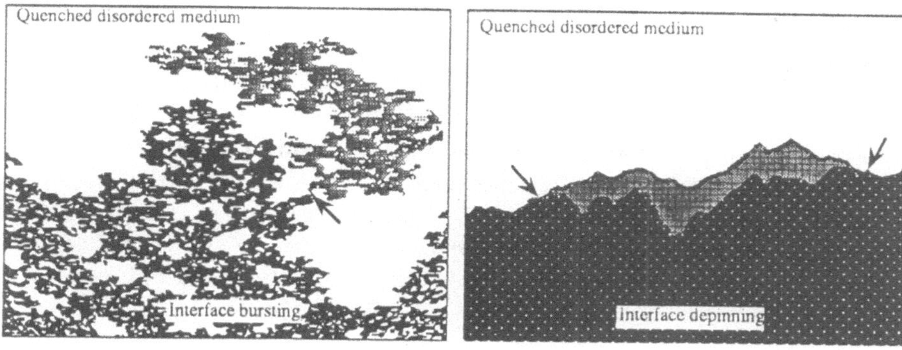

Fig. 1. In the left figure, the system grows by a succession of bursts (gray region), via a very localized breakthrough indicated by the arrow. The interface is self-similar, and the events (bursts) can be considered as independent; in the right figure, the growth can be better seen as a succession of depinning leading to a new interface equilibrium (limit of the gray region). The interface is self-affine, and the successive depinnings are correlated. In both cases the "source" (of injected fluid, etc.) is at the bottom of the figures.

(IF case). This number is, in a region of concentration p,

$$\mathcal{M}_s(p) \cong s^{-1-D_f/D} \mathcal{F}_s\left((p-p_c)s^{\sigma_s}\right) p_f(x)\, C_s(s/s_{\max})$$

where $\sigma_s = D_{\text{red}}/D = 1/(\nu D)$; ν is the critical exponent of the correlation length; C_s is a cutoff function which cuts the values of s larger than $s_{\max}$; $p_f(x)$ is the probability to find the front at coordinate x along the direction of the gradient of concentration. The presence of a cutoff in a region of concentration p is due to existence of the gradient of concentration ∇p, which limits the width of the front to σ_{ft}:

$$s_{\max} \cong s_M \phi\left((p-p_c)/\delta p\right)$$

with $s_M \propto \sigma_{\text{ft}}^D$, and $\delta p = \sigma_{\text{ft}} \nabla p \propto \nabla p^{1/(1+\nu)}$.

It may be needed to know the perimeter fluctuations (case for instance of impedance fluctuations of diffused electrodes). The two results are related via the equality

$$h\, \mathcal{M}_h(p) = s\, \mathcal{M}_s(p)$$

In fact the behavior of $\mathcal{M}_h$, the number of clusters with perimeter h, was first considered [15].

We can use either scaling constraints [15] or a box counting method [16] to determine this behavior.

b) The connection or disconnection of a cluster of size s, occurs at a frequency π_s^0 which depends on the number of red-bonds in a domain of linear size R_s, the main radius of an s-cluster, and

$$\pi_s^0 \cong s^{\sigma_s} \mathcal{P}\left((p-p_c)s^{\sigma_s}\right)$$

This frequency π_s^0 is in fact a frequency per event, or more exactly per unit time when each event takes the same time θ. This was the assumption taken in ref. [15]. However, this is not the case, for instance, in invasion experiments: the time θ_s of an event of size s, depends on the size s.

c) Duration of individual events:

Each event is supposed to take a time θ_s depending on it size and in general it follows a power law in s. Let us take two examples:

- Fluid invasion of a cluster: just after a breakthrough, the pressure sufficient to push the invading fluid though a connecting throat, is larger than all the capillary pressures associated with all the throats of the cluster which is invaded. We can allow that the flux φ through the connecting throat remains constant and that,

$$\theta_s \cong \frac{sV_{\text{pore}}}{\varphi}$$

where V_{pore} is the average volume of a pore.

- Domain walls of ferromagnetic thin films (or in some cases, corrosion patterns): when a blocking spin has been turned over after a slight increase of the applied magnetic field, there is a cluster of spins connected to the blocking spin, such that the energy to reverse these spins remains below the external energy. The growth propagates in this cluster. A rough approximation is to say that the spins at a chemical distance $\ell = 1, 2, 3$, etc. (i.e. which can be reached in at least ℓ steps) from the blocking spin are successively reversed, and that each shell takes the same amount of time θ to be reversed so that

$$\theta_s \cong \theta \ell_{\text{max}}$$

where l_{max} is the average chemical distance of a cluster of size s: $s = (\ell_{\text{max}})^{d_e}$, and

$$\theta_s \cong \theta s^{1/d_e}$$

where d_e is the spreading dimension. For percolating cluster $d_e \cong 1.68$ ($d = 2$), 1.88 ($d = 3$).

So, in general

$$\theta_s \cong \theta s^{\varepsilon}$$

The event frequency per unit time and per cluster of size s is then

$$\pi_s \cong \pi_s^0 \frac{\theta}{\theta_s}$$

d) The number $N_{\text{ev}}(s)$ of events per unit time:

It can be immediately determined from the above quantities after integration over spatial domain (i.e. along coordinate x, the direction of the concentration gradient in a d-dimensional system of size L)

$$N_{\text{ev}}(s) = L^{d-1} \int dx\, \mathcal{M}_s(p)\, \pi_s(p)$$

which can be shown to have the following behavior [15, 18]:

$$N_{\mathrm{ev}}(s) = L^{d-1} s^{-\mathcal{Y}_s - \varepsilon} \, |\nabla p|^{\alpha_N} \, K_s\!\left(\frac{s}{s_M}\right)$$

where $\alpha_N = (D_{\mathrm{f}} - d + 1)\nu/(1+\nu)$ is the exponent of the front surface dependence on the gradient of concentration [27] (see [12] and [26] for more details).

It has been shown when studding the case of invasion percolation [18] that

$$\mathcal{Y}_s = 1 + \frac{D_{\mathrm{f}} - D_{\mathrm{red}}}{D}$$

where $D_{\mathrm{red}} = 1/\nu$ is the exponent of the dependence of the disconnecting bonds (red-bonds) or of the connecting bonds (anti-red bonds) with the size of the window of observation. This expression for $\mathcal{Y}_s$ has been checked numerically in $d = 2$ and is determined exactly in a mean field approach, that is in any $d \geq 6$ (see the discussion in ref. [22]).

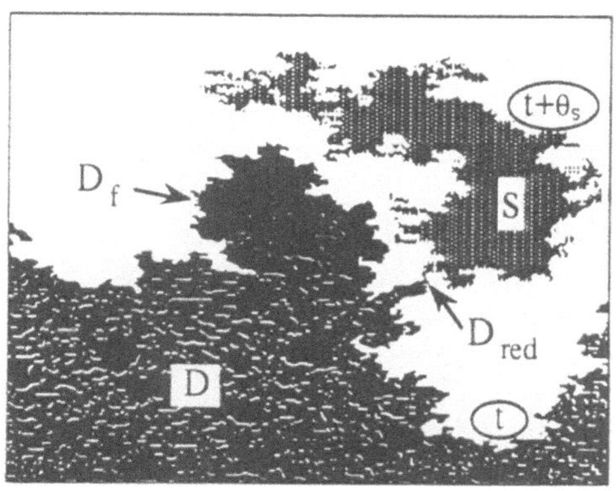

Fig. 2. Shows the various dimensions which appear in the bursting exponent. The burst is suppose to begin at time t, and to finish at $t + \theta_s$. The fractal dimension D_{f} shown here is the one of the external perimeter (and the internal holes have been discarded).

The function $K_s\ (s/s_{\mathrm{max}})$ is a cutoff function which cuts the clusters whose size is larger than $s_M \propto (\sigma_{\mathrm{ft}})^D$ (see [15]).

The fluctuations of the front surface were in fact calculated first [15, 16] and gave for the number $N_{\mathrm{ev}}(h)$ of events with perimeter (or surface in $d = 3$) h the result

$$N_{\mathrm{ev}}(h) = L^{d-1} h^{\mathcal{Y}_h - \varepsilon D/D_{\mathrm{f}}} \, |\nabla p|^{\alpha_N} \, K_h\!\left(\frac{h}{h_M}\right)$$

Here

$$\mathcal{Y}_h = 2 - \frac{D_{\mathrm{red}}}{D_{\mathrm{f}}}$$

and K_h is a cutoff function which cuts the clusters whose perimeter (surface in $d = 3$) is larger than $(\sigma_{\mathrm{ft}})^{D_{\mathrm{f}}}$.

Table 1. The possible values of the fractal dimension D_f, entering in $\mathcal{Y}_s$ or in $\mathcal{Y}_h$.

D_f	$d=2$	$d=3$
External perimeter	$D_h = 7/4$	
Grossman-Aharony perimeter [25]	$D_e = 4/3$	$D \cong 2.52$
Total perimeters	$D = 91/48$ ($D = 1.82$ (invasion))	

3 Noise Generation

To find the noise behavior of fluctuating fronts, we need to know the auto-correlation function of these fluctuations. We shall follow here the lines of the calculations in ref. [15] and [28].

Let $N_s(t)$ be the mass of the cluster connected to the source. $N_s(t)$ is supposed to be a stationary process as t evolves. In the case of fluid invasion $\langle N_s(t)\rangle$ increases with t, due to the injection of fluid. In the case of diffusion, the jump time is much smaller than the diffusion time and $\langle N_s(t)\rangle$ can be considered as constant during the characteristic fluctuation time t_c. Very generally $N_s(t) - \langle N_s(t)\rangle$ can be considered as stationary.

The variance of the increments of N_s is defined by,

$$\langle \Delta N_s(t)^2\rangle = \langle (N_s(t+t_0) - N_s(t))^2\rangle$$

The stationarity is well verified numerically as long as t is not too large (compared to t_c).

Following [15] we find,

$$\langle \Delta N_s(t)^2\rangle = \frac{L^{d-1}}{2}\int dx \sum_s s^2 \mathcal{M}_s[1-\exp(-2\overline{\pi}_s t)]$$

which gives after integration over x, using an effective field approximation,

$$\langle \Delta N_s(t)^2\rangle = L^{d-1}\,|\nabla p|^{-\alpha_N} \sum_s s^{1-D_f/D}[1-\exp(-2\overline{\pi}_s t)]\, K_s\!\left(\frac{s}{s_M}\right)$$

with $\overline{\pi}_s = B s^{D_{\mathrm{red}}/D-\varepsilon}$, while the correlation function is,

$$\langle N_s(t+t_0)N_s(t)\rangle = \frac{L^{d-1}}{2}\,|\nabla p|^{-\alpha_N} \sum_s s^{1-D_f/D}\exp(-2\overline{\pi}_s t)\, K_s\!\left(\frac{s}{s_M}\right)$$

In these expressions the important hypothesis is that successive events are independent, so that the correlation is a sum of exponential functions of time. This independence has been well verified numerically [15]. The spectral density is then a superposition of lorentzian functions. Replacing the sum over s by an integral gives,

$$S(\omega) \cong A \int_{s_m}^{s_M} ds \frac{2Bs^{\rho}}{4B^2 s^{2\sigma} + \omega^2}$$

where
s_m is the lower cutoff of the cluster size ($s_m = 1$),
s_M is the upper cutoff of the cluster size,
$A = L^{d-1} |\nabla p|^{-\alpha_N} K_s(0)$,
$B = \text{const}$,
$\rho = 1 - (D_f - D_{\text{red}})/D - \varepsilon$,
$\sigma = D_{\text{red}}/D - \varepsilon$.

The question is now, does there exist $1/f^{\alpha}$ behavior, with $0 < \alpha < 2$?

There are two cases, depending on the sign of σ, with three regimes in each case:

a) when $0 < \varepsilon < D_{\text{red}}/D$ (i.e. $\sigma > 0$):
i) if $\omega < 2B(s_m)^{\sigma}$, then $S(\omega) \cong$ const.
ii) if $2B(s_m)^{\sigma} < \omega < 2B(s_M)^{\sigma}$, an elementary calculation shows that $S(\omega)$ has the following behavior

$$S(\omega) = C_0 \omega^{(1+\rho)/\sigma - 2} + C_1 + C_2 \omega^2 + \ldots$$

and the condition of existence of $1/f^{\alpha}$ noise is

$$2D < D_f + D_{\text{red}} - \varepsilon D$$

which is never satisfied because $D_f, D_{\text{red}} < D$, and $\varepsilon > 0$.
iii) if $\omega > 2B(s_M)^{\sigma}$, then $S(\omega) \cong \omega^{-2}$, i.e. the noise is Brownian.

b) when $\varepsilon > D_{\text{red}}/D$ (i.e. $\sigma < 0$): It is convenient to make the change $s \to 1/u$, so that we obtain a similar integral for $S(\omega)$ only by changing $s_m \to 1/S_M, s_M \to 1/s_m, \rho \to -\rho - 2, \sigma \to -\sigma$,
i) If $\omega < 2B(s_M)^{\sigma}$, then $S(\omega) \cong$ const.
ii) If $2B(s_M)^{\sigma} < \omega < 2B(s_m)^{\sigma}$, then again

$$S(\omega) = C_0 \omega^{(1+\rho)/\sigma - 2} + C_1 + C_2 \omega^2 + \ldots$$

but the condition of existence for $1/f^{\alpha}$ noise is

$$2D > D_f + D_{\text{red}} - \varepsilon D$$

which is always satisfied.
iii) If $\omega > 2B(s_m)^{\sigma}$, then $S(\omega) \cong \omega^{-2}$, i.e. the noise is Brownian.

To summarize:

- When $\varepsilon < D_{\text{red}}/D$, there never exists a domain with a $1/f^{\alpha}$ noise [28]. At low frequency the spectral density is white noise, while at high frequency it is Brownian noise in $1/f^2$. This is in contradiction with our previous claim [15, 16] that the low frequency regime was $1/f$ noise.

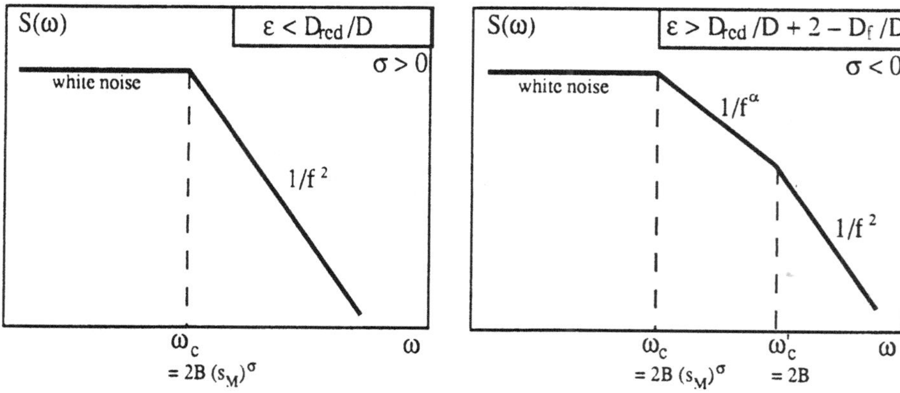

Fig. 3. Spectral density of interface bursting, in two different regimes. For large enough ε, a $1/f$ noise could in principle be observed. In the figure on the right, the condition on ε is such that $\alpha < 2$. Negative values for ε are in general unrealistic.

- When $\varepsilon > D_{\text{red}}/D$ ($D_{\text{red}}/D \cong 0.4$ $(d = 2)$, 0.45 $(d = 3)$).

There exists three behaviors:
  - white noise at very low frequency when $\omega < \omega_{cM}$ with $\omega_{cM} = 2B(s_M)^\sigma$
  - $1/f^\alpha$ noise when $\omega_{cM} < \omega < \omega_{cm}$ with $\omega_{cm} = 2B(s_m)^\sigma = 2B$ and

$$\alpha = 1 + \frac{2D - D_{\text{f}}}{\varepsilon D - D_{\text{red}}}$$

  - $1/f^2$ noise at high frequency when $\omega_{cm} < \omega$.

However $1/f^\alpha$ noise can be observed only when $0 < \alpha < 2$, that is to say when

$$\varepsilon > 2 - \frac{D_{\text{f}} - D_{\text{red}}}{D}$$

This is not case in the examples given in §2c.

4 Rate Equation of the Cluster Distribution

Suppose we want to study the evolution of the number of clusters of size k in a bond percolation problem in which the bonds are removed or added at random at a given rate R. To establish the rate equation it is natural to use the results of section 2, which gives the number $N_{\text{ev}}(i)$ of events of size i per unit time.

Let us first consider a large cluster of size k at the critical concentration p_c, broken into i and j daughters ($i + j + 1 = k$, as one bond disappears). We know that the probability of this event ($i \ll k$) is $N_{\text{ev}}(i) \propto 1 - \mathcal{Y}'_s$, where $\mathcal{Y}'_s$ takes into account the fact that the daughter clusters are obtained by removing a bond belonging to the perimeters (internal and external) of the initial cluster: hence $D_{\text{f}} = D$ and

$$\mathcal{Y}'_s = 2 - \frac{D_{\text{red}}}{D}$$

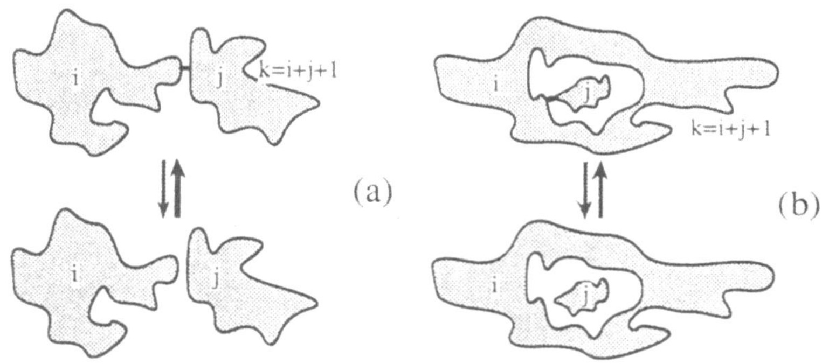

Fig. 4. Connection (black arrow) and disconnection (gray arrow) processes between finite clusters; in figure (a), the process only involves external perimeters ($D_f = D_e$ or D_h), in figure (b), the process involves external and internal perimeters ($D_f = D$) (see the table in section 2 for the definitions of the various D_f).

For scaling and symmetry reasons one expects then that the breaking probability takes the form

$$\frac{i^{-\mathcal{Y}'_s}(k-i)^{-\mathcal{Y}'_s}}{k^{-\mathcal{Y}'_s}}\,\mathcal{S}\left(\frac{i}{k},\frac{k-i}{k}\right)$$

which is such that $\frac{i^{-\mathcal{Y}'_s}(k-i)^{-\mathcal{Y}'_s}}{k^{-\mathcal{Y}'_s}} \to i^{-\mathcal{Y}'_s}$, when $k \to \infty$.
$\mathcal{S}$ is a symmetrical scaling function, with zero degree of homogeneity.

The contribution of this event to $\partial n_k/\partial t$, the rate of removing k-clusters is then proportional to $(i(k-i)/k)^{-\mathcal{Y}'_s}$ times kn_k the probability to be on a k-cluster.

At the critical concentration, Smoluchowski rate equation follows immediately [22]:

$$\frac{\partial n_k}{\partial t} = \frac{1}{2}\sum_{i+j+1=k}(K_{ij}n_in_j - F_{ij}n_{i+j+1}) -$$
$$\sum_{j=0}^{\infty}(K_{kj}n_kn_j - F_{kj}n_{k+j+1}) - K_kn_kn_\infty + F_kn_\infty$$

n_∞ the probability to be on the infinite cluster.

As we just noticed above, the corresponding expressions for $K_{ij}(i \oplus j \Rightarrow i+j+1)$ and $F_{ij}(i \oplus j \Leftarrow i+j+1)$ are obtained (in fig. 4, black and gray arrows respectively), by symmetrization of the preceding results, In the case the bonds are added or removed at random with the same probability, we find at the critical concentration p_c,

$$F_{ij} = (i+j+1)\left(\frac{ij}{i+j+1}\right)^{-\mathcal{Y}_s}\mathcal{S}\left(\frac{i}{i+j+1},\frac{j}{i+j+1}\right)$$

An explicit exact expression for F_{ij} can be obtained in the case of the Bethe lattice with coordination z [29], when $i, j \gg 1$:

$$F_{ij}^{\text{Bethe}} = (i+j+1)\left(\frac{ij}{i+j+1}\right)^{-3/2}\sqrt{\frac{2(z-1)}{\pi(z-2)}} \quad (\text{as } \nu = 1/2,\ D = 4)$$

The coalescence coefficient, K_{ij}, can be obtained using a detailed balance assumption (the equilibrium distribution is that of static percolation):

$$K_{ij} = F_{ij}\frac{n_{i+j+1}^{\text{eq}}}{n_i^{\text{eq}} n_j^{\text{eq}}} = ij\left(\frac{ij}{i+j+1}\right)^{\frac{d}{D}-y_s} S\left(\frac{i}{i+j+1}, \frac{j}{i+j+1}\right)\frac{1}{f(0)}$$

where f is the scaling function of the cluster distribution, $n_s^{\text{eq}}(p) \approx s^{-1-d/D} f[(p-p_c)s^{1/(\nu D)}]$. K_k and F_k have a similar behavior and represent the connection-disconnection probabilities from a front of diffusion (the infinite cluster). The Smoluchowski equation is valid as long as the initial distribution is a random distribution of bonds. Notice here a problem for the case of site dilution, because removing one site can generate more than two clusters [30]. However, this effect seems marginal, and was not observed in the case of fluctuations of diffusion fronts [15, 16], at the precision of our numerical simulations.

An interesting behavior concerns the case when the rate of removing bonds is different from the rate of adding bonds. Then, p becomes time dependent, but the cluster distribution follows adiabatically the percolation distribution if removal and addition of bonds is made at random.

The above equation must first be generalized to arbitrary concentration p. A scaling factor, $u\,[(p-p_c)i^\sigma]\,u[(p-p_c)j^\sigma]/u[(p-p_c)(i+j+1)^\sigma]$ is then expected in F ($\sigma = 1/\nu D$), while in the above equation giving K_{ij}, $1/f(0)$ must be replaced by the p-dependent scaling functions f, which insures the stationary equilibrium constraint.

If τ is the *relative rate* of removing particles, and $R(t)$ the *general time dependent rate* of removing or adding particles, the Smoluchowski coefficients now become

$$\tilde{K}_{ij} = (1-\tau)R(t)K_{ij};\quad \tilde{F}_{ij} = \tau R(t)F_{ij}$$

Such an equation has been studied by Kerstein [29], in the case $\tau = 1$, for the Bethe lattice.

5 Fluctuations of Fronts and Fragmentation

The knowledge of the connection-disconnection behavior of percolation clusters permits not only to establish a Smoluchowski's rate equation of the cluster distribution, but also to determine the fragmentation properties of the percolation clusters [23] (see preceding section).

Recently, Gyure and Edwards [30] presented a scaling theory for the fragmentation of percolation clusters by random bond dilution. They define the ensemble average number a_s of "fragmenting" bonds on a cluster of mass s, and the probability $b_{s',s}$ of obtaining a daughter cluster of mass s' by fragmentation of a cluster of mass s. By "fragmenting" bonds they mean any bond which, when removed, separates the cluster into two pieces.

At the percolation threshold for large enough sizes s and s' the two functions a_s and $b_{s',s}$ are assumed to obey scaling forms

$$a_s \propto s^{\lambda}$$
$$b_{s',s} = s^{-\phi} g\left(\frac{s'}{s}\right)$$

These assumptions were checked analytically on two exactly solvable case, $d = 1$ and the Bethe lattice, and numerically on a 1200×1200 square lattice [30]. The exponent λ is always observed vary close to unity, while the exponent ϕ is found to obey the relationship

$$\phi = 1 + \lambda - \sigma_s$$

where $\sigma_s = 1/(\nu D)$. The numerical evaluation of ϕ in the $d = 2$ square lattice ($\phi = 1.600 \pm 0.006$) agrees well with this relationship.

Now to estimate ϕ, we only need to compare the expression for $b_{s',s}$ with the expression we derived above for the Smoluchowski equation. This immediately leads to,

$$\phi \equiv \mathcal{Y}'_s = 2 - \sigma_s$$

Comparing the two last relations leads to $\lambda = 1$.

This confirms the conjecture of Gyure and Edwards that λ should be equal to unity [23].

6 Dynamical Percolation

When two types of particles diffuse together on a lattice and have different jump probabilities, we are faced with a complicated problem which includes as a limit the case of diffusion of particles in percolating structures (the case when one species has a large jump probability, and the other a negligible one). Such a system is modernized by the so-called "Dynamical Percolation" model introduced by Druger, Ratner and Nitzan [24]. When the jump probabilities are not too different, a mean-field treatment is possible. When the jump probabilities are very different, again a mean-field treatment is possible far from the percolation threshold p'_c of the sites unoccupied by the solver species. Hence the anomalous case is the case of a fast particle diffusing in presence of very slow particles close to the percolation threshold p'_c.

We shall examine here the main features of this anomalous diffusion of the fast particle. We suppose that at time t_n the particle is on a cluster (of empty sites, see fig. 5) of size i_n and radius R_{i_n}. Due to its large diffusion coefficient, its density probability becomes uniform on the i_n-cluster before a particle of the cluster perimeter makes a jump. If at time t_{n+1} the i_n-cluster breaks into two parts i_{n+1} and $(i_n - i_{n+1})$, the particle becomes trapped in one of these pieces, say in the i_{n+1}-cluster. The probability of this event is,

$$F_{i_{n+1}, i_n - i_{n+1}} \frac{i_{n+1}}{i_n} \qquad (i_n > i_{n+1})$$

The factor i_{n+1}/i_n is the probability to fall precisely into the daughter i_{n+1}.

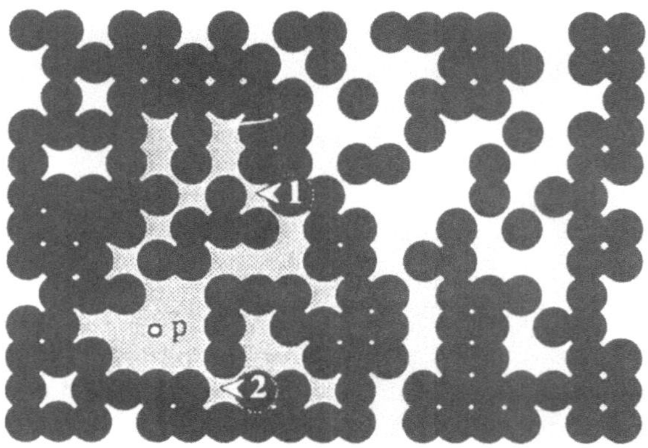

Fig. 5. In the dynamical percolation model two species with different jump probabilities and hard-core exclusion diffuse on a lattice. We consider here a very fast particle (the small white disk p which can visit any connected regions) diffusing in presence of slow particles (large black disks). Very rapidly the distribution probability of the white disk becomes homogeneous inside the allowed cluster (the gray region). If the black particle $n^\circ 1$ jumps, the door is open to the fast particle to diffuse inside a new neighboring cluster of empty sites (here the corresponding cluster percolates); if the black particle $n^\circ 2$ jumps, on the contrary, the cluster is broken into two pieces, and the fast particle p becomes trapped in one of these pieces.

We can roughly localize the particle at the position of the center of mass of the clusters. Hence during the above process we can say that the particle makes a jump of length $|R_{i_n} - R_{i_{n+1}}|$, in an arbitrary direction. The second possible process at time t_{n+1} is the connection of the i_n-cluster to a cluster of size $i_{n+1} - i_n$ leading to an i_{n+1}-cluster. Here again the particle makes a jump of length $|R_{i_n} - R_{i_{n+1}}|$; and the probability of this event is now (the cluster distribution n_i is at equilibrium),

$$K_{i_n, i_{n+1} - i_n} n_{i_{n+1} - i_n} = F_{i_n, i_{n+1} - i_n} \frac{n_{i_{n+1}}}{n_{i_n}} \qquad (i_n < i_{n+1})$$

The probability of a jump of size $|R_{i_n} - R_{i_{n+1}}|$ is then proportional to (the factor S is omitted),

$$i_{n+1} \left(\frac{i_{n+1}(i_n - i_{n+1})}{i_n} \right)^{-y_s} \qquad (\text{when } i_n > i_{n+1})$$

or

$$i_n \left(\frac{i_n}{i_{n+1}}\right)^{d/D} \left(\frac{i_n(i_{n+1} - i_n)}{i_{n+1}}\right)^{-y_s} \qquad \text{(when } i_n < i_{n+1}\text{)}$$

These distributions of probability are dominated by a power law behavior for low values of $|i_n - i_{n+1}|$, and varies like $(1 - i_{\min}/i_{\max})^{-y_f}$, where $i_{\max} = \max\{i_n, i_{n+1}\}$ and $i_{\min} = \min\{i_n, i_{n+1}\}$. The corresponding jump sizes are

$$U = |R_{i_n} - R_{i_{n+1}}| \approx (i_{\max})^{1/D} - (i_{\min})^{1/D} \left\{1 - (i_{\min}/i_{\max})^{1/D}\right\}$$

The probability to have a jump of size U then follows a power law,

$$P(U) \approx U^{-Dy_s}$$

The diffusion process in this extreme case for dynamical percolation is then a Levy flight process. When the situation is not at the threshold p'_c of the empty sites, one expects a Levy flight up to distances smaller than the correlation length ξ of these empty sites, and a Rayleigh flight (Brownian motion) at larger distances. It can be noticed that if the fast particle is an electron jumping from an atom to its neighboring atoms, the clusters to be considered are the clusters of atoms themselves. In the case of many fast particles with only hard core interactions, the problem is the same if we consider only the collective diffusion coefficient of these fast particles.

7 Relations Between Interface Bursting and Interface Depinning [32]

The crossover between invasion percolation and the Eden model has been studied in the middle of the 80ties and it was realized that invasion percolation had anomalous fluctuations which were absent in the Eden model [33]. These anomalous fluctuations are present in systems in which there is a pinning of the interface. This aspect has been studied only recently, and it is not clear at the moment whether all the growths of rough surfaces with quenched disorder belong to the same universality class, depending on the type of experiment performed.

For instance, concerning the roughness of the various self-affine interfaces described in the literature, the situation is far from being clear. From the experimental point of view, the roughness exponent is found equal to $\alpha \cong 0.81$ in fixed rate invasion of a $2d$ Hele-Shaw cell filled with glass beads [38], to $\alpha = 0.73 \pm 0.03$ with a similar system [39], but also between 0.65 and 0.91 [40], and finally equal to $\alpha = 0.65 \pm 0.05$ [14, 41] in imbibition of blotting paper. We shall describe some simple models of growth of pinned interfaces.

In the case of fluid invasion of porous media, Robbins and coworkers [34, 35] pointed out the existence of critical contact angle θ_c, leading to a bursting-depinning transition. They propose a $\{\theta, P\}$ phase diagram (P being the applied pressure) separating static and moving interfaces (but moving slowly enough to avoid viscous effect in the dynamics): when $\theta > \theta_c$, the interface is described by invasion percolation and there exist a critical pressure $P_c(\theta)$ above which bursts appear. When $\theta < \theta_c$, the interface is described by a self-affine structure with a rough interface. In reference [19] the motion of the interface is studied and it is suggested that the expression obtained for $N_{ev}(s)$ (recall section 2) applies if we take $D = d$ (compact growth) and $D_f = d - 1$ (i.e. only the global dimension of the self-affine surface plays a role), while ν remains equal to the correlation length exponent of percolation. The behaviour is checked numerically and agrees well with this choice for D_f and ν, but other values could have even taken, as there is no precise justification for this choice.

The roughness exponent α [32] has been only found numerically [35] and no relation to the above exponents, D, D_f and D_{red}, has been proposed up to now. Martys et al. found $\alpha = H \cong 0.81$. In addition, in the depinning regime ($\theta < \theta_c$), the entire surface progresses coherently, in contrast to the bursting regime where each part of the interface grows almost independently. This behavior is certainly in agreement with the spectral density of the time fluctuating events: in the case of interface bursting, the events are independent and a $1/f^2$ noise is found at

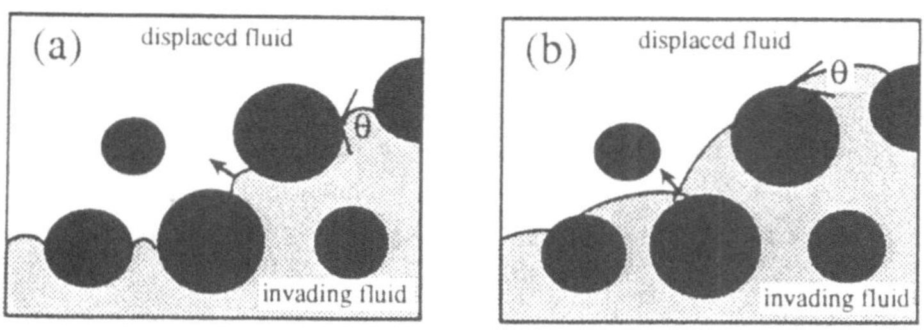

Fig. 6. The invasion of a porous medium (constituted of black spheres of random sizes) by a fluid (in gray) is strongly dependent on the contact angle θ (depending on the mutual wetting properties of the three components, the displaced and invading fluid and the porous medium). In figure (a), the invading fluid is strongly non-wetting, and invasion of two nearest neighboring throats is uncorrelated: "drainage" is modeled by invasion percolation, and only depends on the radii of the throats. In figure (b) the contact angle is now smaller, there exist an overlap during the invasion of nearest neighboring throats, which correlates the evolution of the invading fluid [34].

high frequency. This is not what is observed in the depinning case when $\omega T \gg 1$ (T is the time for the interface to move over a distance ξ and it can be compared to the characteristic time $1/(2Bs_M^\sigma)$ in section 3); the authors found $S(\omega) \propto \omega^{-a}$ with $a \cong 1.55$ [36]; a theory, taking into account the coherent growth, remains to be found to explain this value of a, but this is a very interesting point which will be discussed at the end of this section (see fig. 10).

In another recent paper [37] it has been shown that the same kind of transition could be observed in random three dimensional magnets. The hamiltonian

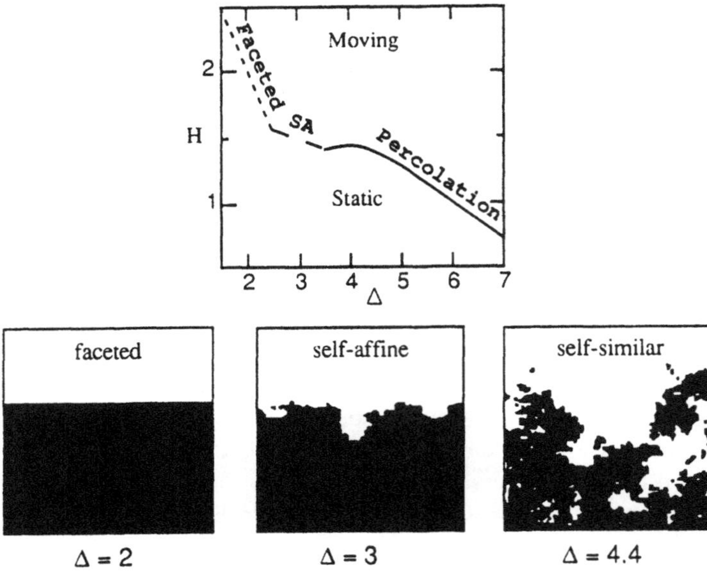

Fig. 7. Growth of interfaces in three-dimensional Ising ferromagnets with random field interactions. The random field distribution has a width Δ; when the randomness is low ($\Delta < 2.42$), the interface is faceted , it becomes self-affine at intermediate randomness ($2.42 < \Delta < 3.41$), and self-similar in a strongly random field ($\Delta > 3.41$). the corresponding phase diagram is shown in the above figure: the drift due to an external field H, has a critical value which separates the moving region from the static one (depinning force) [37].

is that of the standard Ising model with quenched random disorder,

$$\mathcal{H} = -\sum_{\langle i,j \rangle} s_i s_j - \sum_i (H + h_i) s_i$$

where the Ising spin on site i takes the values $S_i = \pm 1$, H is the external magnetic field, and h_i represents the random field, which is described by a uniform random distribution between $-\Delta$ to Δ. Three types of growth nay be found (see fig. 7): percolative structures with interface bursting in strongly disordered systems, self-affine structures with interface depinning in systems with intermediate disorder, and finally faceted growth for weak enough disorder. The roughness exponent in the self affine structures is found around 2/3 (no self-affine growth can be observed in $d = 2$ random magnets).

An interesting model has been proposed [13, 14, 42, 43], in particular for the blotting paper imbibition experiment [14], which is related to directed percolation [31] in a similar manner as interface bursting was related to percolation. In particular it was found that $\alpha = \nu_\perp / \nu_\parallel \cong 0.63$, in close agreement with the experiment. In this directed percolation model, Leschhorn and Tang [44] have investigated the generation of avalanches and their correlations. They found and

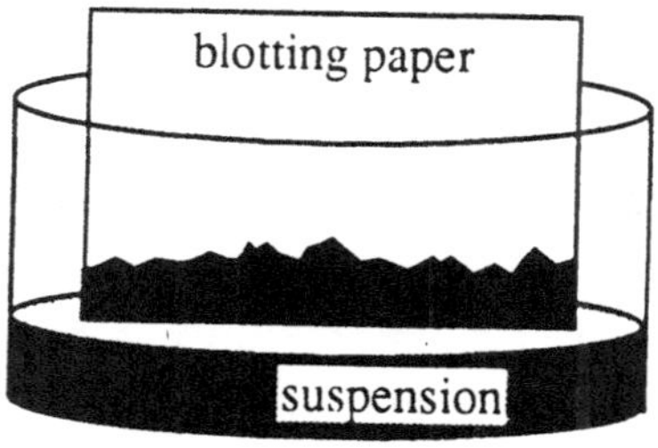

Fig. 8. Experiment of imbibition of a piece of blotting paper by a dye (ink). Capillarity leads to a rough surface, which stops growing when evaporation (and gravity?) balances the incoming flux. The interface profile is digitalized. Very few overhangs can be observed along this interface [14].

avalanche size distribution with a scaling form similar to that of $\mathcal{N}_{\text{ev}}(s)$, i.e.

$$P_{\text{av}}(s) = s^{-\mathcal{K}} \phi\left(\frac{s}{s_0}\right)$$

with $\mathcal{K} = 1.25 \pm 0.05$, and a cutoff size s_0 (the cluster is compact in $d = 1 + 1$ dimension) given by

$$s_0 \propto \xi_{||}(\eta^0_{\text{min}})\, \xi_{\perp}(\eta^0_{\text{min}})$$

where η^0_{min} is the value of the weakest force on the interface.

The value of $\mathcal{K}$ is slightly larger than the value 1.125 obtained when setting $D = d$, $D_{\text{f}} = 1$ and $\nu = 4/3$ in $\mathcal{Y}_s$ [19]. Indeed, the situation is now very different; the system is anisotropic and the events are correlated. The theory relating $\mathcal{K}$ to the other exponents remains to be established.

The introduction of directed percolation can be understood from the simple model described in figure 9. The main idea is the filling of the overhangs. A direct explanation of this constraint leads to the existence of a smoothing force (similar to the one showed in fig. 6b). Nevertheless it must be pointed out that the definition of overhangs supposes the existence of a privileged direction (here the vertical direction). This can be the consequence of the initial horizontal state, but then the growth is supposed not to present large slopes: this also can be a consequence of the gravity field. In any case, few overhangs are effectively observed experimentally. When an overhang is filled then the invasion of new regions of the disordered medium is possible. This can be verified in figures 9 and 10.

In figure 10, we examine the depinning of the interface stabilized along a line of strong pinning defects (directed percolation path). Increasing the driving field (this could correspond to immersing the paper more deeply in the ink), depinning of blocked sites appears. Depending on the order in which the depinning is done, various regions are successively invaded (see fig. 10). We suggest that this correlation could be at the origin of the anomalous exponent (i.e. different from

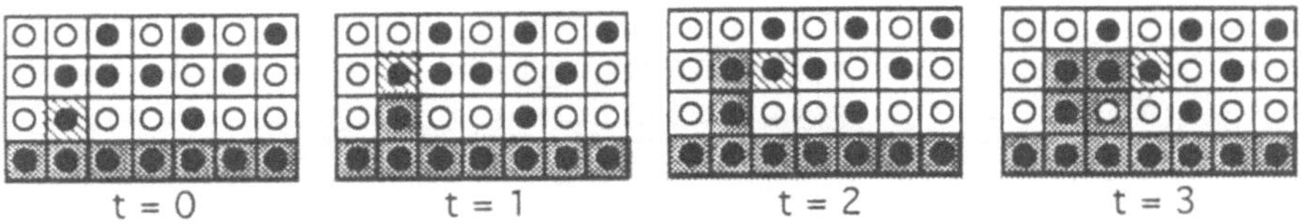

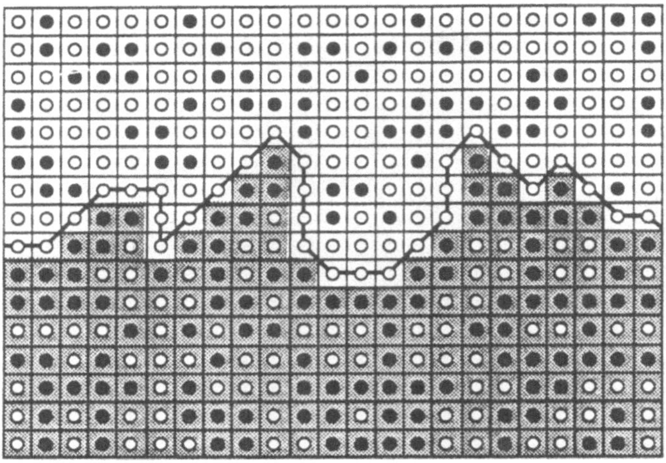

Fig. 9. The model by Buldyrev et al., is a modified invasion percolation process, in which the interface grows with erosion of the overhangs. This is shown in the four upper figures: the unblocked sites are invaded progressively. When an overhang appears ($t = 2$), not only the unblocked site is invaded but also the blocked site below. This may allow the invasion of disconnected clusters of unblocked sites, as can be verified on the figure below. The invasion stops when the interface reaches a directed percolation line of blocked sites (which by definition has no overhangs) [14].

2) in the noise found by Nolle et al. [36] (note however that their model is very different, overhangs are possible), and by Leschhron and Tang [44].

To conclude with this yet very open domain, we shall quote a very recent paper by Roux and Hansen [45]. They examine a simplified version of the Cieplak and Robbins model [29, 33]; The curvature $(y_{i-1} - 2y_i + y_{i+1})$ at any site of the front is supposed to be the relevant physical parameter. It is bounded by a quenched random threshold at any site. The authors also find a roughness exponent around $\alpha = 0.86 \pm 0.03$ (for a system size $L = 512$). It falls to 0.75 when $L = 128$ as observed in ref. [39]. The authors suggest that this is an effective roughness exponent, the real exponent α (which can only be obtained from the spatial derivative of the front or from its power spectrum) would lie between $1 < \alpha < 2$: typically $\alpha \cong 1.2$, the roughness being a roughness of the slopes.

In conclusion, whereas interface bursting seems at present relatively well understood this is not quite the case for interface depinning in spite of very

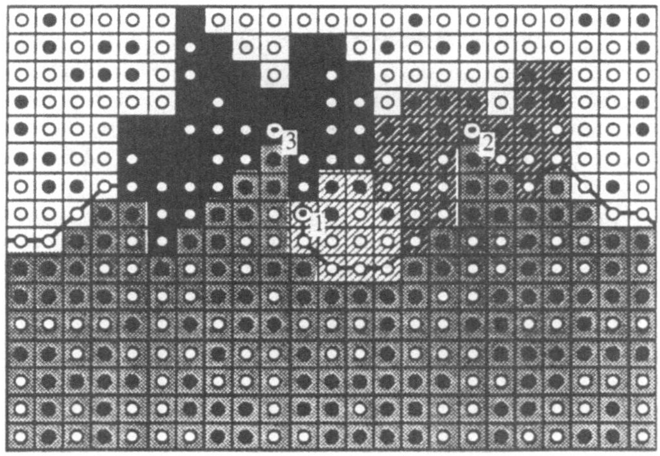

Fig.10. This figure shows the interface depinning in Buldyrev's et al. model [14]. Depending on the position of the depinning sites (1, 2 or 3), different histories are possible, showing the correlated behavior of the growth. Depinning in site 1 leads to invasion of only the region. Depinning in site 2 leads to invasion of not only the region but also the region. The sequence {1,2} is possible, the sequence {2,1} is not. The same result appears with site 3. The growth process is then very different from the case of interface bursting.

important and interesting developments. But even, the treatment of interface bursting remains incomplete in one respect: little is known about the dynamics of fluctuating interfaces in presence of interactions; this case is of great importance in soldering, in intercalation, etc. [46, 47]. Also, and this was out initial purpose when studying fluctuations of diffusion fronts, it would be of real interest to understand the detailed behavior of the fluctuations of impedances between a conducting diffusion front and for example a planar counter electrode. The behavior of an irregular (fractal) electrode is however now partly understood [48].

Acknowledgments

I would like to thank François Devreux, Jacques Neveu, Peter Pfeifer, and Michel Rosso for fruitful discussions. I also acknowledge the organizers of the Vth Max Born Conference for their kind invitation. Laboratoire de Physique de la Matière Condensée is Unité de Recherche D-1254 Associée au Centre National de la Recherche Scientifique.

References

[1] A. H. Thompson, A. J. Katz and R. Raschke, Phys. Rev. Lett. **59** (1987) 29; J. P. Stokes, A. P. Kushnick and M. O. Robbins, Phys. Rev. Lett. **60** (1988) 1386; K. J. Måløy, L. Furuberg, J. Feder and T. Jøssang, Phys. Rev. Lett. **68** (1992) 2161;

[2] Y. Imry and S. Ma, Phys. Rev. Lett. **35** (1975) 1399; G. J. Grinstein, J. Appl. Phys. **55** (1984) 2371; A. Kirilyuk, J. Ferré and D. Renard, Europhys. Lett. **24** (1993) 403;

[3] J. L. Hudson, J. Tabora, K. Krisher and I. G. Krevrekidis, Phys. Lett. A. **179** (1993) 355; D. E. Williams, R. C. Newman, Q. Song and R. G. Kelly, Nature (London) **350** (1991) 216; L. Balázs, L. Nyikos, I. Szabó, and R. Schiller, preprint.

[4] See "Charge Density Waves in Solids", edited by G. Hutiray and J. Solyom, Lecture Notes in Physics, vol. 217 (Springer Verlag, Berlin, 1985).

[5] B. Sapoval, M. Rosso, J. F. Gouyet and J. F. Colonna, Solid State Ionics **18** & **19** (1986) 21.

[6] D. Sornette, Phys. Rev. Lett. **69** (1992) 1287; T. L. Chelidze, Phys. Earth. Planet. Inter. **28** (1982) 93; G. A. Sobolev, Pure Appl. Geophys. **124** (1986) 811.

[7] G. Mackay and N. Jan, J. Phys. A**17** (1984) L

[8] D. Dhar, Phys. Rev. Lett. **64** (1990) 1613.

[9] L. Furuberg, J. Feder, A. Aharony and T. Jøssang, Phys. Rev. Lett. **61** (1988) 2117.

[10] A. H. Thompson, A. J. Katz and C E. Krohn, Advances in Physics, **36** (1987) 625.

[11] P. Bak, C. Tang and K. Wiesenfeld, Phys. Rev. Lett. **59** (1981) 381.

[12] B. Sapoval, M. Rosso and J. F. Gouyet, J. Phys. (Paris) Lett. **46** (1985) L149.

[13] L. H. Tang and H. Leschhorn, Phys. Rev. **45** (1992) R8309.

[14] S. V. Buldyrev, A. L. Barabási, F. Caserta, S. Havlin, H. E. Stanley and T. Vicsek, Phys. Rev. A **45** (1992) R8313; L. A. N. Amaral, A. L. Barabási, S. V. Buldyrev, S. Havlin and H. E. Stanley, Fractals **1** (1993) 818; the rupture lines in paper sheets, seem to have a very similar behavior, see J. Kertész, V. K Horváth and F. Weber, Fractals, **1** (1993) 67.

[15] J. F. Gouyet and Y. Boughaleb, Phys. Rev. B**40** (1989) 4760; B. Sapoval, Philos. Mag. B **59** (1989) 75.

[16] J. F. Gouyet, B. Sapoval, Y. Boughaleb and M. Rosso, Physica A **157** (1989) 620.

[17] S. Roux and E. Guyon, J. Phys. A **22** (1989) 3693.

[18] J. F. Gouyet, Physica A **168** (1990) 581.

[19] N. Martys, M. O. Robbins and M. Cieplak, Phys. Rev. B **44** (1991) 12294.

[20] V. A. Mukhamedov, private communication.

[21] H. J. Hermann, J. Phys. A**26** (1993) L1145; W. R. Roseen and C. K. Mamun, Phys. Rev. B **47** (1993) 11815; S. M. Clifford, J. Geophys. Res. E **98** (1993) 10973; P. Meakin, A. Birovljev, V. Frette, J. Feder and T. Jøssang, Physica A **191** (1992) 227; J. J. E. Herrera, Physica D **57** (1992) 249; L. Furuberg, A. Hansen, E. Hinrichsen, J. Feder and T. Jøssang, Phys. Scr. **38** (1991) 91; R. Pandey and S. Yu, J. Phys. A **24** (1991) 3959; S. Roux and E. Guyon, J. Phys. A **24** (1991) 1611.

[22] J. F. Gouyet, in "Soft Order in Physical Systems", R. Bruisma and Y. Rabin eds., Plenum Pub. Corp., New York, 1993.

[23] J. F. Gouyet, Phys. Rev. B **47** (1993) 5446.

[24] S. D. Druger, A. Nitzan and M. A. Ratner, J. Chem. Phys. **76** (1983) 3133; A. K. Harrison and R. Zwanzig, Phys. Rev. A **32** (1985) 1072.

[25] T. Grossman and A. Aharony, J. Phys. A **20** (1986) L745.

[26] M. Rosso, J. F. Gouyet and B. Sapoval, Phys. Rev. Lett. **57** (1986) 3195; J. F. Gouyet, M. Rosso and B. Sapoval, Phys. Rev. B **37** (1988) 1832.

[27] $|\nabla p|^{\alpha_N}$ is the surface of the front. It can also be written as a function of the width $(\sigma_{\rm ft})^{D_f-d+1}$ and corresponds, as a function of concentration to (x_c is such that $p(x_c) = p_c$)

$$|p(x_c \pm \sigma_{\rm ft}) - p(x_c)|^{-\nu(D_f-d+1)}$$

whose exponent is the exponent ω of ref. [19]. The exponent $\mathcal{Y}_s$ was also found from very different arguments by Martys et al. [19] (and called τ')

[28] J. F. Gouyet, April 1991, unpublished.
[29] A. R. Kerstein, J. Phys. A **22** (1989) 3371.
[30] M. F Gyure and B. F. Edwards, Phys. Rev. Lett. **68** (1992) 2692; B. F. Edwards, M. F. Gyure and M. Ferer, Phys. Rev. A **46** (1992) 6252.
[31] W. Kintzel, in *Percolation Structures and Processes*, G. Deutcher, R. Zallen and J. Adler eds., Annals of the Israel Physical Society, 1983, vol. 5, chap. 18.
[32] T. Vicsek, *Fractal Growth Phenomena*, 2nd ed., part IV (World Scientific Singapore, 1992); J. F. Gouyet, M. Rosso and B. Sapoval, in *Fractals and Disordered Systems*, A. Bundle and S. Havlin eds. (Springer Verlag, Heidelberg, 1992); *Dynamics of Fractal Surfaces*, F. Family and T. Vicsek eds. (World Scientific Singapore, 1991); *Les Houches Workshop*, R. Jullien, J. Kertesz, P. Meakin and D. E. Wolf eds. (Nova Science, New York, 1992); P. Meakin, Phys. Rep. (to appear); T. Halpin-Healey and Y. C. Zhang, Phys. Rev. (to appear).
[33] J. P. Nadal, R. M. Bradley and P. N. Strenski, J. Phys. A **19** (1986) L505.
[34] M. Cieplak and M. O. Robbins, Phys. Rev. Lett. **60** (1988) 2042.
[35] M. Martys, M. Cieplak and M. O. Robbins, Phys. Rev. Lett. **66** (1991) 1058.
[36] C. S. Nolle, B. Koiller, M. Martys and M. O. Robbins, preprint.
[37] H. Ji and M. O. Robbins, Phys. Rev. B **46** (1992) 14519.
[38] V. K. Horvath, F. Family and T. Vicsek, J. Phys. A **24** (1991) L25.
[39] M. A. Rubio, C. A. Edwards, A. Dougherty and J. P. Gollub, Phys. Rev. Lett. **63** (1989) 1685.
[40] S. He, G. Kahanda and P. Z. Wong, Phys. Rev. Lett. **69** (1992) 3731.
[41] S. V. Buldyrev, A. L. Barabási, S. Havlin, J. Kertesz, H. E. Stanley and H. Xenias, Physica A **191** (1992) 220.
[42] K. Sneppen, Phys. Rev. Lett. **69** (1992) 3539; see also Z. Olami, I. Procaccia and R. Zeitak, Phys. Rev. E **49** (1994) 1232.
[43] L. A. N. Amaral, A. L. Barabási, S. V. Buldyrev, S. Havlin and H. E. Stanley, preprint.
[44] H. Leschhorn and L. H. Tang, Phys. Rev. E **49** (1994) 1238.
[45] S. Roux and A. Hansen, J. Phys. I France, **4** (1994) 515.
[46] M. Kolb, J.-F. Gouyet and B. Sapoval, Europhys. Letters **3** (1987) 33; M. Kolb, T. Gobron, J.-F. Gouyet and B. Sapoval, Europhys. Letters **11** (1990) 601.
[47] S. Mazur and S. Reich, J. Phys. Chem., **90** (1986) 1365; R. P. Wool, B.-L. Yuan and O. J. McGarel, Poly. Eng. Sci. **29** (1989) 1340.
[48] See for instance B. Sapoval, "A new approach to the linear and non-linear response of self similar electrodes, in " *Chaos, Solitons and Fractals*", Pergamon Pres., 1994.

Criticality of Self-Avoiding Walks in Fractal Porous Media

Sava Milošević[1] *and Ivan Živić*[2]

[1] Faculty of Physics, University of Belgrade, P.O.Box 368, 11001 Belgrade, Serbia, Yugoslavia
[2] Faculty of Natural Sciences and Mathematics, University of Kragujevac, 34000 Kragujevac, Serbia, Yugoslavia

1 Introduction

The equilibrium properties of a flexible polymer chain in a good solvent enclosed in a quenched random environment, such as a porous medium, appear to be *both* potentially relevant to various experimental situations *and* a controversial theoretical problem for almost two decades. Porous media are often fractals, which means that they display a self–similar distributions of voids (obstacles) over three to four orders of magnitudes in length scale. In a series of recent papers [1-7] we have adopted several families of fractal lattices as models of porous media, and studied the statistics of a polymer chain situated on such lattices. The statistics has been captured by the self–avoiding walk (SAW) model, implying that the appropriate criticality shows up at the critical weight (fugacity) of a single step. In this spirit, we have studied the end–to–end distance critical exponent ν, and the critical exponent γ (associated with the total number of distinct SAWs), in the case of three quite different fractal families: the Sierpinski gasket (SG) family, the plane-filling (PF) family, and the checkerboard (CB) family of fractals. The critical exponents have been calculated by three different methods — the exact renormalization group (RG) method, the Monte Carlo renormalization group (MCRG) technique, and the finite–size scaling (FSS) approach. The obtained results proved valuable in explaining previous theoretical and experimental findings related to the controversial problem of SAWs in various disordered systems.

The self-avoiding walk (SAW) on a lattice is a random walk whose path must not contain self-intersections. Its path corresponds to the shape of a linear flexible polymer chain, and in the case of a good solvent we assume that there is no interaction between two nearest-neaighbour nonconsecutive steps (monomers). In other words, we assume that the polymer is at high temperatures and that its configurations are determined by the excluded volume interactions.

For explaining most of relevant experimental situations, including diffusion of polymers, it turns out that it is important to know the mean square end-to-end distance $\langle R^2 \rangle$, which for large N behaves according to the power law $\langle R_N^2 \rangle \sim N^{2\nu}$, with ν being the related critical exponent. The other pertinent quantity is the number of distinct SAWs (averaged over all possible starting points) that is, for a large number of steps N, governed by the power law $C_N \sim \mu^N N^{\gamma-1}$, where μ is the connectivity constant, and γ is the concomitant critical exponent.

We have calculated the critical exponents ν and γ in the case of several different fractal families. First, we have studied SAWs on the SG family of fractals whose each member can be labelled by an integer b, $2 \le b \le \infty$, so that when $b \to \infty$ both the fractal $d_f = \ln[b(b+1)/2]/\ln b$ and spectral dimension d_s [8,9] approach their Euclidean value 2. Here we briefly recall the fact that each member of the SG fractal family can be constructed in stages. At the initial stage ($r = 1$) of the construction there is an equilateral triangle (generator) that contains b^2 identical smaller triangles of unit side length, out of which only the upper oriented are physically present. The subsequent fractal stages are constructed self-similarly, so that the complete fractal is obtained in the limit $r \to \infty$. In Fig. 1 we have depicted a piece of the $b = 9$ SG fractal lattice, together with a schematic representation of an infinite SAW.

Members of the other two (PF and CB) families of fractals are enumerated by an odd integer b ($3 \le b < \infty$), and are also constructed iteratively. That is, the construction process starts with a generator ($r = 1$) and continues in a self-similar way by successive ($r \to r+1$) enlarging the given stage of construction by a scaling factor b. The particular shapes of the generators for the PF and CB fractals are described and illustrated in references [3] and [4], respectively. As regards the fractal dimensions, it is important to notice that *all* members of the PF family have the same fractal dimension $d_f = 2$, whereas for the CB fractals $d_f = \ln[(b^2+1)/2]/\ln b$ (implying that $d_f \to 2$ when $b \to \infty$).

2 Methods of Calculation of the SAW Critical Exponents

For the first few members of each family of fractals under study, we have calculated the SAW critical exponents ν and γ by applying an exact renormalization group method [10]. The basic idea of the method is illustrated in Fig. 2, in the case of the $b = 2$ SG fractal. One starts with considering an arbitrary stage (r) of the fractal construction and assigns the weight B' to the piece of the SAW path that traverses the r-th stage fractal object. Then, one argues that B' must be some function $f(B)$ of the SAW weight B that corresponds to the next smaller stages ($r-1$) that comprise the given fractal stage of construction. Furthermore, one asserts that $f(B)$, due to the underlying self-similarity, should not depend on r. This means that one can find $f(B)$ by considering the fractal generator ($r = 1$). Thus, in the case depicted in Fig. 2, we can write

$$B' = B^2 + B^3 \,. \tag{1}$$

This is the simplest RG equation, with the nontrivial fixed point $B^* = (\sqrt{5} - 1)/2$. The corresponding eigenvalue 2.3820 determined by

$$\lambda = \left.\frac{\mathrm{d}B'}{\mathrm{d}B}\right|_{B^*} , \tag{2}$$

Fig. 1. A segment of the $b = 9$ SG fractal lattice. The curly line represents a part of an infinite SAW path (or, a part of a flexible polymer chain). The big white triangles (denoted by A through G) can be interpreted as homogeneous pieces (pores), whereas the big black downward oriented triangles can be conceived as obstacles. There are obstacles bigger than those captured within this frame. Theoretically, there are obstacles that are 9^r times bigger (where r is a positive integer) than those depicted here, and, in fact, the smallest downward oriented triangles may be interpreted as stumbling blocks.

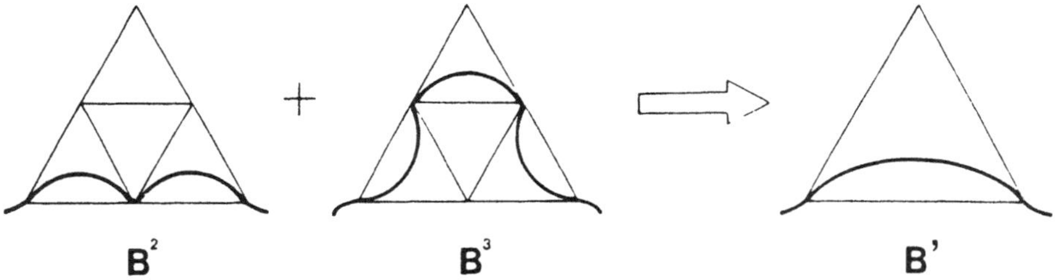

Fig. 2. Schematic representation of the way one obtains the RG equation (1). One should note that here we assume that the SAW walker is not allowed to cross any downward oriented unit triangle, and if he crosses an upward oriented unit triangle he has to do it in one step, and is never allowed to enter the same triangle again. One could stipulate other rules of crossing the upward oriented unit triangles, but it turns out that changes of the rules do not change values of the critical exponents.

and the general formula

$$\nu = \frac{\ln b}{\ln \lambda}, \tag{3}$$

brings about $\nu = 0.7986$, which is definitely larger than the value 3/4 established for the two-dimensional Euclidean lattices. In the case of arbitrary b, the function $f(B)$ is also a polynomial of the form

$$B' = \sum_{N=b}^{b(b+1)/2} a_N B^N, \tag{4}$$

where B is the weight (fugacity) of one SAW step that traverse unit triangle, and a_N is number of all possible SAWs of N steps that traverse the fractal generator ($a_2 = 1$ and $a_3 = 1$, in the case $b = 2$). Hence, the central task in evaluation of ν, for a given b, consists in determining number a_N of all possible SAWs of N steps that traverse the fractal generator. Unfortunately, the exact enumeration of SAWs can be done only for limited initial sequence of the fractal family (for example, to calculate in this way ν for $b = 10$ it would require about 85 days of continuous operating of the IBM 3090 mainframe; this difficulty is common for all families of fractals studied).

In order to find values of ν, we have found it suitable [2] to apply the MCRG method for SAWs on the fractals with $b \geq 10$. Indeed, due to both the inherent self-similarity and the finite ramification of the underlying fractals, this method turned out to work much better than in the case of regular lattices. The starting point of the method consists in treating B', given by (4), as the grand canonical partition function that comprise all possible SAWs that enter and exit the fractal generator at two fixed apexes. Accordingly, one can write

$$\langle N(B) \rangle = \frac{1}{B'} \sum_{N=b}^{b(b+1)/2} N a_N B^N, \tag{5}$$

which appears to be the average number of steps made, at fugacity B, by all possible SAWs that pass the generator. Combining (4) and (5) we get

$$\frac{\mathrm{d}B'}{\mathrm{d}B} = \frac{B'}{B} \langle N(B) \rangle, \tag{6}$$

and comparing the latter with (3), we obtain the equality $\lambda = \langle N(B^*) \rangle$, whereby we find

$$\nu = \frac{\ln b}{\ln \langle N(B^*) \rangle}. \tag{7}$$

This formula enables us to determine the critical exponent ν by calculating $\langle N(B^*) \rangle$ via the constant-fugacity MCRG method [11]. We have undertaken this approach, and thereby we have calculated ν up to $b = 81$ for the SG and CB fractals [2,6], and up to $b = 121$ for the PF fractals [3]. To this must be added that the MCRG approach, when contrasted with the exact RG approach, provided results that deviated at most 0.06% from the corresponding exact findings.

The MCRG approach could not be pushed much further (beyond $b = 100$) using the present day computers, which is somewhat unfortunate since the limit $b \to \infty$ poses an intriguing problem (in view of the fact that the fractal properties, like d_f and d_s, in this limit approach the relevant Euclidean values). In order to asses the asymptotic behaviour of ν, Dhar [12] has applied the FSS method in the case of the SG family of fractals, and we have applied the same method for the PF and the CB families of fractals [3,7]. The essence of the FSS method is the interpretation of the function $B' = f(B)$ as the corner-spin–corner-spin correlation function of the n–vector model (for $n = 0$), whose known asymptotic form (for the large spin distance, in the critical region) provides possibility of finding ν for large b. By this means, it has been established that in all three cases (SG, PF, and CB) ν tends to the Euclidean value 3/4 with the same *negative* correction term $\frac{3}{16} \ln\ln b/\ln b$, when $b \to \infty$.

At the end of this section, we would like to point out that we have been able to calculate the critical exponent γ by applying all three methods (in respective regions of b), that is, the exact RG, MCRG, and FSS method. These calculations are more intricate than in the case of ν, and, for this reason, we address the interested reader to the original references [1,3,5,7,12].

3 Results, Corollary, and Outlook

We present our findings for the SAW critical exponent ν in a way that encapsulates in Fig. 3 the results, obtained via different methods, for the three quite different families of fractals studies. The first striking feature is that in all three cases ν displays qualitatively the same nonmonotonic behaviour as a function of the scaling parameter b, whereas the second (and more) striking feature is that ν, in its decreasing part, crosses the Euclidean value 3/4 for the same finite scaling parameter — $b_h \approx 26$. To our knowledge, this appears to be the first universal element in the critical behaviour of SAWs on fractals. From the practical point of view, this means that there is a borderline in the homogeneity of fractals (measured by the size of the homogeneous parts they are composed of), which are embedded in the two–dimensional Euclidean space. Accordingly, for less homogeneous fractals ν is larger than the Euclidean value 3/4, while for more homogeneous fractals ν lies below 3/4. The first possibility, $\nu \geq 3/4$, seems to be widely accepted, whereas the second, $\nu < 3/4$, is less appreciated, although it has been observed in other investigations, both numerically [13] and experimentally [14].

How can one explain the crossing of the Euclidean value 3/4 for the finite b? This is equivalent to the question — why does ν decrease, with increasing b, when we know exactly that, for small b, it starts with values larger than 3/4? The observed decrease of ν in the region $2 \leq b \leq 100$ imply, according to the formula (7), that $\langle N(B^*)\rangle$ increases faster than b. This increase of $\langle N(B^*)\rangle$ can be explained in the following way. For values of b close to $b = 2$, the dominant number of SAWs are those that quickly traverse the fractal generators, whereas for larger b the walks that contain many rebounds from the generator edges begin

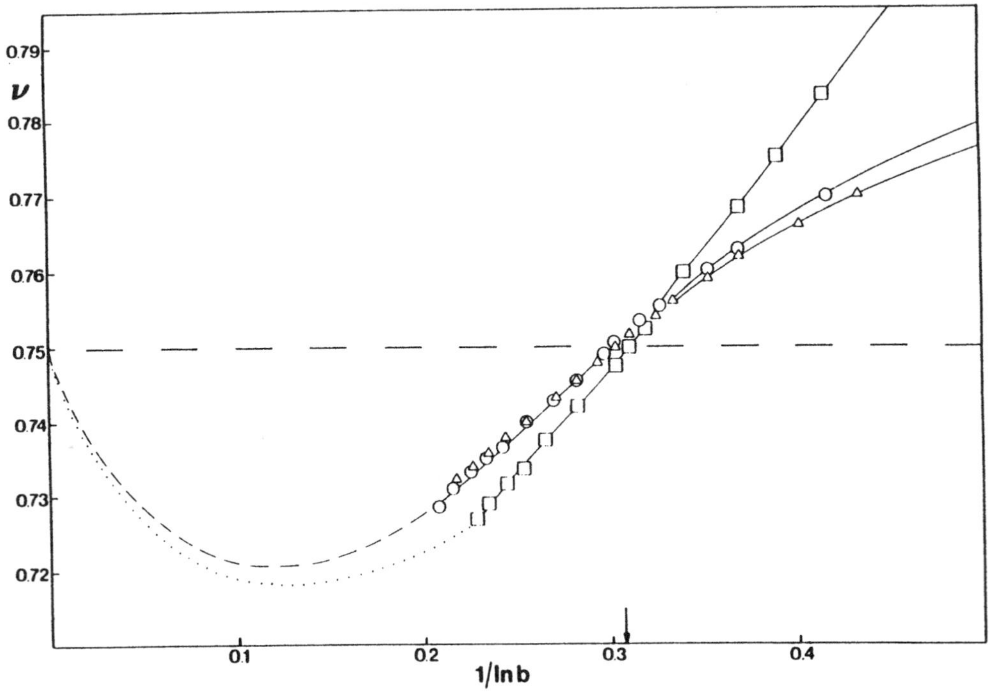

Fig. 3. Data for the SAW critical exponent ν for the SG family of fractals (triangles), PF family (circles), and CB family (squares), plotted as functions of $1/\ln b$. The error bars related to the MCRG data are not depicted in the figure since in all cases they lie well within the corresponding symbols. The dashed horizontal line represents the Euclidean value 3/4. The thin lines, that connect the results, serve merely as guides to the eye, while the dotted and dashed curves represent an extension of the FSS prediction for the CB and PF fractals, respectively. Accordingly, the minimum of ν occurs at $b \approx 1750$ (in the CB case), and at $b \approx 3330$ (in the case of the PF fractals). The small vertical arrow indicates the position of $b_h \approx 26$.

to turn up, which makes the walk look like the Hamiltonian walk (the one that crosses every available unit triangle). This is illustrated in Fig. 1, in which case on can say that the way of traversing the generator B should be characteristic of fractals with small b, whereas the way of running through the generator D should be typical of fractals with larger b. In other words, there is a competition between the excluded volume interaction and the interaction with the walls (of the obstacles, or lattice defects in the case of the PF fractals). They compete in such a way that the former tends to stretch out the SAW path, whereas the latter works on confining most of the SAW path within the homogeneous regions (pores). The same competition causes the minimum of ν, which is expected to take place before the asymptotic approach to the Euclidean value 3/4 from below, in the limit $b \to \infty$ (see Fig. 3). The location of the minimum cannot be detected by the present day computers, but it can be estimated starting with the FSS assumptions [7] (or, less accurately, by some phenomenological arguments [3,6]). As for the meaning of the minimal value of ν that occurs at some b_{min}, we may say that for $b < b_{min}$ the interaction with walls substantially controls the SAW end–to–end distance, while for $b > b_{min}$ the excluded volume interaction begins to preside over.

In Fig. 4 we present our results for the SAW critical exponent γ for the three families of fractals studied. It turns out that γ behaves similarly in all three cases, that is, γ is always larger than the Euclidean value 43/32, and, as a function of the fractal scaling factor b, increases monotonically with b. In the case of the SG family of fractals $\gamma \to 133/32$ when $b \to \infty$ [12], whereas for the other two families (PF and CB) $\gamma \to 103/32$ when $b \to \infty$ [3,7] (however, in all three cases, the first correction term $c \ln\ln b/\ln b$ is the same; with $c = -321/128$, for the SG fractals, and $c = -219/128$, for the PF and CB fractals). Furthermore, it turns out that the particular position of the apparently universal borderline ($b_h \approx 26$; discussed in the foregoing paragraph) is enhanced by the results obtained for γ. Indeed, one can notice from Fig. 4 (and, more precisely, upon comparison of the tables given in [3,5,7]) that for $b > b_h$ (and up to $b = 81$) γ acquires values that are, within the error bars, almost the same for the three families of fractals. To appreciate the sameness of γ (in the interval $b_h < b < 81$), one should observe that for very large b ($b \to \infty$) the difference in γ is about ten times greater than the quite visible difference that occurs for $b < b_h$.

In conclusion, we point out that the universality class pertinent to the SAW criticality on the two–dimensional Euclidean lattices is not retrieved in the limit $b \to \infty$, since γ does not tend to 43/32 (although ν tends to 3/4) when b tends to infinity. In addition, no two fractals with different scaling factors b, belonging to a single family of fractals, have the same criticality, as they have different values for ν, as well as different values for γ. However, the case $b_h \approx 26$ deserves particular attention. Indeed, it springs from our results that the SAW critical exponent ν crosses the value 3/4 at $b_h \approx 26$ *for the three quite different families of fractals*, with γ being almost the same immediately above b_h, that is, for $b_h \approx b \leq 81$. On these grounds, we may conjecture that $b_h \approx 26$ is a universal borderline for fractals in general, which means that one can expect, for instance, that ν is smaller than 3/4 if a fractal is composed of homogeneous patches of linear size larger than b_h. This conjecture should be further scrutinized and used to explain other relevant results. To this end, we quote here two recent examples available for analysis. First, we note that the experimental finding $\nu = 0.68$ should not be surprising as it was obtained in a case of a substratum with very large homogeneous pieces [14]. Secondly, we remark that our surmise can be used to explain certain results related to the controversial problem of SAWs on the critical percolation clusters. More particularly, if one accepts that ν on the backbone of the infinite percolation cluster is definitely larger than 3/4 [15,16] and looks for possible reasons for such a conclusion, then, with the help of our results, it becomes explicable. Namely, the preceding conclusion originated from the MC simulations on backbones (see figure 1 of reference [15]) that appear to

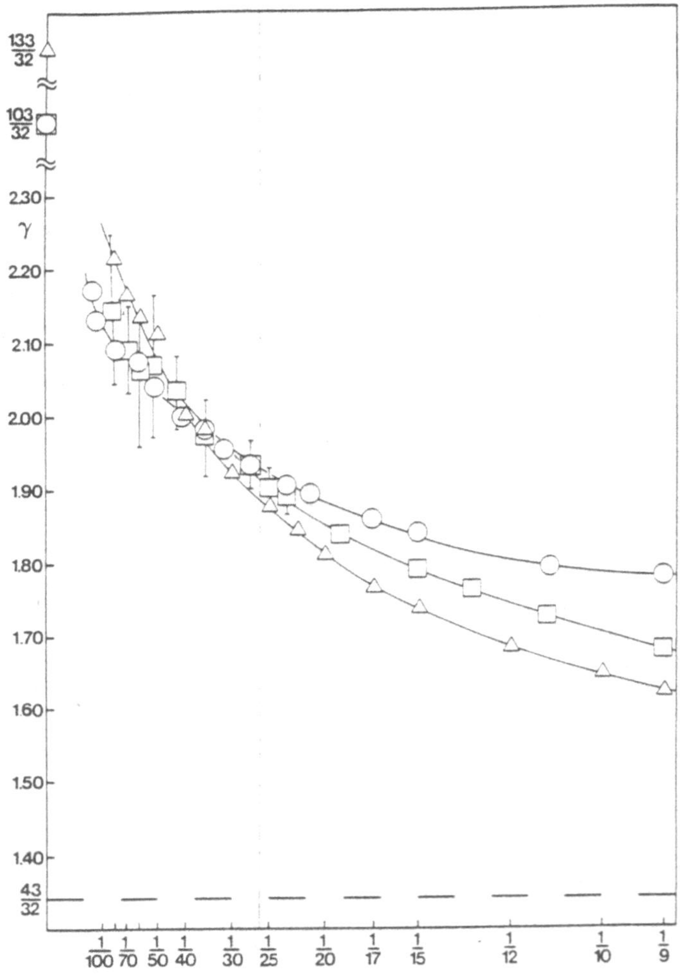

Fig. 4. Data for the SAW critical exponent γ for the SG family of fractals (triangles), PF family (circles), and CB family (squares), plotted as functions of $1/b$. The dashed horizontal line represents the Euclidean value 43/32, while the thin lines that connect the data serve merely as guides to the eye. The error bars related to the MCRG data are visible only in the case of the CB fractals, whereas in the SG and PF case the error bars lie within the corresponding symbols. The thin vertical line indicates the apparently universal borderline $b_h \approx 26$. Finally, one may note that it was necessary to interrupt twice the scale of the vertical axes in order to show the FSS predictions for the asymptotic values of γ.

be finitely ramified fractals with homogeneous parts of the size $b_B \leq 10$, that is, with $b_B < b_h$, and which, on the grounds of our results, should have $\nu > 3/4$. At the end, we expect new evidences that will substantiate our conjecture, as well as new specific verifications that will set the scope of its validity.

References

[1] S. Elezović, M. Knežević, S. Milošević: J. Phys. A: Math. Gen. **20** (1987) 1215;
[2] S. Milošević, I. Živić: J. Phys. A: Math. Gen. **24** (1991) L833;
[3] I. Živić, S. Milošević, H.E. Stanley: Phys. Rev. E **47** (1993) 2430;
[4] S. Elezović–Hadžić, S. Milošević: J. Phys. A: Math. Gen. **25** (1992) 4095;
[5] I. Živić, S. Milošević: J. Phys. A: Math. Gen. **26** (1993) 3393;
[6] S. Milošević, I. Živić: J. Phys. A: Math. Gen. **26** (1993) 7263;
[7] S. Milošević, I. Živić: to be published.
[8] S. Milošević, D. Stassinopoulos, H.E. Stanley: J. Phys. A: Math. Gen. **21** (1988) 1477;
[9] D. Dhar: J. Phys. A: Math. Gen. **21** (1988) 2261;
[10] D. Dhar: J. Math. Phys. **19**, 5 (1978);
[11] S. Redner, S.P. Reynolds: J. Phys. A: Math. Gen. **14** (1981) 2679;
[12] D. Dhar: J. Physique **49** (1988) 397;
[13] D.A. Fábio, R. Aarão, R. Riera: J. Stat. Phys. **71** (1993) 453;
[14] M. Tasserie, A. Hansen, D. Bideau: J. Physique **2** (1992) 2025;
[15] K.Y. Woo, S.B Lee: Phys. Rev. A **44** (1991) 999;
[16] C. Vanderzande, A. Komoda: Phys. Rev. A **45** (1992) R5335;

Fractals in Biology and Medicine

H. E. Stanley[1], *S. V. Buldyrev*[1], *A. L. Goldberger*[3], *S. Havlin*[1,2], *R. N. Mantegna*[1], *S. M. Ossadnik*[1], *C.-K. Peng*[1,3], *F. Sciortino*[1,5], *and M. Simons*[3,4]

[1] Center for Polymer Studies and Department of Physics, Boston University, Boston, MA 02215
[2] Department of Physics, Bar Ilan University, Ramat Gan, ISRAEL
[3] Cardiovascular Division, Harvard Medical School, Beth Israel Hospital, Boston, MA 02215
[4] Department of Biology, MIT, Cambridge, MA 02139,
[5] Dipartimento di Fisica, Universita' "La Sapienza," Piazzale Aldo Moro, 00185 Roma Italy

1 Introduction

The purpose of this talk is to describe some recent progress in applying fractal concepts to systems of relevance to biology and medicine. We review several biological systems characterized by fractal geometry, with a particular focus on the long-range power-law correlations found recently in DNA sequences containing *noncoding* material. Furthermore, we discuss the finding that the exponent α quantifying these long-range correlations ("fractal complexity") is smaller for coding than for noncoding sequences. We also discuss the application of fractal scaling analysis to the dynamics of heartbeat regulation, and report the recent finding that the normal heart is characterized by long-range "anticorrelations" which are absent in the diseased heart.

In the last decade it was realized that some biological systems have no characteristic length or time scale, i.e., they have fractal—or, more generally, self-affine—properties [1,2]. However, the fractal properties in different biological systems, have quite different nature, origin, and appearance. In some cases, it is the geometrical shape of a biological object itself that exhibits obvious fractal features, while in other cases the fractal properties are more "hidden" and can only be perceived if data are studied as a function of time or mapped onto a graph in some special way. After an appropriate mapping, such a graph may resemble a mountain landscape, with jagged ridges of all length scales from very small bumps to enormous peaks. Mathematically, these landscapes can be quantified in terms of fractal concepts such as self-affinity. The main part of the chapter is devoted to the study of such hidden fractal properties that have been recently discovered in DNA sequences and heartbeat activity.

2 Fractal Shapes

In contrast to compact objects, fractal objects have a very large *surface* area. In fact, they are composed almost entirely of "surface." This observation explains why fractals are ubiquitous in biology, where surface phenomena are of crucial importance.

Lungs exemplify this feature (Fig. 1). The surface area of a human lung is as large as a tennis court. The mammalian lung is made up of self-similar branches with many length scales, which is the defining attribute of a fractal surface. The efficiency of the lung is enhanced by this fractal property, since with each breath oxygen and carbon dioxide have to be exchanged at the lung surface. The structure of the bronchial tree has been quantitatively analyzed using fractal concepts [2,4]. In particular, fractal geometry could explain the power law decay of the average diameter of the bronchial tube with the generation number, in contrast to the classical model which predicts an exponential decay [6].

Not only the geometry of the respiratory tree is described by fractal geometry, but also the time-dependent features of inspiration. Specifically, Suki et al. [5] studied airway opening in isolated dog lungs. During constant flow inflations, they found that the lung volume changes in discrete jumps (Fig. 1), and that the probability distribution function of the relative size x of the jumps, $\Pi(x)$, and that of the time intervals t between these jumps, $\Pi(t)$, follow a power law over nearly two decades of x and t with exponents of 1.8 and 2.7, respectively. To interpret these findings, they developed a branching airway model in which airways, labeled ij, are closed with a uniform distribution of opening threshold pressures P. When the "airway opening" pressure P_{ao} exceeds P_{ij} of an airway, that airway opens along with one or both of its daughter branches if $P_{ij} < P_{\text{ao}}$ for the daughters. Thus, the model predicts "avalanches" of airway openings with a wide distribution of sizes, and the statistics of the jumps agree with those $\Pi(x)$ and $\Pi(t)$ measured experimentally. They concluded that power law distributions, arising from avalanches triggered by threshold phenomena, govern the recruitment of terminal airspaces.

A second example is the arterial system which delivers oxygen and nutrients to all the cells of the body. For this purpose blood vessels must have fractal properties [7,8]. The diameter distribution of blood vessels ranging from capillaries to arteries follows a power-law distribution which is one of the main characteristics of fractals. Sernetz et al. [9] have studied the branching patterns of arterial kidney vessels. They analyzed the mass-radius relation and found that it can be characterized by fractal geometry, with fractal dimensions between 2.0 and 2.5. Similarly, the branching of trees and other plants, as well as root systems have a fractal nature [10]. Moreover, the size distribution of plant-supported insects was found to be related to the fractal distribution of the leaves [11].

One of the most remarkable examples of a fractal object is the surface of a cauliflower, where every little head is an "almost" exact reduced copy of the whole head formed by intersecting Fibonacci spirals of smaller heads, which in turn consist of spirals of smaller and smaller heads, up to the fifth order of hierarchy (see Fig. 8.0 in [3]). West and Goldberger were first to describe such a "Fibonacci fractal" in the human lung [2].

Considerable interest in the biological community has also arisen from the possibility that neuron shape can be quantified using fractal concepts. For ex-

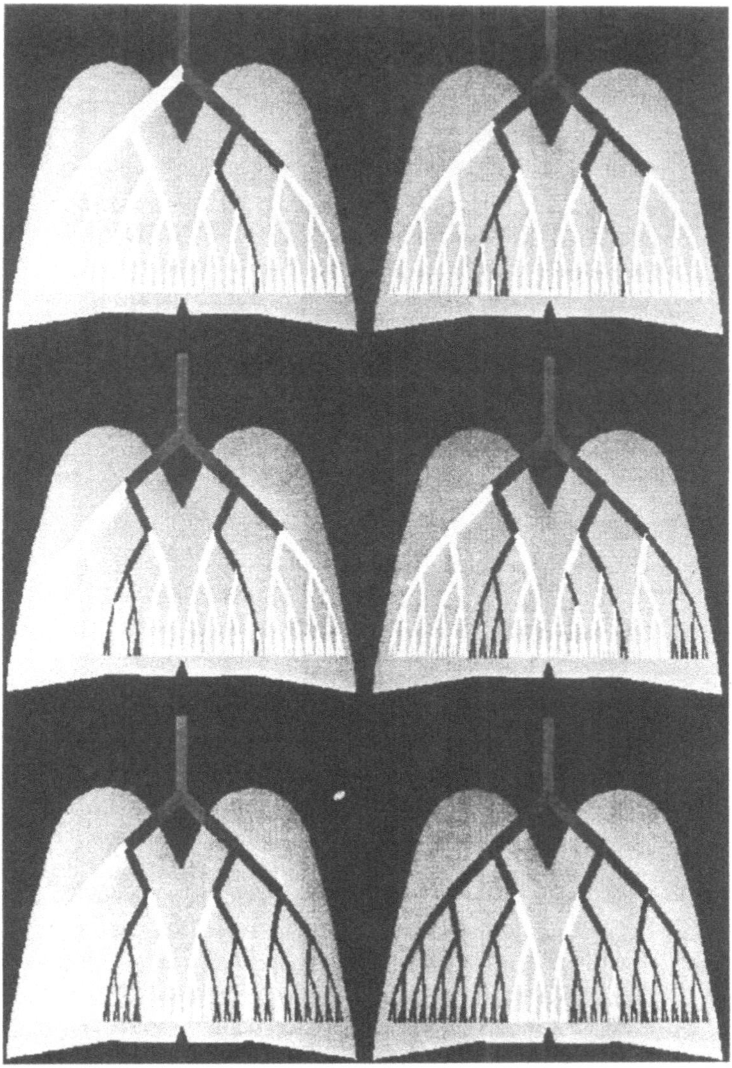

Fig. 1. The dynamic mechanism responsible for filling the lung involves "avalanches" or "bursts" of air that occur in all sizes—instead of an exponential distribution, one finds a power law distribution [5]. The underlying cause of this scale-free distribution of avalanches is the fact that every airway in the lung has its own threshold below which it is not inflated. Shown here is a diagram of the development of avalanches in the airways during airway opening. At first, almost all airways whose threshold value is smaller than the external pressure (red) are closed. Then the airway opening pressure increases until a second threshold is exceeded, and as a result all airways further up the tree whose thresholds are smaller become inflated (green). The airway opening pressure is successively increased until third, fourth, and fifth thresholds are exceeded (yellow, brown, and blue). The last threshold to be exceeded results in filling the airways colored violet; we notice that this last avalanche opens up over 25% of the total lung volume, thereby significantly increasing the total surface area available for gas exchange. After [5]

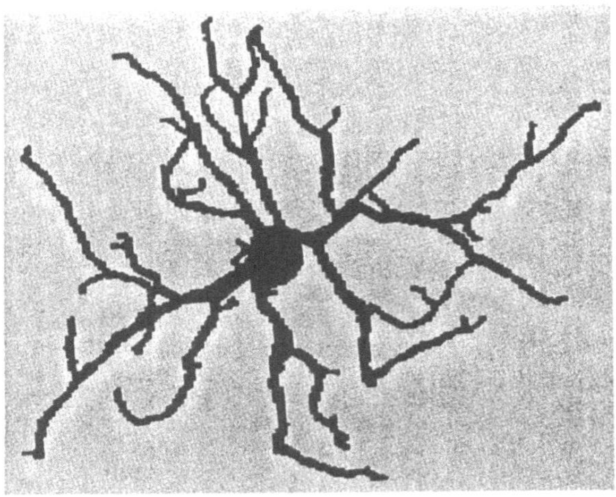

Fig. 2. Photograph of a retinal neuron (nerve cell), the morphology is similar to the DLA archetype. After [13]

ample, Smith et al. [12] studied the fractal features of vertebrate central nervous system neurons in culture and found that the fractal dimension is increased as the neuron becomes more developed. Caserta et al. [13] showed that the shapes of quasi-two-dimensional retinal neurons can be characterized by a fractal dimension d_f. They found for fully developed neurons *in vivo*, $d_f = 1.68 \pm 0.15$, and suggest that the growth mechanism for neurite outgrowth bears a direct analogy with the growth model called *diffusion limited aggregation* (DLA). The branching pattern of retinal vessels in a developed human eye is also similar to DLA [8]. The fractal dimension was estimated to be about 1.7, in good agreement with DLA for the case of two dimensions. For an alternative model for retinal growth see [14].

The DLA-type model governing viscous fingering may also serve to resolve the age-old paradox *"Why doesn't the stomach digest itself?* [15]. Indeed, the concentration of hydrochloric acid in the mammalian stomach after each meal is sufficient to digest the stomach itself, yet the gastric epithelium normally remains undamaged in this harsh environment. One protective factor is gastric mucus, a viscous secretion of specialized cells, which forms a protective layer and acts as a diffusion barrier to acid. Bicarbonate ion secreted by the gastric epithelium is trapped in the mucus gel, establishing a gradient from pH 1-2 at the lumen to pH 6-7 at the cell surface. The puzzle, then, is how hydrochloric acid, secreted at the base of gastric glands by specialized parietal cells, traverses the mucus layer to reach the lumen without acidifying the mucus layer. Bhaskar et al. [15]

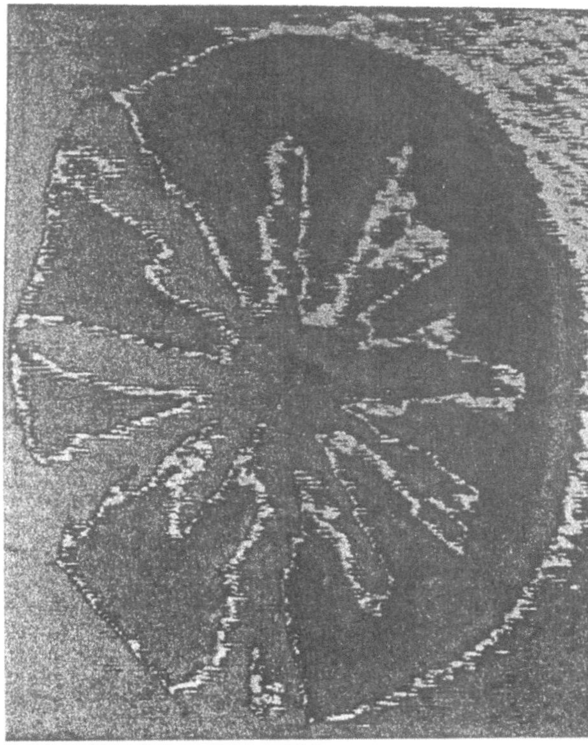

Fig. 3. Viscous fingers reflect the complex interface that develops when one fluid is pumped through another of higher viscosity. Shown is the formation of such viscous fingers or channels when hydrochloric acid is injected into solutions of gastric mucin. These channels may confine the acid and direct it to the lumen, thus protecting the gastric mucosa from acidification and ulceration; when the gastric glands contract, acid is ejected under high enough pressure to form viscous fingers. After [15]

resolved this puzzle by experiments that demonstrate the the possibility that flow of hydrochloric acid through mucus involves viscous fingering—the phenomenon that occurs when a fluid of lower viscosity is injected into a more viscous one (see Fig. 3). Specifically, Bhaskar et al. demonstrated that injection of hydrochloric acid through solutions of pig gastric mucin produces fingering patterns which are strongly dependent on pH, mucin concentration, and acid flow rate. Above pH 4, discrete fingers are observed, while below pH 4, hydrochloric acid neither penetrates the mucin solution nor forms fingers. These *in vitro* results suggest that hydrochloric acid secreted by the gastric gland can penetrate the mucus gel layer (pH 5-7) through narrow fingers, whereas hydrochloric acid in the lumen (pH 2) is prevented from diffusing back to the epithelium by the high viscosity of gastric mucus gel on the luminal side.

Yet another example of DLA-type growth is bacterial colony spread on the surface of agar (gel with nutrient) plates [16] (see Fig. 4). Vicsek et al. [17]

Fig. 4. A typical example of DLA-like colony patterns incubated at 35°C for three weeks after inoculation on the surface of agar plates containing initially 1 g/ℓ of peptone as nutrient. This pattern has a fractal dimension of $d_f \simeq 1.72$. After Matsushita and Fujikawa [16]

studied bacterial colony growth on a strip geometry which results in a self-affine surface (see Fig. 13.19 in [18]). They calculated the roughness exponent α for this surface and found $\alpha = 0.78 \pm 0.07$. The interfacial pattern formation of the growth of bacterial colonies was studied systematically by Ben-Jacob et al. [19]. They demonstrated that bacterial colonies can develop a pattern similar to morphologies in diffusion-limited growth observed in solidification and electrochemical deposition. These include fractal growth, dense-branching growth, compact growth, dendritic growth and chiral growth. The results indicate that the interplay between the micro level (individual bacterium) and the macro level (the colony) play a major role in selecting the observed morphologies similar to those found in nonliving systems.

Another example of fractal interface appears in ecology, in the problem of the territory covered by N diffusing particles [20], see Fig. 5. As seen from the figure, the territory initially grows with the shape of a disk with a relatively smooth surface until it reaches a certain size, at which point the surface becomes increasingly rough. This phenomenon may have been observed by Skellam [21] who plotted contours delineating the advance of the muskrat population and noted that initially the contours were smooth but at later times they became rough (see Fig. 1 in [21]).

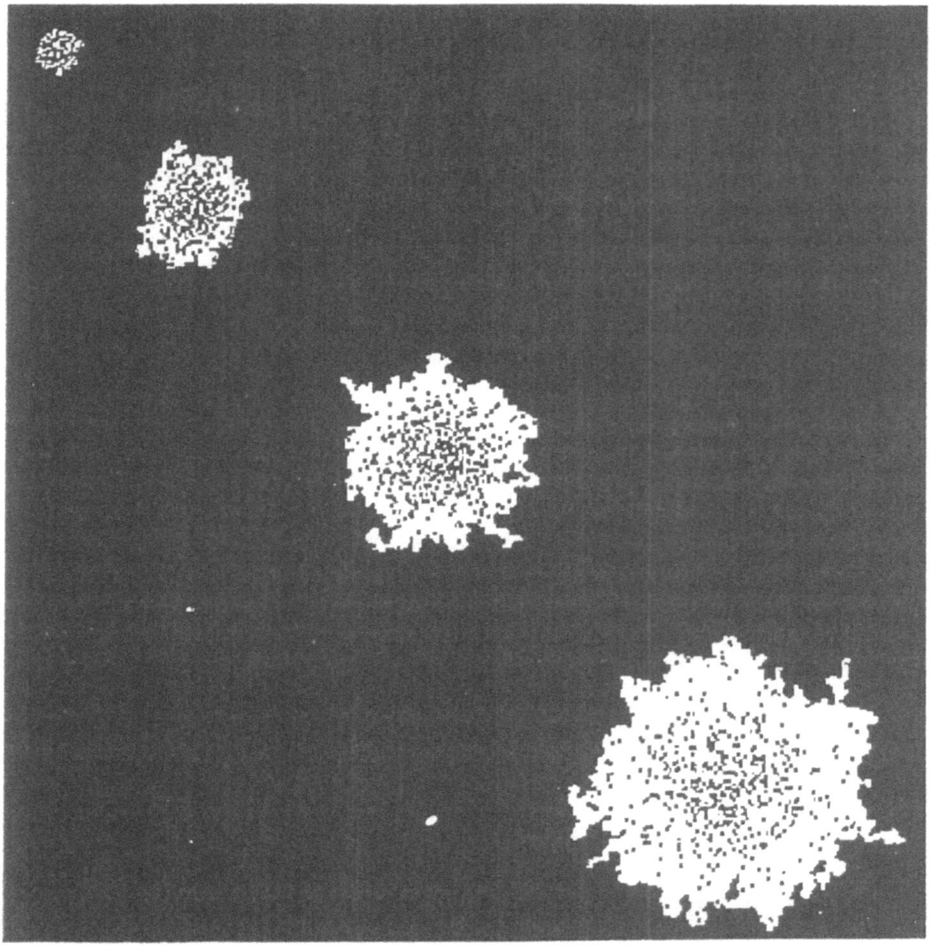

Fig. 5. Snapshots at successive times of the territory covered by N random walkers for the case $N = 500$ for a sequence of times. Note the roughening of the disc surface as time increases. The roughening is characteristic of the experimental findings for the diffusive spread of a population [21]. After [20], courtesy of P. Trunfio

Other biological contexts in which fractal scaling seems to be relevant are the relation between brain size and body weight [22], between bone diameter and bone length [23], between muscle force and muscle mass [23], and between an organism's size and its rate of producing energy and consuming food [24].

3 Long-Range Power Law Correlations

In recent years long-range power-law correlations have been discovered in a remarkably wide variety of systems. Such long-range power-law correlations are a physical fact that in turn gives rise to the increasingly appreciated "fractal geometry of nature" [1,3,18,25–28]. So if fractals are indeed so widespread, it makes sense to anticipate that long-range power-law correlations may be similarly widespread. Indeed, recognizing the ubiquity of long-range power-law correlations can help us in our efforts to understand nature, since as soon as we find power-law correlations we can quantify them with a critical exponent. Quantification of this kind of scaling behavior for apparently unrelated systems allows us to recognize similarities between different systems, leading to underlying unifications that might otherwise have gone unnoticed.

Usually correlations decay exponentially, but there is one major exception: at the critical point [29], the exponential decay of (1a) turns into to a power law decay

$$C_r \sim (1/r)^{d-2+\eta}. \qquad (1)$$

Many systems drive themselves spontaneously toward critical points [30,31]. One of the simplest models exhibiting such "self-organized criticality" is invasion percolation, a generic model that has recently found applicability to describing anomalous behavior of rough interfaces. Instead of occupying all sites with random numbers below a pre-set parameter p, in invasion percolation one "grows" the incipient infinite cluster right at the percolation threshold by the trick of occupying always the perimeter site whose random number is smallest. Thus small clusters are certainly not scale-invariant and in fact contain sites with a wide distribution of random numbers. As the mass of the cluster increases, the cluster becomes closer and closer to being scale invariant or "fractal." Such a system is said to drive itself to a "self-organized critical state" [32].

In the following sections we will attempt to summarize the key findings of some recent work [33–57] suggesting that—under suitable conditions—the sequence of base pairs or "nucleotides" in DNA also displays power-law correlations. The underlying basis of such power law correlations is not understood at present, but this discovery has intriguing implications for molecular evolution and DNA structure, as well as potential practical applications for distinguishing coding and noncoding regions in long nucleotide chains.

4 Information Coding in DNA

The role of genomic DNA sequences in coding for protein structure is well known [58,59]. The human genome contains information for approximately 100,000 different proteins, which define all inheritable features of an individual. The genomic sequence is likely the most sophisticated information database created by nature through the dynamic process of evolution. Equally remarkable is the precise transformation of information (duplication, decoding, etc) that occurs in a relatively short time interval.

The building blocks for coding this information are called *nucleotides*. Each nucleotide contains a phosphate group, a deoxyribose sugar moiety and either a *purine* or a *pyrimidine base*. Two purines and two pyrimidines are found in DNA. The two purines are adenine (A) and guanine (G); the two pyrimidines are cytosine (C) and thymine (T). The nucleotides are linked end to end, by chemical bonds from the phosphate group of one nucleotide to the deoxyribose sugar group of the adjacent nucleotide, forming a long polymer (*polynucleotide*) chain. The information content is encoded in the sequential order of the bases on this chain. Therefore, as far as the information content is concerned, a DNA sequence can be most simply represented as a symbolic sequence of four letters: A, C, G and T, as shown in Fig. 6.

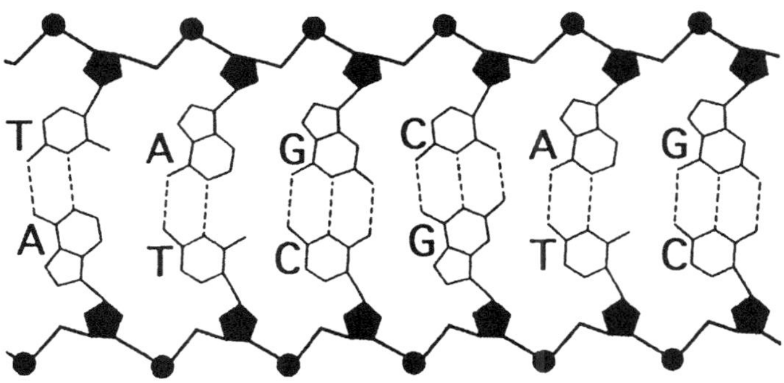

Fig. 6. The base pairing of two "double helix" DNA strands. The two chains of black pentagons and circles represent sugar-phosphate backbones of DNA strands linked by the hydrogen bonds (dashed lines) between complementary base pairs.

In the genomes of high eukaryotic organisms only a small portion of the total genome length is used for protein coding (as low as 5% in the human genome). For example, genes are separated from each other by *intergenic sequences* which are not used for coding proteins and which (especially in mammalian genomes) can be several times longer than genes. Furthermore, in 1977 it was discovered that genes themselves have inclusions which are not used for coding proteins. A gene is transcripted to RNA (pre-mRNA) and then some segments of the pre-mRNA are "spliced out" during the formation of the smaller mRNA molecule. The mRNA then serves as the template for assembling protein. The segments of the chromosomal DNA that are spliced out during the formation of a mature mRNA are called *introns* (for intervening sequences). The coding sequences are called *exons* (for expressive sequences).

The role of introns and intergenomic sequences constituting large portions of the genome remains unknown. Furthermore, only a few quantitative methods are currently available for analyzing information which is possibly encrypted in the noncoding part of the genome.

5 Conventional Statistical Analysis of DNA Sequences

DNA sequences have been analyzed using a variety of models that can basically be considered in two categories. The first types are "local" analyses; they take into account the fact that DNA sequences are produced in sequential order; therefore, the neighboring nucleotides will affect the next attaching nucleotide. This type of analysis, represented by n-step Markov models, can indeed describe some observed short-range correlations in DNA sequences. The second category of analyses is more "global" in nature; they concentrate on the presence of repeated patterns (such as periodic repeats and interspersed base sequence repeats) that are chiefly found in eukaryotic genomic sequences. A typical example of analysis in this category is the Fourier transform, which can identify repeats of certain segments of the same length in nucleotide sequences [58].

However, DNA sequences are more complicated than these two standard types of analysis can describe. Therefore it is crucial to develop new tools for analysis with a view toward uncovering the mechanisms used to code other types of information. Promising techniques for genome studies may be derived from other fields of scientific research, including time-series analysis, statistical mechanics, fractal geometry, and even linguistics.

6 The "DNA Walk"

One interesting question that may be asked by statistical physicists would be whether the sequence of the nucleotides A,C,G, and T behaves like a one-dimensional "ideal gas", where the fluctuations of density of certain particles obey Gaussian law, or if there exist long range correlations in nucleotide content (as in the vicinity of a critical point). These result in domains of all size with different nucleotide concentrations. Such domains of various sizes were known for a long time but their origin and statistical properties remain unexplained. A natural language to describe heterogeneous DNA structure is long-range correlation analysis, borrowed from the theory of critical phenomena [29].

6.1 Graphical Representation

In order to study the scale-invariant long-range correlations of a DNA sequence, we first introduced a graphical representation of DNA sequences, which we term a *fractal landscape* or *DNA walk* [33]. For the conventional one-dimensional random walk model [60,61], a walker moves either "up" $[u(i) = +1]$ or "down" $[u(i) = -1]$ one unit length for each step i of the walk. For the case of an uncorrelated walk, the direction of each step is independent of the previous steps. For the case of a correlated random walk, the direction of each step depends on the history ("memory") of the walker [62–64].

One definition of the DNA walk is that the walker steps "up" $[u(i) = +1]$ if a pyrimidine (C or T) occurs at position i along the DNA chain, while the walker steps "down" $[u(i) = -1]$ if a purine (A or G) occurs at position i (see

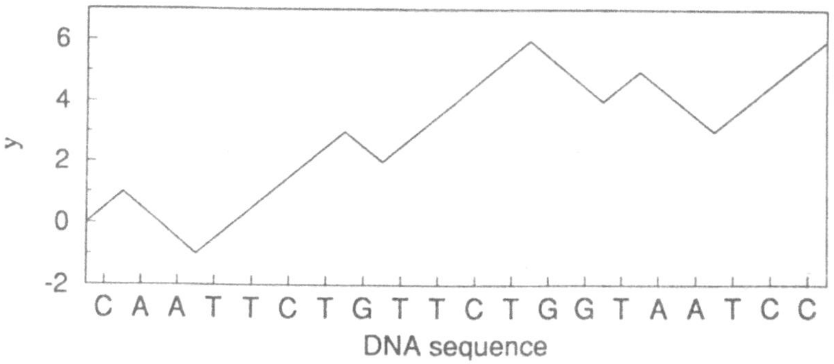

Fig. 7. Schematic illustration showing the definition of the "DNA walk"

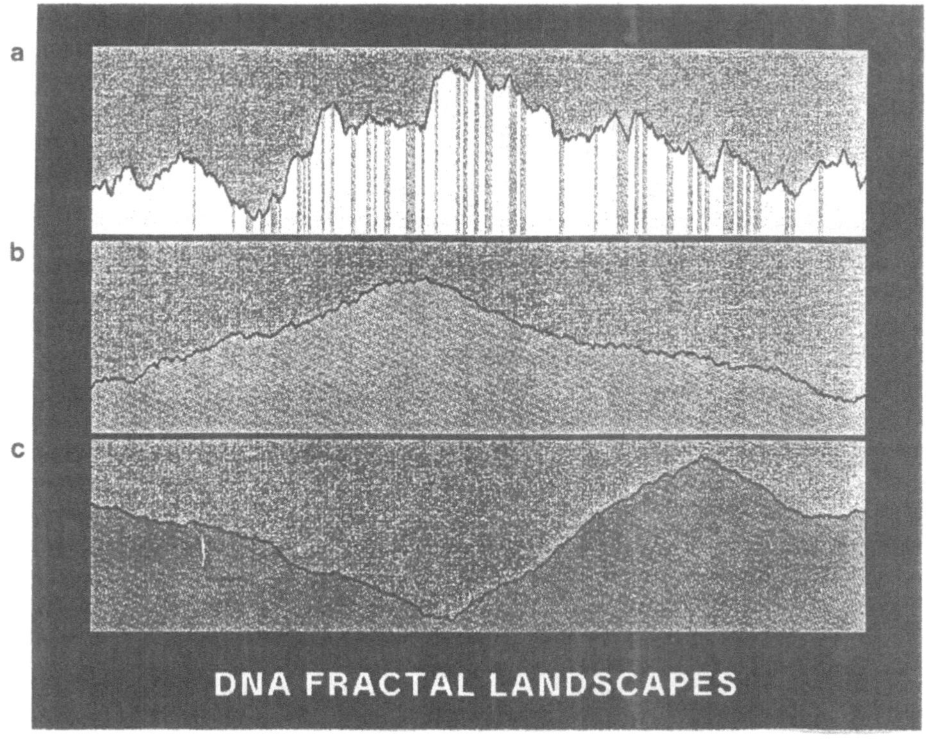

Fig. 8. The DNA walk representations of (a) human β-cardiac myosin heavy chain gene sequence, showing the coding regions as vertical golden bars, (b) the spliced together coding regions, and (c) the bacteriophage lambda DNA which contains only coding regions. Note the more complex fluctuations for (a) compared with the coding sequences (b) and (c). It is found that for almost all coding sequences studied that there appear regions with one strand bias, followed by regions of a different strand bias. In this presentation different step heights for purine and pyrimidine are used in order to align the end point with the starting point. This procedure is for graphical display purposes only (to allow one to visualize the fluctuations more easily) and is not used in any analytic calculations. After [33]

Fig. 7). The question we asked was whether such a walk displays only short-range correlations (as in an n-step Markov chain) or long-range correlations (as in critical phenomena and other scale-free "fractal" phenomena).

There have also been attempts to map DNA sequence onto multi-dimensional DNA walks [34,65]. However, recent work [57] indicates that the original purine-pyrimidine rule provides the most robust results, probably due to the purine-pyrimidine chemical complementarity.

The DNA walk allows one to visualize directly the fluctuations of the purine-pyrimidine content in DNA sequences: Positive slopes on Fig. 8 correspond to high concentration of pyrimidines, while negative slopes correspond to high concentration of purines. Visual observation of DNA walks suggests that the coding sequences and intron-containing noncoding sequences have quite different landscapes. Figure 8a shows a typical example of a gene that contains a significant fraction of base pairs that do *not* code for amino acids. Figure 8b shows the DNA walk for a sequence formed by splicing together the coding regions of the DNA sequence of this same gene (i.e., the cDNA). Figure 8c displays the DNA walk for a typical sequence with only coding regions. Landscapes for intron-containing sequences show very jagged contours which consist of patches of all length scales, reminiscent of the disordered state of matter near critical point. On the other hand, coding sequences typically consist of a few lengthy regions of different strand bias, resembling domains in the system in the ferromagnet state. These observations can be tested by rigorous statistical analysis. Figure 8 naturally motivates a quantification of these fluctuations by calculating the "net displacement" of the walker after ℓ steps, which is the sum of the unit steps $u(i)$ for each step i. Thus $y(\ell) \equiv \sum_{i=1}^{\ell} u(i)$.

6.2 Correlations and Fluctuations

An important statistical quantity characterizing any walk [60,61] is the root mean square fluctuation $F(\ell)$ about the average of the displacement; $F(\ell)$ is defined in terms of the difference between the average of the square and the square of the average,

$$F^2(\ell) \equiv \overline{[\Delta y(\ell) - \overline{\Delta y(\ell)}]^2} = \overline{[\Delta y(\ell)]^2} - \overline{\Delta y(\ell)}^2, \tag{2}$$

of a quantity $\Delta y(\ell)$ defined by $\Delta y(\ell) \equiv y(\ell_0+\ell) - y(\ell_0)$ (see also Chaps. 1 and 5). Here the bars indicate an *average* over all positions ℓ_0 in the gene. Operationally, this is equivalent to (a) using calipers preset for a fixed distance ℓ, (b) moving the beginning point sequentially from $\ell_o = 1$ to $\ell_o = 2, \cdots$ and (c) calculating the quantity $\Delta y(\ell)$ (and its square) for each value of ℓ_o, and (d) averaging all of the calculated quantities to obtain $F^2(\ell)$.

The mean square fluctuation is related to the auto-correlation function

$$C(\ell) \equiv \overline{u(\ell_0)u(\ell_0 + \ell)} - \overline{u(\ell_0)}^2 \tag{3}$$

through the relation

$$F^2(\ell) = \sum_{i=1}^{\ell} \sum_{j=1}^{\ell} C(j-i). \tag{4}$$

The calculation of $F(\ell)$ can distinguish three possible types of behavior.

1. If the base pair sequence were random, then $C(\ell)$ would be zero on average [except $C(0) = 1$], so $F(\ell) \sim \ell^{1/2}$ (as expected for a *normal* random walk).
2. If there were local correlations extending up to a characteristic range R (such as in Markov chains), then $C(\ell) \sim \exp(-\ell/R)$; nonetheless *the asymptotic* ($\ell \gg R$) *behavior* $F(\ell) \sim \ell^{1/2}$ *would be unchanged from the purely random case.*
3. If there is no characteristic length (i.e., if the correlation were "infinite-range"), then the scaling property of $C(\ell)$ would not be exponential, but would most likely to be a power law function, and the fluctuations will also be described by a power law

$$F(\ell) \sim \ell^{\alpha} \tag{5}$$

with $\alpha \neq 1/2$.

Figure 8a shows a typical example of a gene that contains a significant fraction of base pairs that do *not* code for amino acids. It is immediately apparent that the DNA walk has an extremely jagged contour which corresponds to long-range correlations. Figure 9 shows double logarithmic plots of the mean square fluctuation function $F(\ell)$ as a function of the linear distance ℓ along the DNA chain for a typical intron-containing gene.

The fact that the data for intron-containing and intergenic (i.e., noncoding) sequences are linear on this double logarithmic plot confirms that $F(\ell) \sim \ell^{\alpha}$. A least-squares fit produces a straight line with slope α substantially larger than the prediction for an uncorrelated walk, $\alpha = 1/2$, thus providing direct experimental evidence for the presence of long-range correlations.

On the other hand, the dependence of $F(\ell)$ for coding sequences is not linear on the log-log plot: its slope undergoes a crossover from 0.5 for small ℓ to 1 for large ℓ. However, if a single patch is analyzed separately, the log-log plot of $F(\ell)$ is again a straight line with the slope close to 0.5. This suggests that within a large patch the coding sequence is almost uncorrelated.

It is known that functional proteins usually form a single compact three-dimensional conformation that corresponds to the global energy minimum in the conformational space. Recently, Shakhnovich and Gutin [66] found that in order to have such a minimum it is sufficient that an amino acid sequence forms an uncorrelated random sequence. The finding of Peng et al. [33] of the lack of long range correlations in the coding nucleotide sequences provides more evidence for this hypothesis, since there exist almost one-to-one correspondence between amino acid sequences and their nucleotide codes. Furthermore, this finding may also indicate that the *lack* of long range correlations in the amino acid sequences is, in fact, a necessary condition for a functional biologically active protein.

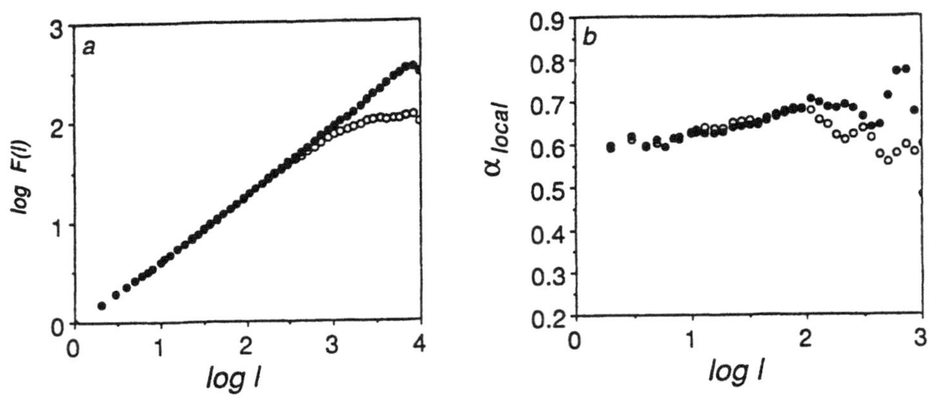

Fig. 9. (a) Double logarithmic plots of the mean square fluctuation function $F(\ell)$ as a function of the linear distance ℓ along the DNA chain for the rat embryonic skeletal myosin heavy chain gene (o) and its "intron-spliced sequence" (•). (b) The corresponding local slopes, α_{local}, based on pairs of successive data points of part (a). We see that the values of α are roughly constant. For this specific gene, the sequence with exons removed has an even broader scaling regime than the DNA sequence of the entire gene, indicated by the fact that part (a) is linear up to 10,000 nucleotides. After [42]

7 Differences Between Correlation Properties of Coding and Noncoding Regions

The initial report [33] on long-range (scale-invariant) correlations only in noncoding DNA sequences has generated contradicting responses. Some [34,35, 38,39] support our initial finding, while some [35,40,44,50] disagree. However, the conclusions of Refs. [36] and [35,40,44,50] are inconsistent *with one another* in that [35] and [50] doubt the existence of long-range correlations (even in noncoding sequences) while [36] and [40,44] conclude that even coding regions display long-range correlations ($\alpha > 1/2$). Prabhu and Claverie [40] claim that their analysis of the putative *coding* regions of the yeast chromosome III [67] produces a *wide range of exponent values*, some larger than 0.5. The source of these contradicting claims may arise from the fact that, in addition to normal statistical fluctuations expected for analysis of rather short sequences, coding regions typically consist of only a few lengthy regions of alternating strand bias. Hence conventional scaling analyses cannot be applied reliably to the entire sequence but only to sub-sequences.

Peng et al. [56] have recently applied the "bridge method" to DNA, and have also developed a similar method specifically adapted to handle problems associated with non-stationary sequences which they term *detrended fluctuation analysis* (DFA).

The idea of the DFA method is to compute the dependence of the standard error of a linear interpolation of a DNA walk $F_d(\ell)$ on the size of the interpolation segment ℓ. The method takes into account differences in local nucleotide content and may be applied to the entire sequence which has lengthy patches. In contrast with the original $F(\ell)$ function, which has spurious crossovers even for ℓ much smaller than a typical patch size, the detrended function $F_d(\ell)$ shows linear behavior on the log-log plot for all length scales up to the characteristic patch size, which is of the order of a thousand nucleotides in the coding sequences. For ℓ close to the characteristic patch size the log-log plot of $F_d(\ell)$ has an abrupt change in its slope.

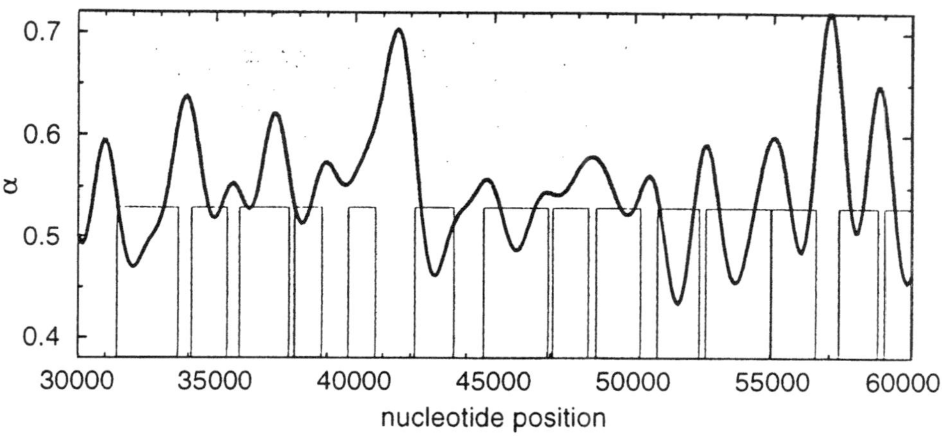

Fig. 10. Analysis of section of Yeast Chromosome III using the sliding box *Coding Sequence Finder* "CSF" algorithm. The value of the long-range correlation exponent α is shown as a function of position along the DNA chain. In this figure, the results for about 10% of the DNA are shown (from base pair #30,000 to base pair #60,000). Shown as vertical bars are the putative genes and open reading frames; denoted by the letter "G" are those genes that have been more firmly identified (March 1993 version of *GenBank*). Note that the local value of α displays **minima** where genes are suspected, while between the genes α displays **maxima**. This behavior corresponds to the fact that the DNA sequence of genes lacks long-range correlations ($\alpha = 0.5$ in the idealized limit), while the DNA sequence in between genes possesses long-range correlations ($\alpha \approx 0.6$). After [57]

The DFA method clearly supports the difference between coding and noncoding sequences, showing that the coding sequences are less correlated than noncoding sequences for the length scales less than 1000, which is close to characteristic patch size in the coding regions. One source of this difference is the tandem repeats (sequences such as AAAAAA...), which are quite frequent in noncoding sequences and absent in the coding sequences.

To provide an "unbiased" test of the thesis that noncoding regions possess but coding regions lack long-range correlations, Ossadnik et al. [57] analyzed several artificial uncorrelated and correlated "control sequences" of size 10^5 nucleotides using the GRAIL neural net algorithm [68]. The GRAIL algorithm identified about 60 putative exons in the uncorrelated sequences, but only about 5 putative exons in the correlated sequences.

Using the DFA method, we can measure the local value of the correlation exponent α along the sequence (see Fig 10) and find that the local minima of α as a function of a nucleotide position usually correspond to noncoding regions, while the local maxima correspond to noncoding regions. Statistical analysis using the DFA technique of the nucleotide sequence data for yeast chromosome III (315,338 nucleotides) shows that the probability that the observed correspondence between the positions of minima and coding regions is due to random coincidence is less than 0.0014. Thus, this method—which we called the "coding sequence finder" (CSF) algorithm—can be used for finding coding regions in the newly sequenced DNA, a potentially important application of DNA walk analysis.

8 Long-Range Correlations and Evolution

What is the biological meaning of the finding of long-range correlations in DNA? If two nucleotides whose positions differ by 1000 base pairs were uncorrelated, then there might be no meaning. However, the finding that they are correlated suggests some underlying organizational property. The long-range correlations in DNA sequences are of interest because they may be an indirect clue to its three-dimensional structure [45,54] or a reflection of certain scale-invariant properties of long polymer chains [53,55]. In any case, the statistically meaningful long-range "scale-invariant" (see Fig. 11) correlations in noncoding regions and their absence in coding regions will need to be accounted for by future explanations of global properties in gene organization and evolution.

Molecular evolutionary relationships are usually inferred from comparison of coding sequences, conservation of intron/exon structure of related sequences, analysis of nucleotide substitutions, and construction of phylogenetic trees [69]. The changes observed are conventionally interpreted with respect to nucleotide sequence composition (mutations, deletions, substitutions, alternative splicing, transpositions, etc.) rather than overall genomic organization.

Very recently, Buldyrev et al. [55] sought to assess the utility of DNA correlation analysis as a complementary method of studying gene evolution. In particular, they studied the changes in "fractal complexity" of nucleotide organization of a single gene family with evolution. A recent study by Voss [36] reported that the correlation exponent derived from Fourier analysis was lowest for sequences from organelles, but paradoxically higher for invertebrates than vertebrates. However, this analysis must be interpreted with caution since it was based on pooled data from different gene families rather than from the quantitative examination of any single gene family (see also [70,71]).

The hypothesis that the fractal complexity of genes from higher animals is greater than that of lower animals, using single gene family analysis was tested

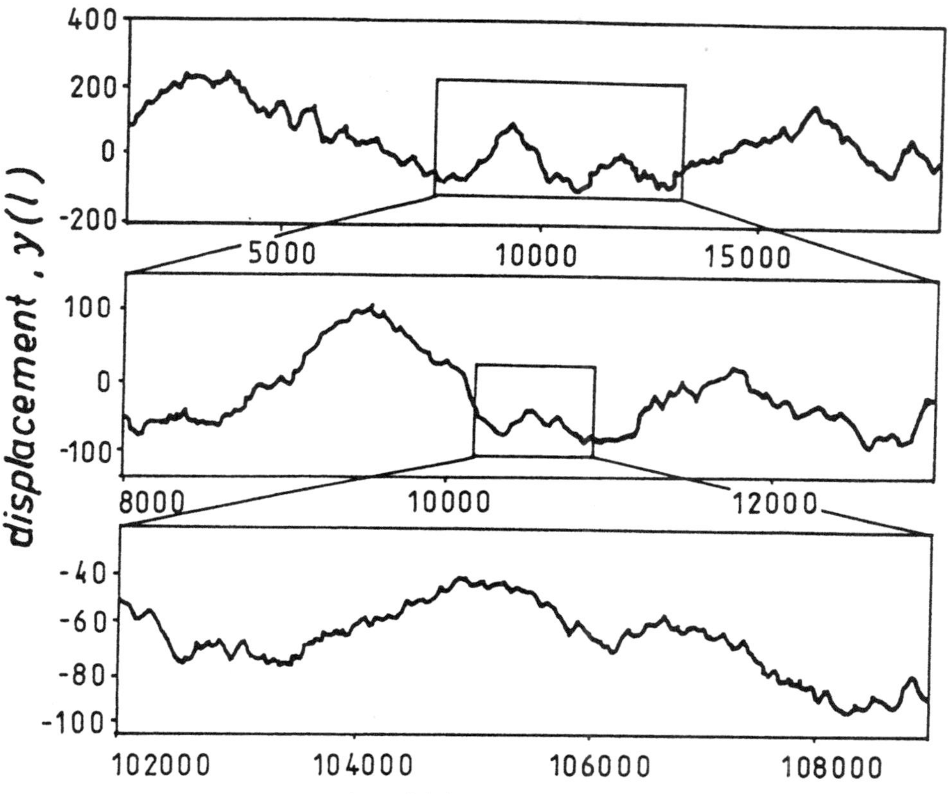

Fig. 11. The DNA walk representation for the rat embryonic skeletal myosin heavy chain gene ($\alpha = 0.63$). At the top the entire sequence is shown. In the middle the solid box shown in the top is magnified. At the bottom the solid box shown in the middle is magnified. The statistical self-similarity of these plots is consistent with the existence of a scale-free or fractal phenomenon which we call a fractal landscape. Note that one must magnify the segment by different factors along the ℓ (horizontal) direction and the y (vertical) direction; since F has the same units (dimension) as y, these magnification factors M_ℓ and M_y (along ℓ and y directions respectively) are related to the scaling exponent α by the simple relation $\alpha = \log(M_y)/\log(M_\ell)$ [e.g., from top to middle, $\log(M_y)/\log(M_\ell) = \log(2.07)/\log(3.2) = 0.63$]

in [55]. This analysis focuses on the genome sequences from the conventional (Type II) myosin heavy chain (MHC) family. Such a choice limits potential bias that may arise secondary to non-uniform evolutionary pressures and differences in nucleotide content between unrelated genes. The MHC gene family was chosen because of the availability of completely sequenced genes from a phylogenetically diverse group of organisms, and the fact that their relatively long sequences are well-suited to statistical analysis.

The landscape produced by DNA walk analysis reveals that each MHC cDNA consists of two roughly equal parts with significant differences in nucleotide content (Fig. 12). The first part that codes for the heavy meromyosin or "head" of the protein molecule has a slight excess of purines (52% purines and 48% pyrimidines); the second part that codes for the light meromyosin or "tail" has about 63% purines and 37% pyrimidines. The *absolute nucleotide* contents are not shown in the graphical representation of Fig. 12a because we subtract the average slope from the landscape to make relative fluctuations around the average more visible. Indeed, one can easily see from Fig. 12a that the relative concentration of pyrimidines in the first part ("uphill" region) of the myosin cDNA is much higher than in the second ("downhill" region).

The landscapes of Fig. 12 show that the coding sequences of myosins remain practically unchanged with evolution, while the entire gene sequences become more heterogeneous and complex. The quantitative measurements of the exponent α by DFA method confirm this visual observation showing that for all coding sequences of MHC family $\alpha \approx 0.5$. In contrast, for entire genes of MHC family, the value of α monotonically increases from lower eukaryotes to invertebrates and from invertebrates to vertebrates [55]. A stochastic model of random deletions and insertions of DNA portions was developed in [55] to explain this finding; see also [72–77].

Two major theories have been advanced to explain the origin and evolution of introns. One suggests that precursor genes consisted entirely of coding sequences and introns were inserted later in the course of evolution to help facilitate development of new structures in response to selective pressure, perhaps, by means of "exon shuffling" [78]. The alternative theory suggests that precursor genes were highly segmented and subsequently organisms not requiring extensive adaptation or new development or, perhaps, facing the high energetic costs of replicating unnecessary sequences, lost their introns [79,80]. Support for these hypotheses has remained largely conjectural; no models have been brought forward to support either process. The landscape analysis of the MHC gene family and the stochastic model [53,55] here are more consistent with the former view.

9 Other Biological Systems with Long-Range Correlations

The catalog of systems in which power law correlations appear has grown rapidly in recent years [32,81,82]; see also Chap. 2. What do we anticipate for other biological systems? Generally speaking, when "entropy wins over energy"—i.e., randomness dominates the behavior—we find power laws and scale invariance.

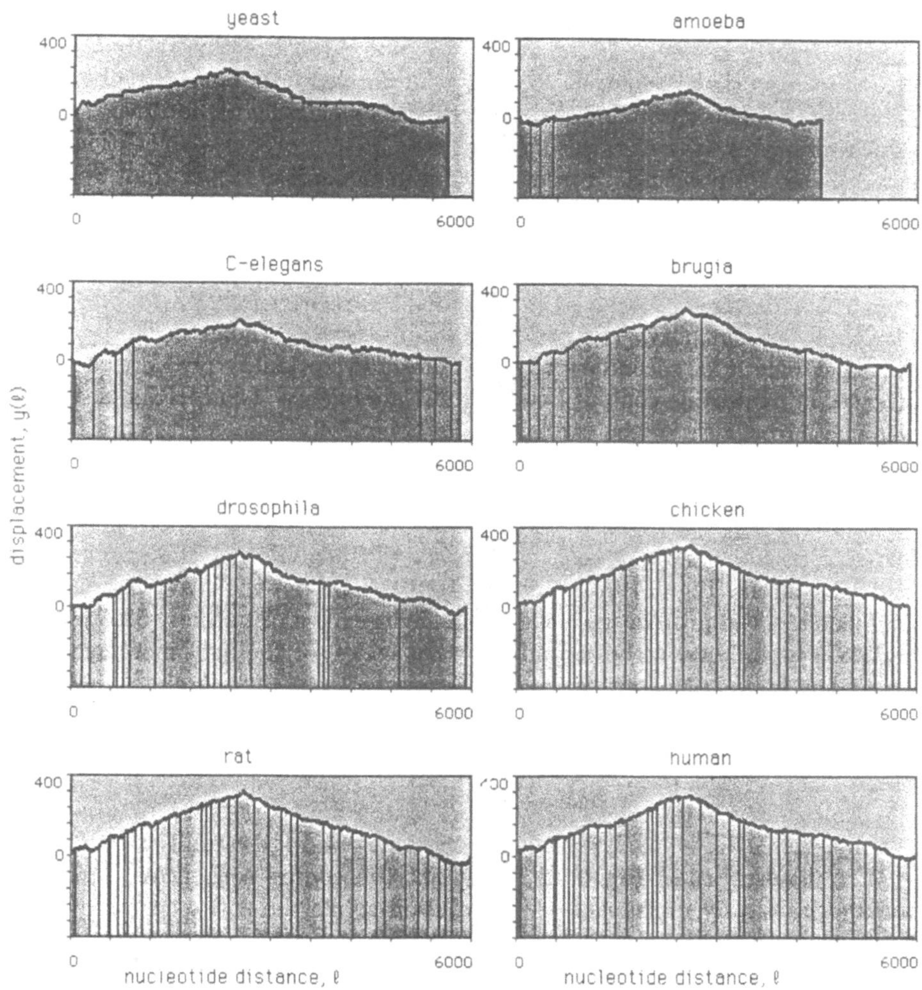

Fig. 12. The DNA walk representations of 8 cDNA sequences from the MHC family. DNA landscapes are plotted so that the end points have the same vertical displacement as the starting points [33]. The graphs are for yeast, amoeba, worms: *C. elegans*, *Brugia malayi*, drosophila, chicken, rat and human (from top to bottom, left to right).

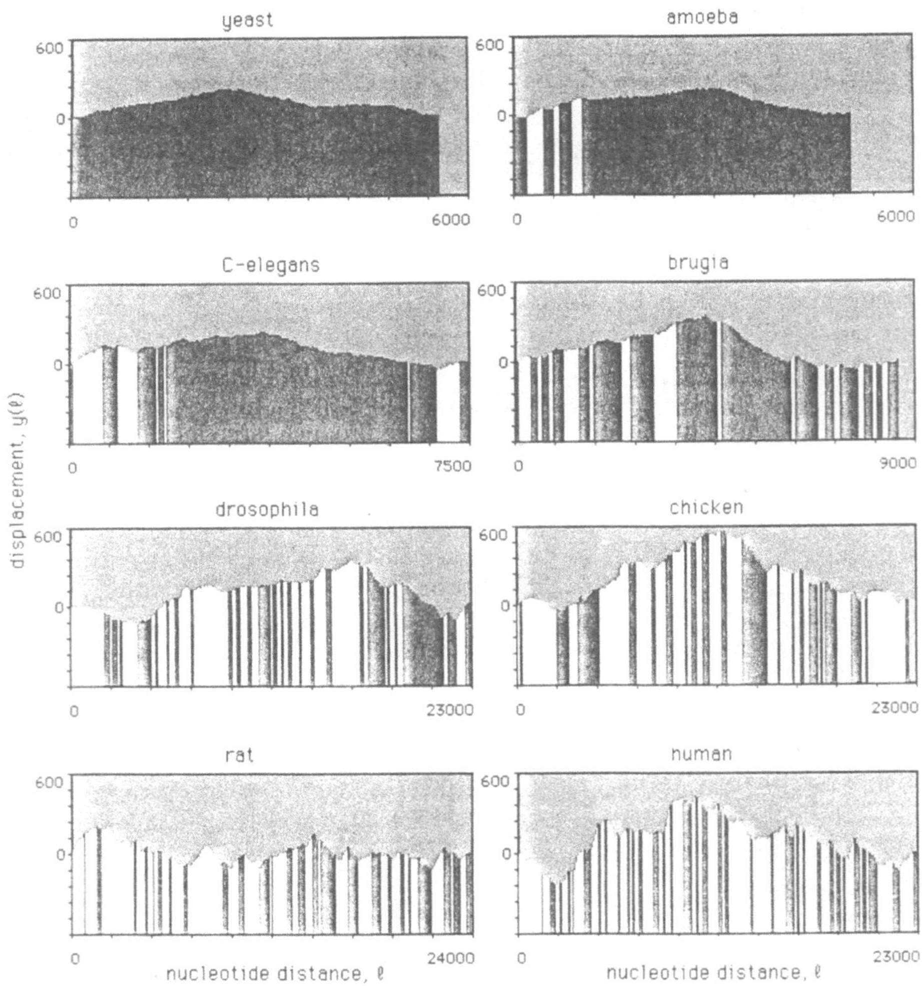

Fig. 13. Same as Fig. 12, for the full gene. The shaded areas denote coding regions of the genes. The DNA walks for the genes show increasing "complexity" with evolution. In contrast, the cDNA walks all show remarkably similar crossover patterns due to sequential "up-hill" and "down-hill" slopes representing different purine/pyrimidine strand biases in the regions coding for the head and tail of the MHC molecule, respectively. After [55]

The absence of characteristic length (or time) scales may confer important biological advantages, related to adaptability of response [2]. Biological systems sometimes are described in language that makes one think of a Swiss watch. Such mechanistic or "Rube Goldberg" descriptions must in some sense be incomplete, since it is only some appropriately-chosen averages that appear to behave in a regular fashion. The trajectory of each individual biological molecule is of necessity random—albeit correlated. Thus one might hope that recent advances in understanding "correlated randomness" [83,62–64] could be relevant to biological phenomena.

9.1 The Human Heartbeat

Traditionally, clinicians describe the normal electrical activity of the heart as "regular sinus rhythm." However, cardiac interbeat intervals fluctuate in a complex, apparently erratic manner in healthy subjects even at rest. Analysis of heart rate variability has focused primarily on short time oscillations associated with breathing (0.15–0.40 Hz) and blood pressure control (~ 0.1 Hz) [80]. Fourier analysis of longer heart rate data sets from healthy individuals typically reveals a $1/f$-like spectrum for frequencies < 0.1 Hz [84–87].

Peng et al. [88] recently studied scale-invariant properties of the human heartbeat time series, the output of a complicated integrative control system. The analysis is based on the digitized electrocardiograms of beat-to-beat heart rate fluctuations over very long time intervals (up to 24 h $\approx 10^5$ beats) recorded with an ambulatory monitor. The time series obtained by plotting the sequential intervals between beat n and beat $n+1$, denoted by $B(n)$, typically reveals a complex type of variability. The mechanism underlying such fluctuations is related to competing neuroautonomic inputs. Parasympathetic (vagal) stimulation decreases the firing rate of pacemaker cells in the heart's sinus node; sympathetic stimulation has the opposite effect. The nonlinear interaction (competition) between these two branches of the involuntary nervous system is the postulated mechanism for much of the erratic heart rate variability recorded in healthy subjects, although non-autonomic factors may also be important.

To study these dynamics over large time scales, the time series is passed through a digital filter that removes fluctuations of frequencies > 0.005 beat^{-1}, and plot the result, denoted by $B_L(n)$, in Fig. 13. One observes a more complex pattern of fluctuations for a representative healthy adult (Fig. 13a) compared to the "smoother" pattern of interbeat intervals for a subject with severe heart disease (Fig. 13b). These heartbeat time series produce a contour reminiscent of the irregular landscapes that have been widely studied in physical systems.

To quantitatively characterize such a "landscape," Peng et al. introduce a mean fluctuation function $F(n)$, defined as

$$F(n) \equiv \overline{|B_L(n'+n) - B_L(n')|}, \tag{6}$$

where the bar denotes an average over all values of n'. Since $F(n)$ measures the average difference between two interbeat intervals separated by a time lag n, $F(n)$ quantifies the magnitude of the fluctuation over different time scales n.

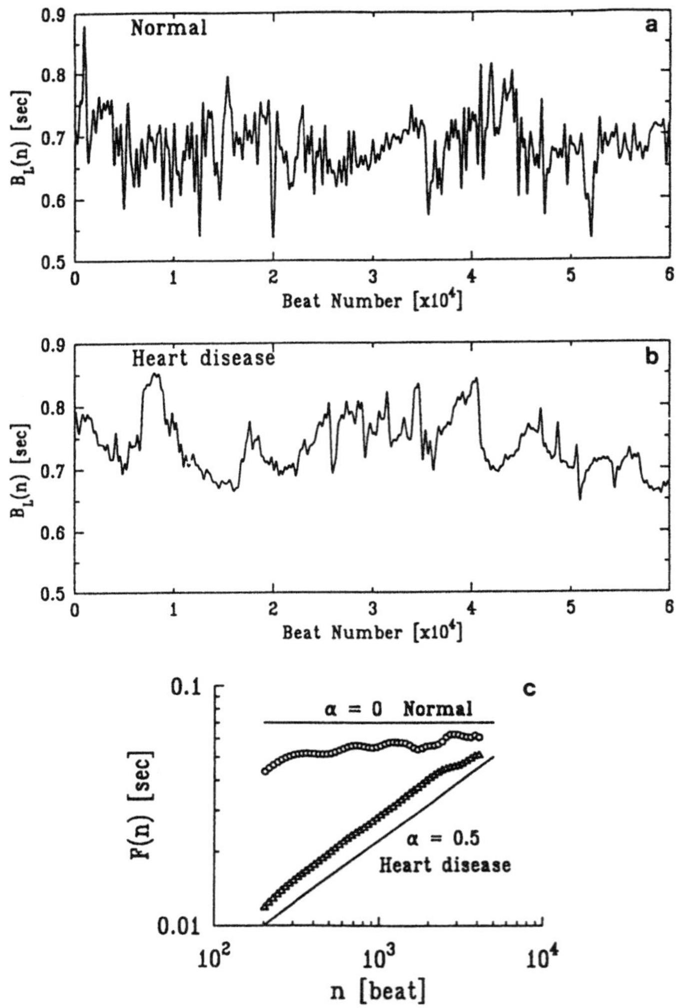

Fig. 14. The interbeat interval $B_L(n)$ after low-pass filtering for (a) a healthy subject and (b) a patient with severe cardiac disease (dilated cardiomyopathy). The healthy heartbeat time series shows more complex fluctuations compared to the diseased heart rate fluctuation pattern that is close to random walk ("brown") noise. (c) Log-log plot of $F(n)$ vs n. The circles represent $F(n)$ calculated from data in (a) and the triangles from data in (b). The two best-fit lines have slope $\alpha = 0.07$ and $\alpha = 0.49$ (fit from 200 to 4000 beats). The two lines with slopes $\alpha = 0$ and $\alpha = 0.5$ correspond to "$1/f$ noise" and "brown noise," respectively. We observe that $F(n)$ saturates for large n (of the order of 5000 beats), because the heartbeat interval are subjected to physiological constraints that cannot be arbitrarily large or small. The low-pass filter removes all Fourier components for $f \geq f_c$. The results shown here correspond to $f_c = 0.005$ beat^{-1}, but similar findings are obtained for other choices of $f_c \leq 0.005$. This cut-off frequency f_c is selected to remove components of heart rate variability associated with physiologic respiration or pathologic Cheyne-Stokes breathing as well as oscillations associated with baroreflex activation (Mayer waves). After [88]

Figure 13c is a log-log plot of $F(n)$ vs n for the data in Figs. 13a and 13b. This plot is approximately linear over a broad physiologically-relevant time scale (200 – 4000 beats) implying that

$$F(n) \sim n^{\alpha}. \tag{7}$$

It is found that the scaling exponent α is markedly different for the healthy and diseased states: for the healthy heartbeat data, α is close to 0, while α is close to 0.5 for the diseased case. Note that $\alpha = 0.5$ corresponds to a random walk (a Brownian motion), thus the low-frequency heartbeat fluctuations for a diseased state can be interpreted as a stochastic process, in which the heartbeat intervals $I(n) \equiv B(n+1) - B(n)$ are uncorrelated for $n \geq 200$.

To investigate these dynamical differences, it is helpful to study further the correlation properties of the time series. It is useful to study $I(n)$ because it is the appropriate variable for the aforementioned reason. Since $I(n)$ is stationary, one can apply standard spectral analysis techniques [3] (see also Chap. 1). Figures 14a and 14b show the power spectra $S_I(f)$, the square of the Fourier transform amplitudes for $I(n)$, derived from the same data sets (without filtering) used in Fig. 13. The fact that the log-log plot of $S_I(f)$ vs f is linear implies

$$S_I(f) \sim \frac{1}{f^{\beta}}. \tag{8}$$

The exponent β is related to α by $\beta = 2\alpha - 1$ [62]. Furthermore, β can serve as an indicator of the presence and type of correlations:

1. If $\beta = 0$, there is no correlation in the time series $I(n)$ ("white noise").
2. If $0 < \beta < 1$, then $I(n)$ is correlated such that positive values of I are likely to be close (in time) to each other, and the same is true for negative I values.
3. If $-1 < \beta < 0$, then $I(n)$ is also correlated; however, the values of I are organized such that positive and negative values are more likely to alternate in time ("anti-correlation") [62].

For the diseased data set, we observe a flat spectrum ($\beta \approx 0$) in the low frequency region (Fig. 14b) confirming that $I(n)$ are not correlated over long time scales (low frequencies). Therefore, $I(n)$, the first derivative of $B(n)$, can be interpreted as being analogous to the *velocity* of a random walker, which is uncorrelated on long time scales, while $B(n)$—corresponding to the *position* of the random walker—are correlated. However, this correlation is of a trivial nature since it is simply due to the summation of uncorrelated random variables.

In contrast, for the data set from the healthy subject (Fig. 14a), we obtain $\beta \approx -1$, indicating *nontrivial* long-range correlations in $B(n)$—these correlations are not the consequence of summation over random variables or artifacts of non-stationarity. Furthermore, the "anti-correlation" properties of $I(n)$ indicated by the negative β value are consistent with a nonlinear feedback system that "kicks" the heart rate away from extremes. This tendency, however, does not only operate on a beat-to-beat basis (local effect) but on a wide range of time

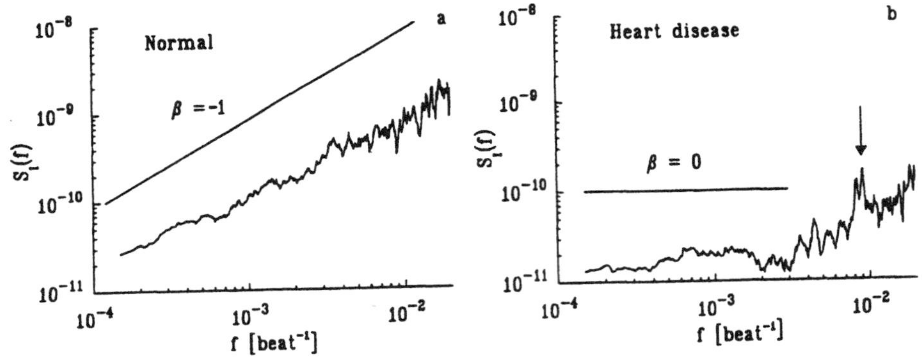

Fig. 15. The power spectrum $S_I(f)$ for the interbeat interval increment sequences over $\sim$ 24 hours for the same subjects in Fig. 13. (a) Data from a healthy adult. The best-fit line for the low frequency region has a slope $\beta = -0.93$. The heart rate spectrum is plotted as a function of "inverse beat number" (beat^{-1}) rather than frequency (time^{-1}) to obviate the need to interpolate data points. The spectral data are smoothed by averaging over 50 values. (b) Data from a patient with severe heart failure. The best-fit line has slope 0.14 for the low frequency region, $f < f_c = 0.005$ beat^{-1}. The appearance of a pathologic, characteristic time scale is associated with a spectral peak (arrow) at about 10^{-2} beat^{-1} (corresponding to Cheyne-Stokes respiration). After [88]

scales. To our knowledge, this is the first explicit description of long-range anti-correlations in a fundamental biological variable, namely the interbeat interval increments.

Furthermore, the histogram for the heartbeat intervals increments is found to be well-described by a Lévy stable distribution. For a group of subjects with severe heart disease, it is found that the distribution is unchanged, but the long-range correlations vanish. Therefore, the different scaling behavior in health and disease must relate to the underlying dynamics of the heartbeat. Applications of this analysis may lead to new diagnostics for patients at high risk of cardiac disease and sudden death.

9.2 Physiological Implications

The finding of nontrivial long-range correlations in healthy heart rate dynamics is consistent with the observation of long-range correlations in other biological systems that do not have a characteristic scale of time or length. Such behavior may be adaptive for at least two reasons. (i) The long-range correlations serve as an organizing principle for highly complex, nonlinear processes that generate fluctuations on a wide range of time scales. (ii) The lack of a characteristic scale helps prevent excessive *mode-locking* that would restrict the functional responsiveness of the organism. Support for these related conjectures is provided by observations from severe diseased states such as heart failure where the breakdown of long-range correlations is often accompanied by the emergence of a dominant frequency mode (e.g., the Cheyne-Stokes frequency). Analogous transitions to highly periodic regimes have been observed in a wide range of other disease states including certain malignancies, sudden cardiac death, epilepsy and fetal distress syndromes.

The complete breakdown of normal long-range correlations in any physiological system could theoretically lead to three possible diseased states: (i) a random walk (brown noise), (ii) highly periodic behavior, or (iii) completely uncorrelated behavior (white noise). Cases (i) and (ii) both indicate only "trivial" long-range correlations of the types observed in severe heart failure. Case (iii) may correspond to certain cardiac arrhythmias such as fibrillation. More subtle or intermittent degradation of long-range correlation properties may provide an early warning of incipient pathology. Finally, we note that the long-range correlations present in the healthy heartbeat indicate that the neuroautonomic control mechanism actually drives the system away from a single steady state. Therefore, the classical theory of homeostasis, according to which stable physiological processes seek to maintain "constancy" [89], should be extended to account for this dynamical, far from equilibrium, behavior.

9.3 Human Writings

Long-range correlations have been found recently in human writings [81]. A novel, a piece of music or a computer program can be regarded as a one-dimensional string of symbols. These strings can be mapped to a one-dimensional random walk model similar to the DNA walk (Sect. 6) allowing calculation of the correlation exponent α using (4a). Values of α between 0.6 and 0.9 were found for various texts.

An interesting fractal feature of languages was found in 1949 by Zipf [90]. He observed that the frequency of words as a function of the word order decays as a power law (with a power close to -1) for more than four orders of magnitude. A theory for this empirical finding, based on assumptions of coding words in the brain, was given by Mandelbrot [1,91]. A related interesting statistical measure of short-range correlations in languages and in general series sequences is the entropy and redundancy defined by Shannon [92].

9.4 Dynamics of Membrane Channel Openings

Ions, such as potassium and sodium, cannot cross the lipid cell membrane. They can, however, enter or exit the cell through ion channel proteins that are distributed on the cell membrane. These proteins spontaneously fluctuate between open or closed states. Liebovitch [93] found that the histograms of the open and closed duration times of some channels are self-similar and behave as power laws. This approach may provide new models for the ion channel gating mechanisms.

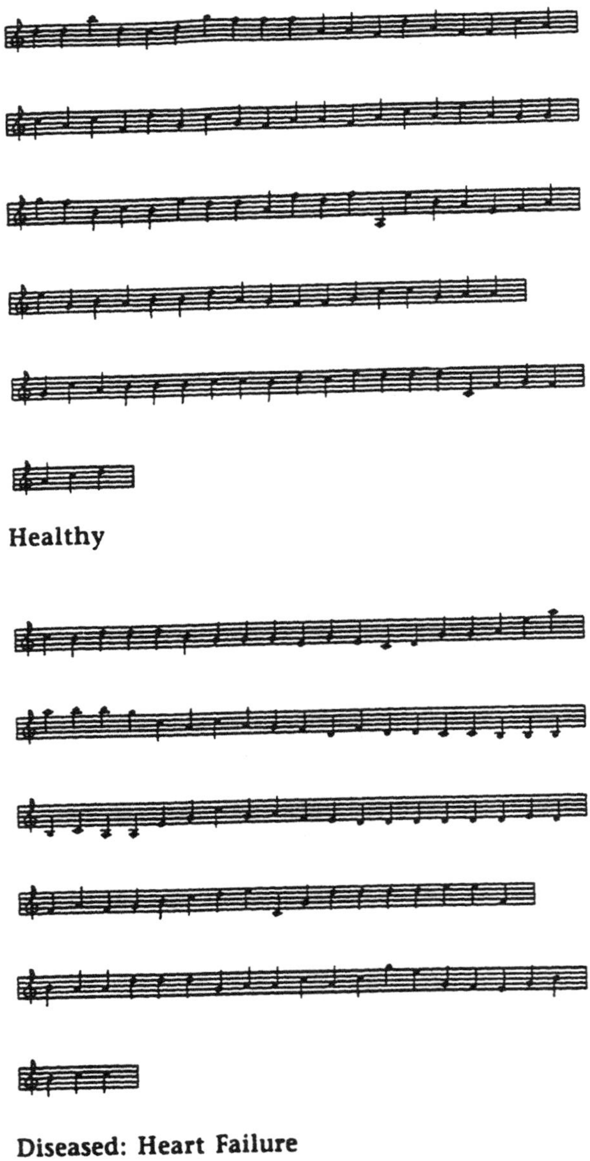

Fig. 16. Musical mapping of two heartbeat times series, derived from normal (top) and pathologic (bottom) data sets. The original heart beat time series were obtained from 24 hour recordings consisting of about 10^6 heartbeats. The heartbeat time series were then low-pass filtered to remove fluctuations > 0.05 (beat^{-1}), roughly equivalent to averaging every 200 beats. The pattern of fluctuations in the normal is more complex than that of the "music" generated from the abnormal data sets. Musical compositions based on these times series are available on cassette by request along with the "scores"; contact Zachary D. Goldberger (e-mail: ary "at" astro.bih.harvard.edu). There is a nominal charge for copying and mailing. After [97], courtesy of Z.D. Goldberger

9.5 Fractal Music and the Heartbeat

Fourier analysis of instantaneous fluctuations in amplitude as well as inter-note intervals for certain classical music pieces (e.g., Bach's First Brandenburg Concerto) reveals a $1/f$ distribution over a broad frequency range [94-96]. Voss and Clarke [95] used an algorithm for generating $1/f$-noise to "compose" music.

Based on the observation of different scaling patterns for healthy and pathologic heartbeat time series [88], it was very recently postulated [97] that (i) actual biological rhythms such as the heartbeat might serve as a more natural template for musical compositions than artificially-generated noises, and (ii) audibly appreciable differences between the note series of healthy and diseased hearts could potentially serve as the basis for a clinically useful diagnostic test. Accordingly, a computer program was devised to map heart rate fluctuations onto intervals of the diatonic musical scale [97]. As anticipated, the *normal* ($1/f$-like) heartbeat obtained from the low pass filtered time series reported in [88] generated a more variable (complex) type of music than that generated by the *abnormal* times series (Fig. 15).

The "musicality" of these transcriptions is intriguing and supports speculations about the brain's possible role as a translator/manipulator of biological $1/f$-like noise into aesthetically pleasing art works. Current investigations are aimed at extending these preliminary observations by (i) comparing the "musicality" of note sequences generated by natural (biological) vs. artificial (computer simulated) correlated and uncorrelated noises, and (ii) using heartbeat time series as the template for simultaneously generating fluctuations in musical rhythm and intensity, not only pitch.

9.6 Fractal Approach to Biological Evolution

Fossil data suggest that evolution of biological species takes place as intermittent bursts of activity, separated by relatively long periods of quiescence [98]. Recently Bak and Sneppen [99] suggested that these spontaneous catastrophic extinctions may be related to the power law distribution of avalanches of growth observed in a model of self-organized criticality (SOC). Such SOC models are reminiscent of recent surface growth models based on the concept of directed invasion percolation [100].

Acknowledgements.

We are grateful to R. Bansil, A.-L. Barabasi, K.R. Bhaskar, F. Caserta, G. Daccord, W. Eldred, P. Garik, Z.D. Goldberger, Z. Hantos, J.M. Hausdorff, R.E. Hausman, P. Ivanov, T.J. LaMont, H. Larralde, M.E. Matsa, J. Mietus, A. Neer, J. Nittmann, F. Pétak, I. Rabin, F. Sciortino, A. Shehter, M.H.R. Stanley, B. Suki, P. Trunfio, M. Ukleja, and G.H. Weiss for major contributions to those results reviewed here that represent collaborative research efforts. We also wish to thank C. Cantor, C. DeLisi, M. Frank-Kamenetskii, A.Yu. Grosberg, G. Huber, I. Labat, L. Liebovitch, G.S. Michaels, P. Munson, R. Nossal, R. Nussinov,

R.D. Rosenberg, J.J. Schwartz, M. Schwartz, E.I. Shakhnovich, M.F. Shlesinger, N. Shworak, and E.N. Trifonov for valuable discussions. Partial support was provided to to SVB and HES by the National Science Foundation, to ALG by the G. Harold and Leila Y. Mathers Charitable Foundation, the National Heart, Lung and Blood Institute and the National Aeronautics and Space Administration, to SH by Israel-USA Binational Science foundation, and to C-KP by an NIH/NIMH Postdoctoral NRSA Fellowship.

References

[1] B.B. Mandelbrot: *The Fractal Geometry of Nature* (W.H. Freeman, San Francisco 1982)

[2] B.J. West, A.L. Goldberger: J. Appl. Physiol., **60**, 189 (1986);
B.J. West, A.L. Goldberger: Am. Sci., **75**, 354 (1987);
A.L. Goldberger, B.J. West: Yale J. Biol. Med. **60**, 421 (1987);
A.L. Goldberger, D.R. Rigney, B.J. West: Sci. Am. **262**, 42 (1990);
B.J. West, M.F. Shlesinger: Am. Sci. **78**, 40 (1990);
B.J. West: *Fractal Physiology and Chaos in Medicine* (World Scientific, Singapore 1990);
B.J. West, W. Deering: Phys. Reports **xx, xxx** (1994)

[3] A. Bunde, S. Havlin, eds.: *Fractals and Disordered Systems* (Springer-Verlag, Berlin 1991)
A. Bunde, S. Havlin, eds.: *Fractals in Science* (Springer-Verlag, Berlin 1994)

[4] M.F. Shlesinger, B.J. West: Phys. Rev. Lett. **67**, 2106 (1991)

[5] B. Suki, A.-L. Barabasi, Z. Hantos, F. Peták, H.E. Stanley: Nature **368** 615 (1994)

[6] E.R. Weibel, D.M. Gomez: Science **137**, 577 (1962)

[7] A.A. Tsonis, P.A. Tsonis: Perspectives in Biology and Medicine **30**, 355 (1987)

[8] F. Family, B.R. Masters, D.E. Platt: Physica D **38**, 98 (1989);
B.R. Masters, F. Family, D.E. Platt: Biophys. J. (Suppl.) **55**, 575 (1989);
B.R. Masters, D.E. Platt: Invest. Ophthalmol. Vis. Sci. (Suppl.) **30**, 391 (1989)

[9] M. Sernetz, J. Wübbeke, P. Wlczek: Physica A **191**, 13 (1992)

[10] H. Takayasu: *Fractals in the Physical Sciences* (Manchester University Press, Manchester 1990)

[11] D.R. Morse: Nature **314**, 731 (1985)

[12] T.G. Smith, W.B. Marks, G.D. Lange, W.H. Sheriff Jr., E.A. Neale: J. Neuroscience Methods **27**, 173 (1989)

[13] H.E. Stanley: Bull. Am. Phys. Soc. **34**, 716 (1989);
F. Caserta, H.E. Stanley, W.D. Eldred, G. Daccord, R.E. Hausman, J. Nittmann: Phys. Rev. Lett. **64**, 95 (1990);
F. Caserta, R.E. Hausman, W.D. Eldred, H.E. Stanley, C. Kimmel: Neurosci. Letters **136**, 198 (1992);
F. Caserta, W.D. Eldred, E. Fernandez, R.E. Hausman, L.R. Stanford, S.V. Buldyrev, S. Schwarzer, H.E. Stanley: J. Neurosci. Methods (in press)

[14] D.R. Kayser, L.K. Aberle, R.D. Pochy, L. Lam: Physica A **191**, 17 (1992)

[15] K.R. Bhaskar, B.S. Turner, P. Garik, J.D. Bradley, R. Bansil, H.E. Stanley, J.T. LaMont: Nature **360**, 458 (1992)

[16] T. Matsuyama, M. Sugawa, Y. Nakagawa: FEMS Microb. Lett. **61**, 243 (1989); H. Fujikawa, M. Matsushita: J. Phys. Soc. Japan **58**, 387 (1989); M. Matsushita, H. Fujikawa: Physica A **168**, 498 (1990)
[17] T. Vicsek, M. Cserző, V.K. Horváth: Physica A **167**, 315 (1990); S. Matsuura, S. Miyazima: Physica A **191**, 30 (1992)
[18] T. Vicsek: *Fractal Growth Phenomena, Second Edition* (World Scientific, Singapore 1992)
[19] E. Ben-Jacob, H. Shmueli, O. Shochet, A Tenenbaum: Physica A **187**, 378 (1992); E. Ben-Jacob, A. Tenenbaum, O. Shochet, O. Avidan: Physica A **202**, 1 (1994)
[20] H. Larralde, P. Trunfio, S. Havlin, H.E. Stanley, G.H. Weiss: Nature **355**, 423 (1992); M.F. Shlesinger: Nature **355**, 396 (1992); H. Larralde, P. Trunfio, S. Havlin, H.E. Stanley, G.H. Weiss: Phys. Rev. A **45**, 7128 (1992); S. Havlin, H. Larralde, P. Trunfio, J.E. Kiefer, H.E. Stanley, G.H. Weiss: Phys. Rev. A **46**, R-1717 (1992)
[21] J.G. Skellam: Biometrika **38**, 196 (1951)
[22] P.H. Harvey, J.R. Krebs: Science **249**, 140 (1990)
[23] R.H. Peters: *The Ecological Implications of Body Size* (Cambridge University Press, Cambridge 1983); J.E.I. Hokkanen: J. Theor. Biol. **499**, 499 (1956); A.A. Biewener: Science **250**, 1097 (1990)
[24] M. Sernetz, B. Gelleri, J. Hofmann: J. Theor. Biol. **187**, 209 (1985); R.R. Strathmann: Science **250**, 1091 (1990)
[25] J. Feder: *Fractals* (Plenum, NY, 1988)
[26] D. Stauffer, H.E. Stanley: *From Newton to Mandelbrot: A Primer in Theoretical Physics* (Springer-Verlag, Heidelberg & N.Y. 1990)
[27] E. Guyon, H.E. Stanley: *Les Formes Fractales* (Palais de la Découverte, Paris 1991); **English translation:** *Fractal Forms* (Elsevier North Holland, Amsterdam 1991)
[28] H.E. Stanley, N. Ostrowsky, eds.: *On Growth and Form: Fractal and Non-Fractal Patterns in Physics*, Proceedings 1985 Cargèse NATO ASI, Series E: Applied Sciences (Martinus Nijhoff, Dordrecht 1985)
[29] H.E. Stanley: *Introduction to Phase Transitions and Critical Phenomena* (Oxford University Press, London 1971)
[30] F. Zernike: Physica **7**, 565 (1940)
[31] H.E. Stanley, N. Ostrowsky, eds.: *Correlations and Connectivity: Geometric Aspects of Physics, Chemistry and Biology*, Proceedings 1990 Cargèse Nato ASI, Series E: Applied Sciences (Kluwer, Dordrecht 1990)
[32] P. Bak, C. Tang, K. Wiesenfeld: Phys. Rev. Lett. **59**, 381 (1987); P. Bak, C. Tang, K. Wiesenfeld: Phys. Rev. A **38**, 3645 (1988)
[33] C.-K. Peng, S.V. Buldyrev, A.L. Goldberger, S. Havlin, F. Sciortino, M. Simons, H.E. Stanley: Nature **356**, 168 (1992)
[34] W. Li, K. Kaneko: Europhys. Lett. **17**, 655 (1992)
[35] S. Nee: Nature **357**, 450 (1992)
[36] R. Voss: Phys. Rev. Lett. **68**, 3805 (1992)
[37] J. Maddox: Nature **358**, 103 (1992)
[38] P.J. Munson, R.C. Taylor, G.S. Michaels: Nature **360**, 636 (1992)
[39] I. Amato: Science **257**, 747 (1992)
[40] V.V. Prabhu, J.-M. Claverie: Nature **357**, 782 (1992)

[41] P. Yam: Sci. Am. **267[3]**, 23 (1992)
[42] C.-K. Peng, S.V. Buldyrev, A.L. Goldberger, S. Havlin, F. Sciortino, M. Simons, H.E. Stanley: Physica A **191**, 25 (1992)
[43] H.E. Stanley, S.V. Buldyrev, A.L. Goldberger, J.M. Hausdorff, S. Havlin, J. Mietus, C.-K. Peng, F. Sciortino, M. Simons: Physica A **191**, 1 (1992)
[44] C.A. Chatzidimitriou-Dreismann, D. Larhammar: Nature **361**, 212 (1993); D. Larhammar, C.A. Chatzidimitriou-Dreismann: Nucleic Acids Res. **21**, 5167 (1993)
[45] A.Yu. Grosberg, Y. Rabin, S. Havlin, A. Neer: Biofisika (Russia) **38**, 75 (1993); A.Yu. Grosberg, Y. Rabin, S. Havlin, A. Neer: Europhys. Lett. **23**, 373 (1993)
[46] J.-L. Sikorav, G.M. Church: J. Mol. Biol. **222**, 1085 (1991)
[47] A.Yu. Grosberg, A.R. Khokhlov: *Statistical Physics of Macromolecules* (Nauka Publishers, Moscow 1989)
[48] A.Yu. Grosberg, S.K. Nechaev, E.I. Shakhnovich: J. Physique **49**, 2095 (1988)
[49] K.W. Adolph, in: *Chromosome Structure and Function*, ed. by M.S. Risley (Van Nostrand, New York 1986); M. Takahashi: J. Theor. Biol. **141**, 117 (1989)
[50] S. Karlin, V. Brendel: Science **259**, 677 (1993)
[51] C.-K. Peng, S.V. Buldyrev, A.L. Goldberger, S. Havlin, M. Simons, H.E. Stanley: Phys. Rev. E **47**, 3730 (1993)
[52] N. Shnerb, E. Eisenberg: Phys. Rev. E **49**, R1005 (1994)
[53] S.V. Buldyrev, A.L. Goldberger, S. Havlin, C.-K. Peng, M. Simons, H.E. Stanley: Phys. Rev. E **47**, 4514 (1993)
[54] A.S. Borovik, A.Yu. Grosberg, M.D. Frank-Kamenetskii: preprint
[55] S.V. Buldyrev, A.L. Goldberger, S. Havlin, C.-K. Peng, H.E. Stanley, M.H.R. Stanley, M. Simons: Biophys. J. **65**, 2673 (1993)
[56] C.-K. Peng, S.V. Buldyrev, S. Havlin, M. Simons, H.E. Stanley, A.L. Goldberger: Phys. Rev. E **49**, 1685 (1994)
[57] S.M. Ossadnik, S.V. Buldyrev, A.L. Goldberger, S. Havlin, R.N. Mantegna, C.-K. Peng, M. Simons, H.E. Stanley: Biophys. J. **67**, 64 (1994) H.E. Stanley, S.V. Buldyrev, A.L. Goldberger, S. Havlin, C.-K.Peng, M. Simons: [Proceedings of Internat'l Conf. on Condensed Matter Physics, Bar-Ilan], Physica A **200**, 4 (1993); H.E. Stanley, S.V. Buldyrev, A.L. Goldberger, S. Havlin, S.M. Ossadnik, C.-K. Peng, M. Simons: Fractals **1**, 283-301 (1993)
[58] S. Tavaré, B.W. Giddings, in: *Mathematical Methods for DNA Sequences*, Eds. M.S. Waterman (CRC Press, Boca Raton 1989), pp. 117-132
[59] J.D. Watson, M. Gilman, J. Witkowski, M. Zoller: *Recombinant DNA* (Scientific American Books, New York 1992).
[60] E.W. Montroll, M.F. Shlesinger: "The Wonderful World of Random Walks" in: *Nonequilibrium Phenomena II. From Stochastics to Hydrodynamics*, ed. by J.L. Lebowitz, E.W. Montroll (North-Holland, Amsterdam 1984), pp. 1–121
[61] G.H. Weiss: *Random Walks* (North-Holland, Amsterdam 1994)
[62] S. Havlin, R. Selinger, M. Schwartz, H.E. Stanley, A. Bunde: Phys. Rev. Lett. **61**, 1438 (1988); S. Havlin, M. Schwartz, R. Blumberg Selinger, A. Bunde, H.E. Stanley: Phys. Rev. A **40**, 1717 (1989); R.B. Selinger, S. Havlin, F. Leyvraz, M. Schwartz, H.E. Stanley: Phys. Rev. A **40**, 6755 (1989)

[63] C.-K. Peng, S. Havlin, M. Schwartz, H.E. Stanley, G.H. Weiss: Physica A **178**, 401 (1991);
C.-K. Peng, S. Havlin, M. Schwartz, H.E. Stanley: Phys. Rev. A **44**, 2239 (1991)
[64] M. Araujo, S. Havlin, G.H. Weiss, H.E. Stanley: Phys. Rev. A **43**, 5207 (1991);
S. Havlin, S.V. Buldyrev, H.E. Stanley, G.H. Weiss: J. Phys. A **24**, L925 (1991);
S. Prakash, S. Havlin, M. Schwartz, H.E. Stanley: Phys. Rev. A **46**, R1724 (1992)
[65] C.L. Berthelsen, J.A. Glazier, M.H. Skolnick: Phys. Rev. A **45**, 8902 (1992)
[66] E.I. Shakhnovich, A.M. Gutin: Nature **346**, 773 (1990)
[67] S.G. Oliver et al.: Nature **357**, 38 (1992)
[68] E.C. Uberbacher, R.J. Mural: Proc. Natl. Acad. Sci. USA **88**, 11261 (1991)
[69] W.-H. Li, D. Graur: *Fundamentals of Molecular Evolution* (Sinauer Associates, Sunderland MA 1991)
[70] H.E. Stanley, S.V. Buldyrev, A.L. Goldberger, S. Havlin, C-K Peng, M. Simons: Physica A **200**, 4 (1993)
[71] S.V. Buldyrev, A. Goldberger, S. Havlin, C.-K. Peng, F. Sciortino, M. Simons, H.E. Stanley: Phys. Rev. Lett. **71**, 1776 (1993);
R.F. Voss, Phys. Rev. Lett. **71**, 1777 (1993)
[72] W. Li: International Journal of Bifurcation and Chaos **2**, 137 (1992)
[73] J. Jurka: J. Mol. Evol. **29**, 496 (1989)
[74] R.H. Hwu, J.W. Roberts, E.H. Davidson, R.J. Britten: Proc. Natl. Acad. Sci. USA. **83**, 3875 (1986)
[75] J. Jurka, T. Walichiewicz, A. Milosavljevic: J. Mol. Evol. **35**, 286 (1992)
[76] J. Des Cloizeaux: J. Physique (Paris) **41**, 223 (1980);
P.G. de Gennes: *Scaling Concepts in Polymer Physics* (Cornell Univ. Press, Ithaca 1979)
[77] G.F. Joyce: Nature **338**, 217 (1989)
[78] W. Gilbert: Nature **271**, 501 (1978)
[79] P. Hagerman: Ann. Rev. Biochem. **59**, 755 (1990)
[80] W.F. Doolitle, in: *Intervening Sequences in Evolution and Development*, ed. by E. Stone, R. Schwartz (Oxford University Press, NY 1990), p. 42
[81] A. Schenkel, J. Zhang, Y-C. Zhang: Fractals **1**, 47 (1993);
M. Amit, Y. Shmerler, E. Eisenberg, M. Abraham, N. Shnerb: Fractals **2**, xxx (1994)
[82] R.N. Mantegna: Physica A **179**, 23 (1991)
[83] H.E. Stanley, N. Ostrowsky, eds.: *Random Fluctuations and Pattern Growth: Experiments and Models*, Proceedings 1988 NATO ASI, Cargèse (Kluwer Academic Publishers, Dordrecht 1988)
[84] R.I. Kitney, O. Rompelman: *The Study of Heart-Rate Variability* (Oxford University Press, London 1980);
S. Akselrod, D. Gordon, F.A. Ubel, D.C. Shannon, A.C. Barger, R.J. Cohen: Science **213**, 220 (1981)
[85] M. Kobayashi, T. Musha: IEEE Trans. Biomed. Eng. **29**, 456 (1982)
[86] A.L. Goldberger, D.R. Rigney, J. Mietus, E.M. Antman, S. Greenwald: Experientia **44**, 983 (1988)
[87] J.P. Saul, P. Albrecht, D. Berger, R.J. Cohen: *Computers in Cardiology* (IEEE Computer Society Press, Washington, D.C. 1987), pp. 419–422;
D.T. Kaplan, M. Talajic: Chaos **1**, 251 (1991)
[88] C.-K. Peng, J.E. Mietus, J.M. Hausdorff, S. Havlin, H.E. Stanley, A.L. Goldberger: Phys. Rev. Lett. **70**, 1343 (1993);
C.K. Peng, Ph.D. Thesis, Boston University (1993);

C.K. Peng, S.V. Buldyrev, J.M. Hausdorff, S. Havlin, J.E. Mietus, M. Simons, H.E. Stanley, A.L. Goldberger, in: *Fractals in Biology and Medicine*, ed. by G.A. Losa, T.F. Nonnenmacher, E.R. Weibel (Birkhauser Verlag, Boston 1994)

[89] W.B. Cannon: Physiol. Rev. **9**, 399 (1929)

[90] G.K. Zipf: *Human Behavior and the Principle of "Least Effort"* (Addison-Wesley, New York 1949)

[91] L. Brillouin: *Science and Information Theory* (Academic Press, New York 1956)

[92] C.E. Shannon: Bell Systems Tech. J. **80**, 50 (1951)

[93] L.S. Liebovitch, J. Freichbarg, J.P. Koniarek: Math. Biosci. **89**, 36 (1987); L.S. Liebovitch: Biophys. J. **55**, 373 (1989)

[94] R.V. Voss, J. Clarke: Nature **238**, 317 (1975)

[95] R.V. Voss, J. Clarke: J. Acous. Soc. Am. **63**, 258 (1978)

[96] M. Schroeder: *Fractals Chaos, Power Laws: Minutes from an Infinite Paradise* (W. H. Freeman, New York 1991)

[97] H.E. Stanley, S.V. Buldyrev, A.L. Goldberger, Z.D. Goldberger, S. Havlin, S.M. Ossadnik, C.-K. Peng, M. Simons: Physica A **205**, 214–253 (1994)

[98] N. Eldredge, S. J. Gould, in: *Models in Paleobiology*, edited by T. J. M. Schopf (Freeman and Cooper Inc., San Francisco 1972); S. J. Gould, N. Eldredge: *Nature* **366**, 223 (1993)

[99] P. Bak, K. Sneppen: Phys. Rev. Lett. **71**, 4083 (1993)

[100] S. Havlin, A.-L. Barabási, S.V. Buldyrev, C.-K. Peng, M. Schwartz, H.E. Stanley, T. Vicsek, in: *Growth Patterns in Physical Sciences and Biology* [Proc. NATO Advanced Research Workshop, Granada, Spain, October 1991], J.M. Garcia-Ruiz, E. Louis, P. Meakin, L. Sander, eds. (Plenum, New York 1993), pp. 85–98; L.-H. Tang, H. Leschhorn: Phys. Rev. A **46**, R-8309 (1992); S.V. Buldyrev, A.-L. Barabási, F. Caserta, S. Havlin, H.E. Stanley, T. Vicsek: Phys. Rev. A **46**, R-8313 (1992); A.-L. Barabási, S. V. Buldyrev, S. Havlin, G. Huber, H. E. Stanley, T. Vicsek, in: R. Jullien, J. Kertész, P. Meakin, D. E. Wolf (eds.), *Surface Disordering: Growth, Roughening, and Phase Transitions* [Proc. of the Les Houches Workshop 1992] (Nova Science, New York 1992), pp. 193–204

Diffusion of Lattice Gases in Disordered Lattices

Klaus W. Kehr and Thomas Wichmann

Institut für Festkörperforschung, Forschungszentrum Jülich, D–52425 Jülich, Germany

1 Abstract

A survey of recent simulations and analytical work on collective diffusion of lattice gases in lattices with site-energy disorder is given. In one-dimensional disordered lattices, exact results have been obtained in the limit of small and large particle concentrations. Simulation results at intermediate concentrations show that the time-dependent site-occupancy correlations are not yet satisfactorily understood. In higher-dimensional disordered lattices, an effective-medium theory, which uses mean-field transition rates, gives an improvement over the phenomenological description and qualitative agreement with the simulations. The main effects derive from saturation of deep trapping sites. One conspicuous example is the existence of a coefficient of collective diffusion in a situation where no single-particle diffusion coefficient exists.

2 Introduction

This paper is concerned with the coefficient of collective diffusion of lattice gases in lattices with site-energy disorder. A survey of recent simulations and of analytical derivations of this transport coefficient in lattices of different dimensions will be given. The understanding of transport processes in lattice gases in disordered lattices is important for several reasons. There are various disordered materials where the diffusivity or mobility of many particles, that cannot simultaneously occupy the same sites, is of interest. Examples are hydrogen in disordered alloys [1] and metglasses [2], and particles on imperfect or disordered surfaces [3]. From a more fundamental point of view, interesting problems arise from the correlations between the occupancies of different sites in nonequilibrium situations, which determine the transport properties in disordered lattices. We will address this problem, which is not yet solved even for the one-dimensional disordered chain.

In lattice gases one considers two quite different diffusion coefficients, namely the coefficient of collective or chemical diffusion, and the coefficient of tagged-particle diffusion. The coefficient of collective diffusion is defined through Fick's law, i. e. , it describes the decay of density disturbances towards the equilibrium density, on a mesoscopic scale. The tagged-particle diffusion coefficient follows from the asymptotic mean-square displacement of individual particles. Diffusion of tagged particles is theoretically more complicated to understand than collective diffusion, due to additional backward correlations that appear in the random walk of individual particles. Hence, we will concentrate in this paper on the coefficient of collective diffusion, where the identity of the particles is disregarded.

3 Model; Mean-Field Transition Rates

The model that we consider is the usual lattice-gas model where particles occupy the sites of d-dimensional hypercubic lattices ($d = 1,2,3,...$). The overall concentration c of the particles is defined as the quotient of the number of particles and the number of sites. Multiple occupancy of the sites is excluded, otherwise the particles have no further interactions. Disorder is introduced by assigning random, quenched transition rates Γ_{ij} where Γ_{ij} is the transition rate of a particle from site i to a neighbor site j, if the site j is empty.

Collective diffusion is completely described by the probability $P(l,t)$ of finding a particle at lattice site l at time t, with appropriate initial conditions. The quantity $P(l,t)$ obeys a master equation; for the following argument it suffices to consider only that part of the master equation for $P(1,t)$ that involves two neighboring sites 1 and 2. We designate with $\partial_{12}/\partial t$ the time derivative of $P(1,t)$, which results from transitions between 1 and 2. It is given by

$$\frac{\partial_{12}}{\partial t}P(1,t) = \Gamma_{21}[P(2,t) - P(2,1,t)] - \Gamma_{12}[P(1,t) - P(1,2,t)] \ . \qquad (1)$$

P(1,2,t) is the joint probability of finding a particle at site 1 and another particle at site 2 at time t. The terms in the square brackets represent the probabilities that a particle is present at site 2 and no particle at site 1, and a particle at site 1 and no particle at site 2, respectively. Only under these conditions transitions from 2 to 1, or from 1 to 2, respectively, are allowed. If the transition rates *between* the sites 1 and 2 are symmetric, $\Gamma_{12} = \Gamma_{21}$, the joint probabilities cancel each other in (1) and we have

$$\frac{\partial_{12}}{\partial t}P(1,t) = \Gamma_{21}[P(2,t) - P(1,t)] \ . \qquad (2)$$

I. e. , the problem has been reduced to a single-particle problem [4]. The consequences are that the coefficient of collective diffusion is given for the ideal hypercubic lattices with a uniform transition rate Γ by $D_{\text{coll}} = \Gamma$ (with the lattice constant $a = 1$), and that D_{coll} is given in the random-barrier model and in the model with randomly blocked sites by the diffusion coefficient $D_{\text{s.p.}}$ of single, independent particles [5]. Hence no concentration dependence of D_{coll} is present in these cases.

The simplest nontrivial case where collective diffusion is influenced by the time-dependent joint probabilities and hence $D_{\text{coll}} \neq D_{\text{s.p.}}$ is then the case of site-energy disorder. This model is abstractly defined by specifying disordered rates Γ_i which originate from the sites i, and are independent of the terminal sites. Concretely, one may consider lattices where the sites have random energies $E_i \leq 0$, which are selected from a common distribution $\nu(E)$. The transition rates then follow from an Arrhenius law,

$$\Gamma_i = \Gamma_0 \exp(E_i/k_B T) \qquad E_i \leq 0 \ . \tag{3}$$

The distribution $\nu(E)$ of the random energies may be Gaussian, exponential, or of other forms. One particularly simple case is the dichotomic distribution, where two types of sites occur, namely "free" sites with probability $1 - c_t$, with energy $E_i = 0$ and transition rate Γ_0, and "trap" sites with probability c_t, energy $E_i = E_t$, and transition rate $\Gamma^< = \Gamma_0 \exp(E_t/k_B T)$.

The single-particle diffusion coefficient of the random trap model is exactly known in all dimensions [6],

$$D_{\text{s.p.}} = \{\Gamma^{-1}\}^{-1} \tag{4}$$

where the braces indicate an average over the distribution of the random transition rates $\rho(\Gamma)$. Of course, the application of (4) requires the existence of the first inverse moment of $\rho(\Gamma)$. The general formula will be exemplified for the dichotomic random-trap model,

$$D_{\text{s.p.}} = \left(\frac{c_t}{\Gamma^<} + \frac{1-c_t}{\Gamma_0}\right)^{-1} \ . \tag{5}$$

Again, this formula holds for arbitrary dimensions.

It is evident that the joint probabilities do not cancel each other in the part of the master equation (1) when the transition rate Γ_{21} is replaced by Γ_2 and Γ_{12} by Γ_1, respectively, as appropriate for the random-trap model. The formulation of a master equation for the joint probability $P(1,2,t)$, etc. , leads to an intractable hierarchy of equations. The simplest way to break this hierarchy is to factorize the two-particle joint probability. Since we are interested in the behavior of small deviations from equilibrium we also linearize the product. Thus we make the ansatz

$$P(1,2,t) \approx P(1)P(2) + P(1)\delta P(2,t) + P(2)\delta P(1,t) \tag{6}$$

where, e. g. , $\delta P(1,t) = P(1,t) - P(1)$ is the deviation from the equilibrium occupation probability. We will refer to this ansatz as the "mean-field approximation".

The factorization is exact for the equilibrium state (but *not* for stationary states) and the stationary solution of the master equation is obtained when the condition of detailed balance holds,

$$\Gamma_1 P(1)[1 - P(2)] = \Gamma_2 P(2)[1 - P(1)] \ . \tag{7}$$

Insertion of the approximation (6) into the part of the master equation (1) leads to a master equation with mean-field transition rates,

$$\frac{\partial_{12}}{\partial t}\delta P(1,t) = \Gamma_{21}^{\mathrm{MF}}\delta P(2,t) - \Gamma_{12}^{\mathrm{MF}}\delta P(1,t) , \tag{8}$$

where

$$\Gamma_{21}^{\mathrm{MF}} = \Gamma_2 - (\Gamma_2 - \Gamma_1)P(1) . \tag{9}$$

The same reasoning can be made for any pair of adjacent sites i, j.

The mean-field transition rates can be brought into another form by applying the condition of detailed balance,

$$\Gamma_{21}^{\mathrm{MF}} = \Gamma_1 \frac{P(1)}{P(2)} . \tag{10}$$

Different but related mean-field transition rates were introduced by Gartner and Pitis [7], they are symmetric.

It is instructive to consider the mean-field transition rates in the limiting cases

$$\Gamma_{21}^{\mathrm{MF}} = \begin{cases} \Gamma_2 & c \to 0 \\ \Gamma_1 & c \to 1 . \end{cases} \tag{11}$$

The first case corresponds to the trivial single-particle limit, while the second case represents the diffusion of isolated vacancies, as we will discuss explicitly for the linear chain.

4 Linear Chains

4.1 Limit of Large Concentrations

We will first consider lattice gases on disordered linear chains in the limit of large concentrations, $c \to 1$. To this end we consider the dichotomic chain in the almost filled situation, as sketched in Fig. 1(a). The single vacancy sees a potential as indicated in Fig. 1(b). In this potential the transition rates *into* the sites are specified, i. e. $\Gamma_{i-1,i} = \Gamma_{i+1,i}$, in accordance with the result of the mean-field approximation (11). We note that this approximation is exact in this limit, since it correctly describes the physical process. The case when the transition rates into the sites are specified has been called "in-site randomness" by van den Broeck [8].

It remains to derive the diffusion coefficient for the case of in-site randomness. This can be conveniently done by using an exact formula for the mean first-passage time of a particle on a finite disordered linear chain of length L with the reflecting starting site 0 and the absorbing terminal site L [9]. The result is

$$D = \{\Gamma\}^{-1}\{\Gamma^{-1}\}^{-2} . \tag{12}$$

As before, the braces designate the average over the distribution of the transition rates. As an example, for the dichotomic random-trap model

$$D = [c_t \Gamma^< + (1-c_t)\Gamma_0]^{-1} \left[\frac{c_t}{\Gamma^<} + \frac{1-c_t}{\Gamma_0}\right]^{-2} . \tag{13}$$

This formula gives the diffusion coefficient of single vacancies in lattice gases on disordered dichotomic linear chains with a concentration c_t of trap sites, in the limit $c \to 1$.

Figure 2 demonstrats that the numerical simulations of the diffusion coefficient agree completely with the formula given above. Hence the correctness of (13) has been verified. The curve shows a shallow minimum around the value $c_t \approx 2/3$. Indeed, (13) has a minimum as a function of c_t at $c_t = 2/3$ for $\Gamma^< \ll \Gamma$, and in this limit

$$D = \frac{27}{4}\frac{(\Gamma^<)^2}{\Gamma_0} . \tag{14}$$

This result may be compared with the result for the ordered chain with $c_t = 2/3$ where two trap sites are followed by a free site. The potential felt by the vacancy is a periodic succesion of the double mountain that is recognized in Fig. 1(b), and the diffusion coefficient of the periodic arrangement is, in the limit $\Gamma^< \ll \Gamma_0$,

$$D = 9\frac{(\Gamma^<)^2}{\Gamma_0} . \tag{15}$$

At first glance it may seem paradoxical that a single vacancy diffuses more slowly in a lattice gas on a disordered lattice with a concentration of about 2/3 of deep sites, compared to the situation with only deep sites where $D_{\text{coll}} = \Gamma^<$. The discussion given above demonstrates, by transcription of the problem to the effective mean-field transition rates, that this is indeed the case.

As already said, the diffusion coefficient of single vacancies in the case of $c \to 1$, represents the coefficient of collective diffusion of lattice gases, in this limit, on disordered chains.

4.2 Arbitrary Concentrations

Using the mean-field transition rates (10) the coefficient of collective diffusion can be derived for arbitrary concentrations of the lattice gas by using the exact expression for the mean first-passage time [9]. The result is

$$D = \{P^2\Gamma\}^{-1}\{(P\Gamma)^{-1}\}^{-2} . \tag{16}$$

It is easily examined that the limiting cases of small and large particle concentrations follow correctly. The limiting case (12) of $c \to 1$ is obtained directly by replacing P by 1. The low-concentration limit (4) is obtained by observing that in this limit $P \sim \Gamma^{-1}$, as is evident from the condition of detailed balance.

A completely equivalent result to (16) was obtained by Gartner and Pitis [7], although their formula looks rather different. These authors also calculated the first correction to the mean-field approximation.

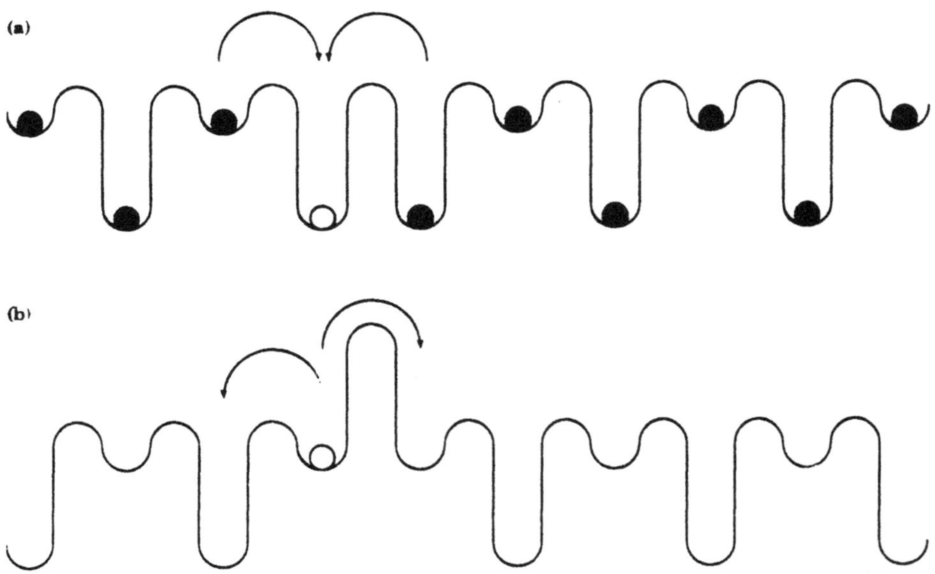

Fig. 1. (a) Dichotomic chain with only a single vacancy (open circle) left. All other lattice sites are occupied by particles (full circles). Arrows indicate remaining transition possibilities of the particles. (b) Potential as seen from the single vacancy of (a). Arrows show possibile transitions of the vacancy.

Numerical simulations of the collective diffusion were done by preparing the system with a uniform density plus a cosine profile,

$$c(x) = c + \delta c \cos(kx) \tag{17}$$

where $k = 2\pi/\lambda$ with λ the wavelength of the density profile and x is a discrete coordinate along the chain. The initial amplitude δc was chosen small compared to c and the decay with time was observed during the dynamics of the particles,

$$\delta c(t) = \delta c \exp(-D_{\mathrm{coll}} k^2 t) \ . \tag{18}$$

We used the vectorized computer code that was described in [10]. Figures 3 and 4 give a comparison of the simulation results with the mean-field approximation and the first correction to it. The observed concentration dependence of the diffusion coefficient is smoother than predicted by theory. In the limiting cases $c \to 0$ and $c \to 1$ theory and simulations agree, the deviations of some data points for $c_{\mathrm{t}} = 0.2$ and $c \to 1$ are due to difficulties in determining D_{coll} in this region.

The discrepancies between theory and simulations are most pronounced around $c \approx c_{\mathrm{t}}$. The first correction to the mean-field theory improves the situation, but there remains a considerable difference. It is fitting to point out that the ordered situation for $c_{\mathrm{t}} = 1/2$ and arbitrary concentrations c was completely solved in an early paper by Richards [11]. Also in this case the largest difference between

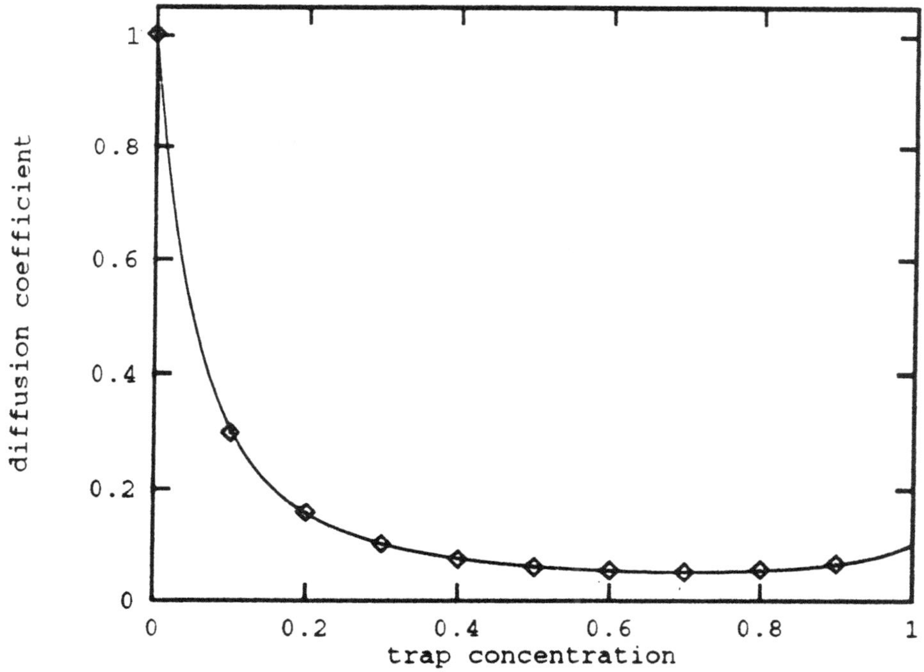

Fig. 2. Diffusion coefficient of single vacancies on a linear chain with "in-site randomness" as a function of the trap concentration c_t. The full curve represents the theoretical expression (13) with the ratio of transition rates $\Gamma^{<}/\Gamma_0 = 0.1$ and the symbols ($\diamond$) are results of numerical simulations.

the exact result and the mean-field result occurs for $c = c_t$. We conclude that the proper treatment of the time-dependent particle correlations is not yet understood in the disordered situation.

5 Collective Diffusion in Higher-Dimensional Disordered Lattices

We now direct our attention to the coefficient of collective diffusion of lattice-gas particles in lattices with random site-energies in two and three dimensions. The diffusion of many particles in lattices with trapping sites that can be saturated by particles was of continued interest to metal physicists in the past. We mention the early work of Oriani [12]; see [13] for further references. Kirchheim et al. [14] have introduced the model of a Gaussian distribution of site energies for the interpretation of solubility and diffusivity data of hydrogen in metglasses. Diffusion in the presence of trapping sites was also of interest to surface physicists. Several approximate derivations of the diffusion coefficient were given in the literature. They are based on the decomposition of the coefficient of collective diffusion in the product of a kinetic factor and a thermodynamic factor,

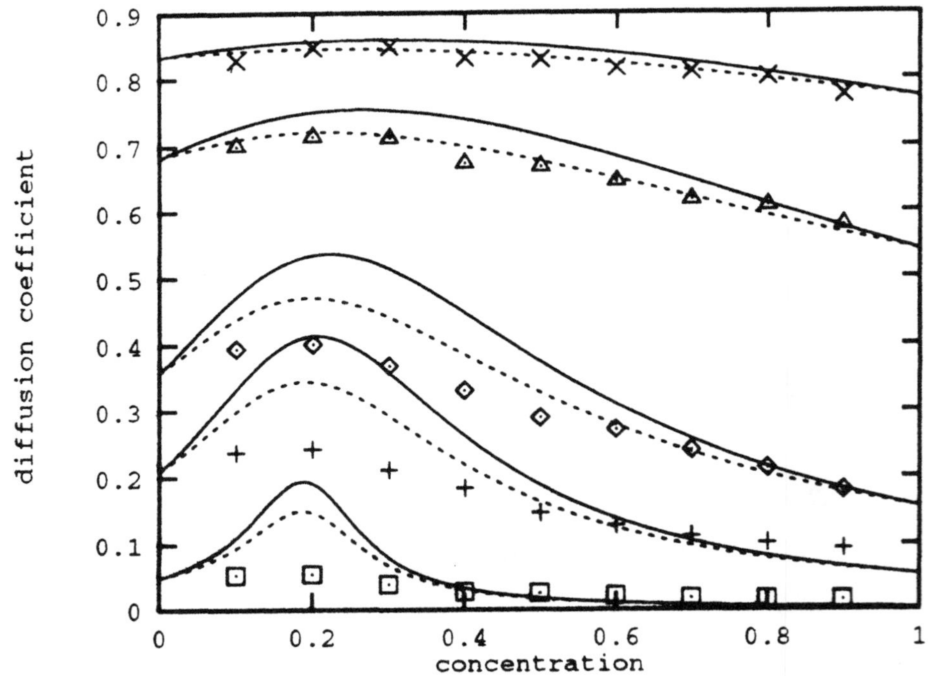

Fig. 3. Coefficient of collective diffusion as a function of the concentration of the lattice gas, for the dichotomic random-trap model with the trap concentration $c_t = 0.2$. The different symbols represent simulation results with the ratios $\Gamma^{<}/\Gamma_0$ of the transition rates: ($\times$) 0.5; ($\triangle$) 0.3; ($\diamond$) 0.1; (+) 0.05; ($\square$) 0.01. The full curves represent the mean-field result and the dashed curves include the first correction obtained by Gartner and Pitis [7].

$D_{\text{coll}} = Bc\frac{\partial\mu}{\partial c}$. While the thermodynamic factor $c\frac{\partial\mu}{\partial c}$ can be calculated exactly from equilibrium statistical mechanics for the model of random site-energies, assumptions are necessary for the kinetic factor B. We cite an explicit expression for the coefficient of collective diffusion which represents the specialization of a recent formula of Salomons [15] to the dichotomic model,

$$D_{\text{coll}} = D_0 \frac{(1-c)[(1-c_t)c_1 + c_t c_2 \Gamma^{<}/\Gamma_0]}{(1-c_t)c_1(1-c_1) + c_t c_2(1-c_2)} \quad . \tag{19}$$

D_0 is the diffusion coefficient in the ideal lattice, c_1 and c_2 are the concentrations of the lattice-gas particles on the free and trap sites, respectively. These concentrations are explicitly given by

$$c_1 = \frac{1}{2(1-c_t)} \left[c - c_t - \frac{1}{e^{-\beta E_t} - 1} \right. \\ \left. + \sqrt{(c-c_t)^2 + 2\frac{c + c_t - 2cc_t}{e^{-\beta E_t} - 1} + \left(\frac{1}{e^{-\beta E_t} - 1}\right)^2} \right] \tag{20}$$

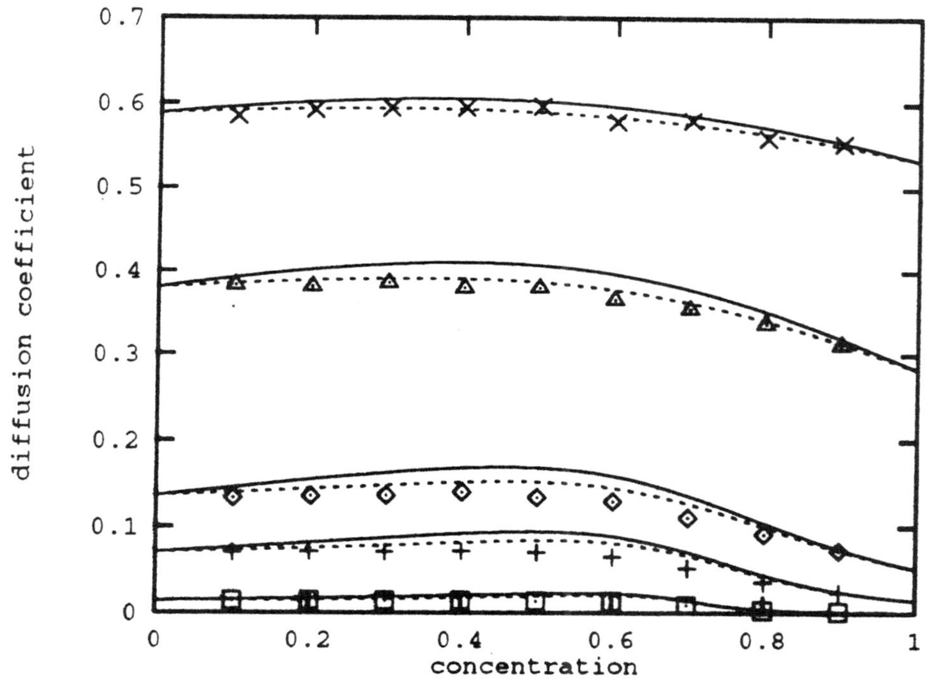

Fig. 4. Coefficient of collective diffusion as a function of the concentration of the lattice gas, for the dichotomic random-trap model with the trap concentration $c_t = 0.7$. The different symbols and curves have the same meaning as in Fig. 3.

$$c_2 = \frac{c - c_1(1 - c_t)}{c_t} \quad . \tag{21}$$

We will refer to the above expression (19) as the phenomenological theory. A comparison of the formula (19) with the simulation data will be made in conjunction with the results of the effective-medium theory.

It is important to examine the limiting cases for D_{coll}. First, in the limit of very small particle concentration $c \to 0$ one obtains the correct single-particle diffusivity of (5), $D_{\text{coll}}^{\text{phen}}(c \to 0) = D_{\text{s.p.}}$. In the limit of $c \to 1$ and $\Gamma^< \ll \Gamma_0$ one finds

$$D_{\text{coll}}^{\text{phen}}(c \to 1, \Gamma^< \ll \Gamma_0) = D_0(1 - c_t) \ . \tag{22}$$

Comparison of this result with known results from percolation theory shows that it cannot be correct, except in infinite dimensions. If very deep traps are present and if they are saturated by filling nearly all sites of the system, the trapping sites act as more or less permanent blocking sites to the diffusion of the remaining particles.

The diffusive behavior of the particles on the free sites now depends crucially on the concentration of the blocking sites. If their concentration is so small that the lattice of free sites percolates, diffusion of the particles on the free sites is possible. Further, the transition rates between the free sites are symmetric, hence the coefficient of collective diffusion coincides with the coefficient of single-particle diffusion, as was discussed in Sect. 3. The coefficient of single-particle diffusion in lattices with smaller concentrations of randomly blocked sites was considered by Tahir-Kheli [16] and Ernst et al. [17]. They found

$$D = D_0(1 - \frac{c_t}{f} + \ldots) \tag{23}$$

where f is the correlation factor for tagged-particle diffusion in ideal lattices, in the limit $c \rightarrow 1$. For instance, in the simple-cubic lattice $f = 0.653\ldots$. Numerical investigations [18] showed that (23) is a good approximation for c_t not too close to $1-p_c$ where p_c is the percolation threshold for free sites. It was also verified in [18] that the single-particle diffusion coefficient (23) coincides with the collective diffusivity. Clearly (23) is incompatible with (22), except for $f = 1$, which is the correct correlation factor in infinite dimensions. Our numerical simulations, which are given in the next section, agree with (23), not with (22). Hence they confirm the physically reasonable picture of diffusion in the presence of blocked sites, for $c_t < 1 - p_c$.

No exact expression can be given for the opposite case, $c_t > 1-p_c$. Diffusion ceases to exist for the completely immobile situation, that is for $\Gamma^{<}/\Gamma_0 \rightarrow 0$. For $\Gamma^{<}/\Gamma_0 \ll 1$ diffusion is only possible through the dynamical processes that particles may escape from trap sites, i. e. , one expects $D_{\text{coll}} \sim \Gamma^{<}$. This is in contradiction to (22) where $D_{\text{coll}} \sim D_0 = \Gamma_0$.

6 Effective-Medium Theory

To improve the phenomenological theory, which was discussed in the preceding section, we have developed an effective-medium theory for collective diffusion in lattice gases in disordered lattices in arbitrary dimensions. A detailed description will be given elsewhere [19], hence only a brief characterization of the approximation and of some results will be given.

The effective-medium theory for the collective diffusion is based on previous work on single-particle diffusion in disordered lattices with random distributions of the bonds. That work can be regarded as an extension of the work of Kirkpatrick on the conductivity of disordered lattices [20] to time-dependent hopping problems, see the review [6]. In the single-bond approximation and in the static limit, the self-consistency condition for the effective-medium transition rate Γ_{EMT} of the random-barrier model reads

$$\left\{ \frac{\Gamma_{\text{EMT}} - \Gamma_{\text{sym}}}{(d-1)\Gamma_{\text{EMT}} + \Gamma_{\text{sym}}} \right\} = 0 \tag{24}$$

where d is the dimensionality of the lattice. The average in (24) is over the distribution of the transition rates Γ_{sym}. The diffusion coefficient is then given by $D = \Gamma_{EMT}$. The application of the effective-medium theory to lattice gases requires first the reduction to a single-particle problem. This is accomplished by the introduction of mean-field transition rates. In addition, the single-particle transition rates have to be symmetric, in order to apply the single-bond formulation of the random-barrier model. The mean-field transition rates which were introduced in Sect. 3 are not symmetric. Symmetric mean-field transition rates were introduced in Ref. [7], which will be used for the further derivations. Explicitly we use

$$\Gamma_{ij}^{\mathrm{sym}} = \frac{\Gamma_i P(i)[1-P(j)]}{\{P(1-P)\}} \quad . \tag{25}$$

The symmetry of these rates is obvious from the condition of detailed balance, (7), and the factor in the denominator has been added for proper normalization.

For $d = 1$ $\Gamma_{\mathrm{EMT}} = \{\Gamma_{\mathrm{sym}}^{-1}\}^{-1}$ is obtained. This corresponds to the result of Ref. [7] for linear chains, which is equivalent to the result (16) given in Sect. 4. In $d = 2$ and 3 the self-consistency condition can be solved explicitly for the dichotomic distribution. For other distributions such as the Gaussian one a numerical evaluation of (24) is necessary, which poses no problem.

Figures 5 and 6 show the results of the effective-medium approximation for the dichotomic random-trap model for two trap concentrations, together with the simulation results and the phenomenological expression (19). We first discuss the case of small trap concentration c_{t}. For small particle concentrations $c < c_{\mathrm{t}}$ both expressions give a description of the simulation data, with better agreement of the effective-medium theory. For small c_{t} there is a steep rise of the diffusion coefficient with particle concentration. This is of experimental relevance: the result confirms the intuitive picture that traps can be saturated by particles, thus allowing the remaining particles to perform diffusion processes more effectively. For larger particle concentration $c > c_{\mathrm{t}}$ the effective-medium theory and the phenomenological theory disagree with the simulation data. On the other hand, the simulation data for $\Gamma^{<} \ll \Gamma_0$ extrapolate nicely to the value that is predicted by (23).

For the large trap concentration $c_{\mathrm{t}} = 0.75$ the effective-medium theory gives a better description of the data than the phenomenological expression for $c < c_{\mathrm{t}}$. For $c > c_{\mathrm{t}}$ the phenomenological expression is in complete disagreement with the simulation data, as expected from the above discussion. Also the effective-medium theory does not reproduce the numerical data, but it gives at least a rough qualitative description.

Finally we point out that the phenomenological expression (19) is obtained as the $d \to \infty$ limit of the effective-medium theory, as can be derived from (24) . We thus observe that the phenomenological derivations can be justified in the limit of infinite dimensions of the underlying lattices. The difficulty of the phenomenological theories resides in the kinetic factor. Concretely, particles in deep traps create backward correlation in the diffusion steps of the other particles which are difficult to correctly take into account. In the limit of high dimensions, the mobile particles have so many possibilities for jumps that these backward correlations become irrelevant. Hence it is plausible that the phenomenological expression is regained in this limit.

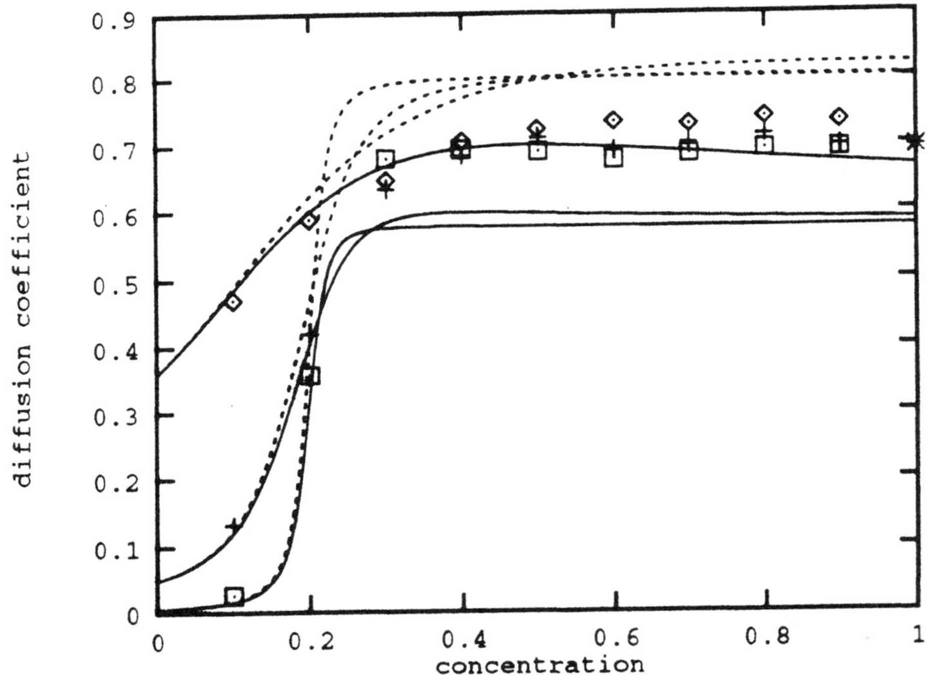

Fig. 5. Coefficient of collective diffusion as a function of the concentration of the lattice gas, for the dichotomic random-trap model with the trap concentration $c_t = 0.2$. The different symbols represent simulation results with the ratios of the transition rates $\Gamma^{<}/\Gamma_0$: ($\diamond$) 0.1; (+) 0.01; ($\square$) 0.001. The star indicates the limit (23). The full curves represent the effective-medium approximation and the dashed curves the phenomenological expression (19).

7 Collective Diffusion in the Absence of a Coefficient of Single-Particle Diffusion

In this section we will show that collective diffusion is possible even in the case where no diffusion coefficient of single, independent particles exists. The phenomenon to be discussed is caused by the saturation of those sites that are responsible for the vanishing of the single-particle diffusion coefficient, by a finite concentration of particles.

We consider an exponential distribution of site energies,

$$\nu(E) = \frac{1}{\sigma} \exp(\frac{E}{\sigma}) \qquad E \leq 0 \tag{26}$$

where σ characterizes the width of the distribution. When combined with the Arrhenius law (3) for the transition rates of particles from a site with energy E,

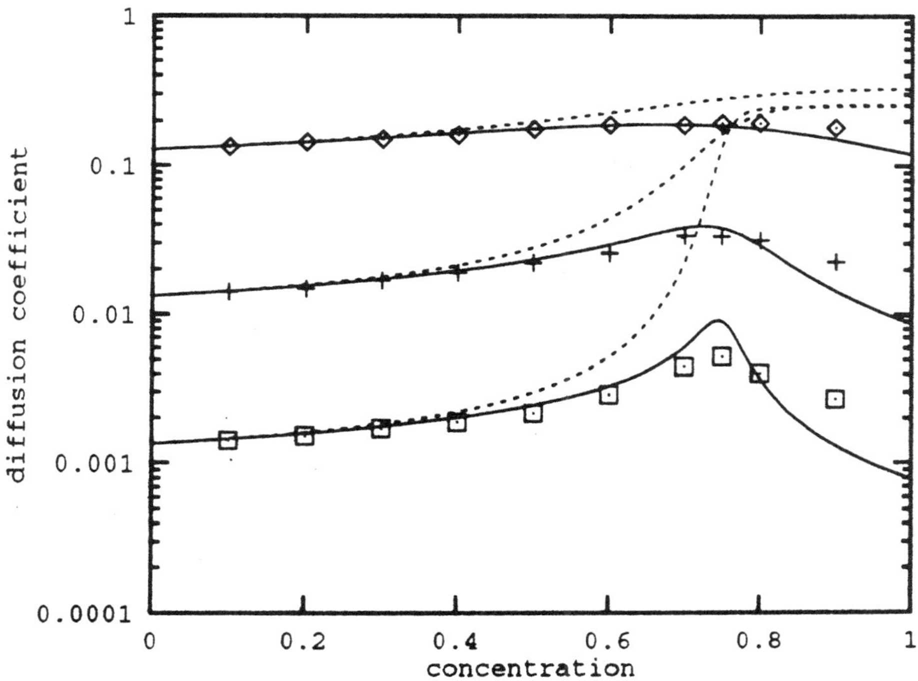

Fig. 6. Coefficient of collective diffusion as a function of the concentration of the lattice gas, for the dichotomic random-trap model with the trap concentration $c_t = 0.75$. The different symbols and curves have the same meaning as in Fig. 5.

we have the following distribution of transition rates in our model,

$$\rho(\Gamma) = \frac{\alpha}{\Gamma_0} \left(\frac{\Gamma}{\Gamma_0} \right)^{\alpha - 1} \tag{27}$$

with the parameter $\alpha = k_B T / \sigma$. The diffusion coefficient of a single, independent particle follows from (4),

$$D_{\text{s.p.}}^{-1} = \int_0^{\Gamma_0} d\Gamma \frac{1}{\Gamma} \rho(\Gamma) \ . \tag{28}$$

The integral on the right-hand side diverges for $\alpha \leq 1$; this means that no coefficient of single-particle diffusion exists for temperatures smaller than the critical value σ / k_B. This fact leads to the phenomenon of dispersive transport [21]. Another wellknown consequence of the model is subdiffusive behavior of the mean-square displacement of independent particles [22],

$$< r^2(t) > \sim t^{2/d_W} \tag{29}$$

where $d_W = 2/\alpha$ for $d \geq 2$.

We now introduce many particles in a lattice where the random site energies are given by the exponential distribution (26). The equilibrium distribution over the energies is then the Fermi-Dirac distribution,

$$f(E) = [1 + \exp[\beta(E-\mu)]]^{-1}, \tag{30}$$

to be multiplied with the distribution of site energies (26). The chemical potential follows from the given equilibrium concentration

$$c = \int_{-\infty}^{0} dE \nu(E) f(E) \ . \tag{31}$$

We introduce a small disturbance of the equilibrium distribution, as formulated in (17), where now x is a discrete coordinate in one of the lattice directions. The investigations have been made in 2 and 3 dimensions. We observe [23] that the initial amplitute decays diffusively, in accordance with (18).

The results of the numerical simulations are shown in Fig. 7, together with the results of the effective-medium theory for this model. One sees that the diffusion coefficient $D_{\text{coll}}(c)$ extrapolates to the correct finite value for $c \to 0$ and $\alpha > 1$, and to zero for $c \to 0$ and $\alpha \leq 1$. There is good agreement between the effective-medium theory and the simulations at smaller concentrations and rough qualitative agreement at larger concentrations.

For small particle concentrations we can give the following qualitative picture [23]. Let us replace the actual distribution of the particles at finite temperature, according to the Fermi-Dirac expression, by a sharp distribution up to a "Fermi-level" E^*. This pseudo Fermi-level is determined by

$$\int_{-\infty}^{E^*} dE \nu(E) = c \ , \tag{32}$$

i. e. , $E^* = \sigma \ln c$; note that $E^* \leq 0$. At small concentrations one may replace the many-particle diffusion problem by the problem of diffusion of few particles that are thermally excited above the pseudo-Fermi level E^*. Since the concentration of blocked sites is small for small particle concentration, we may estimate the coefficient of collective diffusion by

$$D_{\text{coll}}^{-1} \approx D_{\text{s.p.}}^{-1} = \int_{E^*}^{0} dE \nu(E) \frac{1}{\Gamma(E)} \ . \tag{33}$$

Evaluation of the integral gives

$$D = \frac{\Gamma_0(\frac{1}{\alpha}-1)}{c^{-(\frac{1}{\alpha}-1)}-1} \ . \tag{34}$$

This expression is plotted in Fig. 8 where also the numerical data and the effective-medium result from Fig. 7 are reproduced. We note that the qualitative expression approaches the effective-medium result for $c \to 0$. Expression (34) diverges for $c \to 1$, which is clearly outside of the range of the validity of the argument.

The small-concentration limit of (34) reads

$$D \approx \Gamma_0(\frac{1}{\alpha} - 1)c^{(\frac{1}{\alpha}-1)} \ . \tag{35}$$

I. e. , we predict a power-law behavior of the collective diffusivity in the model with exponentially distributed site energies, for $c \to 0$. Fig. 8 demonstrates that the data are consistent with this prediction. However, there are too few data to speak of a quantitative verification.

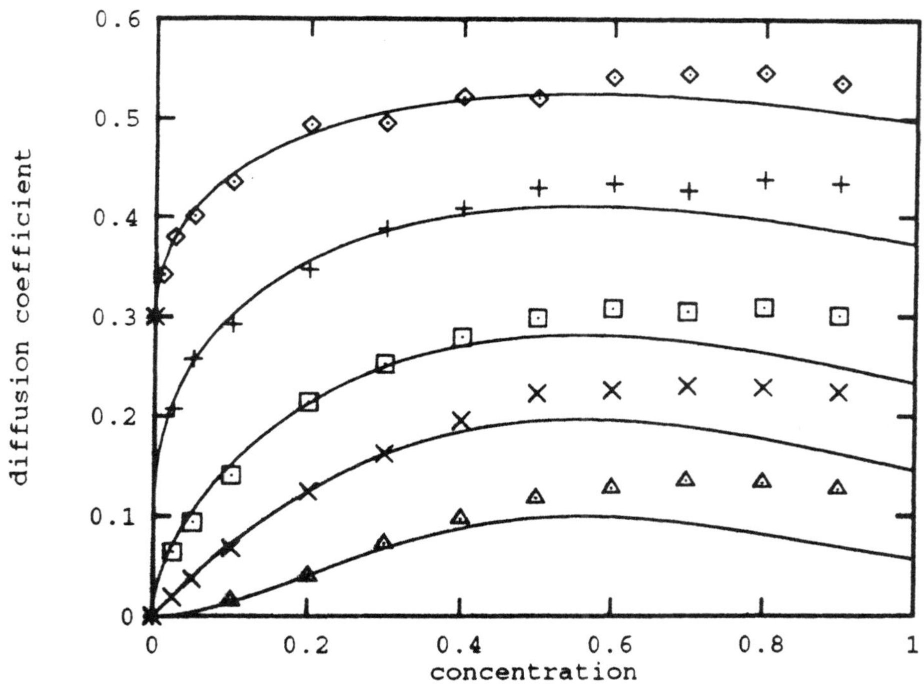

Fig. 7. Coefficient of collective diffusion as a function of the concentration of the lattice gas, for an exponential distribution of site energies. The different symbols represent simulation results of the parameters α: ($\diamond$) 10/7; (+) 1; ($\square$) 2/3; ($\times$) 1/2; ($\triangle$) 1/3. Stars indicate the limit of single-particle diffusion. The full curves represent the effective-medium approximation.

8 Conclusions

We point out that the coefficient of collective diffusion of lattice gases in lattices with site-energy disorder is exactly known in some limiting cases only. These are

1. the trivial limit of small concentrations, $c \to 0$

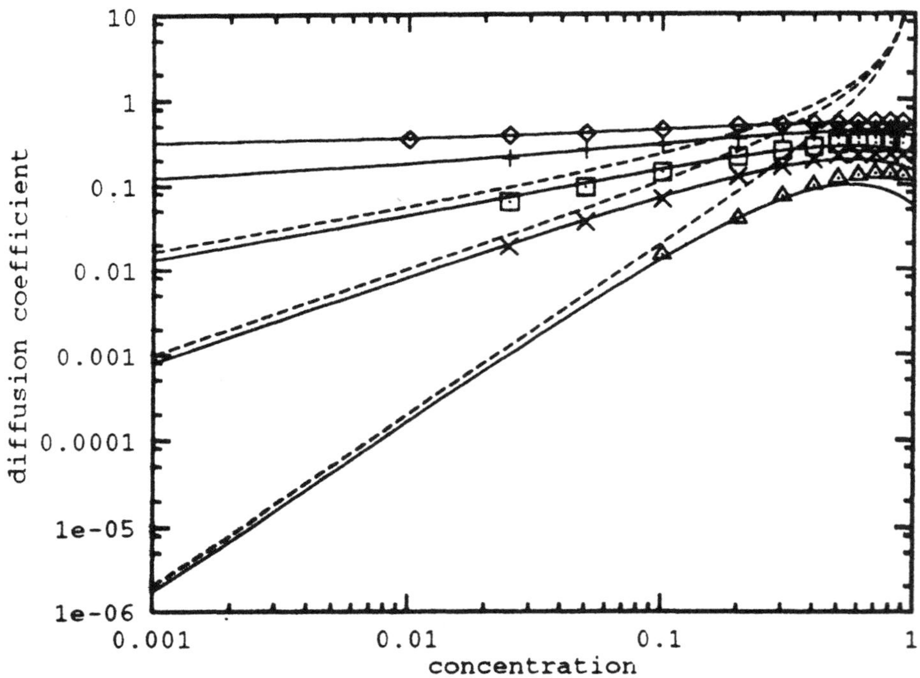

Fig. 8. Coefficient of collective diffusion as a function of the concentration of the lattice gas, for an exponential distribution of site energies. The scales are logarithmic. The different symbols and the full curves have the same meaning as in Fig. 7. The dashed curves show the qualitative approximation (34).

2. in linear chains, the limit of small vacancy concentrations, $c \to 1$
3. for $d \geq 2$, the limit of saturated deep traps, $c \to 1$ and $\Gamma^{<} \ll \Gamma_0$, when their concentration is small, $c_t \ll 1 - p_c$
4. the limit of infinite-dimensional lattices.

Comparison of the mean-field approximation and its first correction with numerical simulations has demonstrated that the effect of correlations in the site occupancies of different sites in nonequilibrium situations is not yet understood for arbitrary concentrations of the lattice-gas particles. The discrepancies are most pronounced for $c \approx c_t$. Their proper treatment is certainly an open theoretical problem.

In higher-dimensional lattices, the phenomenological theory provides a good description for small particle concentrations, $c \ll c_t$. The effective-medium theory constitutes an improvement for $c < c_t$ and it provides a qualitative overall description for all concentrations. Hence we conclude that the problem of determining D_{coll} is now clarified for practical purposes.

Collective diffusion may strongly increase with increasing particle concentration when deep traps are saturated by a fraction of the particles, thus allowing the remaining particles to diffuse uninpeded by trapping events. One extreme situation is provided by the exponential distribution of the site energies where the coefficient of single-particle diffusion vanishes for low temperatures. We have seen that a coefficient of collective diffusion may exist at finite concentrations, due to the saturation effects of the traps.

Finally we should also point out some obvious omissions in the work presented. Tagged-particle diffusion, which is also interesting, has been omitted by intention. Some data on tagged-particel diffusion have been given in [24]. Most severe is the neglect of interactions of the lattice-gas particles beyond the site-exclusion. Experimental systems have interacting particles and there has already been considerable work on such systems, for instance work devoted to the problem of universal response in disordered superionic conductors [25, 26]. Our hope was to contribute to the derivation of the coefficient of collective diffusion in rather idealized systems; apparently this is already a difficult task.

References

[1] See, for instance, R.C. Brouwer, J. Rector, N. Koeman, and R. Griessen, Phys. Rev. B **40** (1989) 3546

[2] R. Kirchheim, Progr. Mat. Sci. **32** (1988) 262

[3] for referencies, see C. H. Mak, H. C. Anderson, and S. M. George, J. Chem. Phys. **88** (1988) 4052

[4] R. Kutner, Phys. Lett. A **81** (1981) 239

[5] R. Kutner and K. W. Kehr, Phil. Mag. A **48** (1983) 199

[6] For a review, see J. W. Haus and K. W. Kehr, Phys. Rep. **150** (1987) 263

[7] P. Gartner and R. Pitis, Phys. Rev. B **45** (1992) 7739

[8] C. Van den Broeck, in: Noise and nonlinear phenomena in nuclear systems, eds. J. L. Munoz-Cobo and F. C. Difilippo (Plenum, New York, 1989) p. 3

[9] K. W. Kehr, O. Paetzold, and T. Wichmann, Phys. Lett. A **182** (1993) 135

[10] O. Paetzold, Comput. Phys. Commu. **64** (1991) 1

[11] P. M. Richards, Phys. Rev. B **16** (1977) 1393

[12] R. A. Oriani, Acta metall. **18** (1970) 147

[13] K. W. Kehr and O. Paetzold, Defect and Diffusion Forum **95-98** (1993) 1167

[14] R. Kirchheim, F. Sommer and G. Schluckebier, Acta metall. **30** (1982) 1059

[15] E. Salomons, J. Phys. C **21** (1988) 5953

[16] R. A. Tahir-Kheli, Phys. Ref. B **28** (1983) 3049

[17] M. H. Ernst, T. M. Nieuwenhuizen, and P. F. J. van Velthoven, J. Phys. A **20** (1987) 5335

[18] M. Braun and K.W. Kehr, Phil.Mag.A **61** (1990) 855

[19] T. Wichmann and K. W. Kehr, in preparation

[20] S. Kirkpatrick, Rev. Mod. Phys. **45** (1973) 574

[21] H. Scher, M. F. Shlesinger, and J. T. Bendler, Phys. Today **44** (1) (1991) 26

[22] S. Havlin, B. L. Trus, and G.H. Weiss, J. Phys. A. **19** (1986) L817

[23] T. Wichmann, K. G. Wang, and K. W. Kehr, J. Phys. A **27** (1994) L263

[24] K. W. Kehr and O. Paetzold, Physica A **190** (1992) 1

[25] D. Knoedler, and W. Dieterich, Physica A **191** (1992) 426

[26] W. Dieterich, D. Knoedler, and P. Pendzig, J. Non-cryst. Solids (in press)

Single–ion Random Walk on a Lattice in an Attractive Coulomb Cutoff Potential

R.Kutner[1*], *D.Knödler*[1], *P.Pendzig*[1], *R.Przeniosło*[2] *and W.Dieterich*[1]

[1] Fakultät für Physik, Universität Konstanz, Postfach 5560, D–78434 Konstanz, Germany
[2] Department of Physics, Warsaw University, PL–00681 Warsaw, Hoża 69, Poland

1 Introduction

The general concept where charge carriers perform a random walk on a regular or disordered lattice under the influence of a Coulomb force has found wide applications in a variety of fields, when a formulation in terms of molecularly defined quantities is required. This approach allows one to study systematically on microscopic and macroscopic scales quantities like the time– and frequency–dependent diffusion coefficient, the static and dynamic conductivity, the mobility, dielectric permittivity as well as different types of effective escape and reaction rates. The fields of application can be the dispersive transport in ionically conducting solid electrolytes [1], diffusional dissociation and recombination, e.g. photodissociation [2], diffusion–controlled chemical and encounter reactions [3, 4, 5] diffusion–limited aggregation [6, 7], gelation [8], and segregation [9], as well as energy transfer processes between molecules, e.g. in biophysics [10].

The traditional, continuum Onsager approach [11, 12] is based on the Smoluchowski equation [13, 14] extended by Kramers [15, 16] based on the Fokker–Planck and Kramers equations [17]. The above concepts are only valid asymptotically and on macroscopic length scales. So far they are the main theoretical tools for the study of static and dynamic properties.

The problem, studied here by means of random walks, can be generally classified as a thermally activated escape or dissociation process. In particular, the work is motivated by the "universal dynamic response" of low–conductivity materials, characterized by approximate power–laws of the frequency–dependent conductivity and the related dielectric permittivity [1, 18, 19, 20, 21, 22, 23, 24]. Whether dispersive transport is mainly determined by single–ion motion or by collective phenomena is not yet completely understood.

We systematically study, by standard Monte Carlo simulation [25], and by exact enumeration procedure [26, 27], c.f. Sect.3, the inhomogeneous random walk of a single ion in a periodic attractive Coulomb cutoff potential (c.f. Sect.2) for

---

* Permanent address: Department of Physics, Warsaw University, PL–00681 Warsaw, Hoża 69, Poland

short, intermediate and long-time scales, as a function of period length and inverse temperature $\beta(= 1/k_BT)$ (c.f. Sect.4). Our approach provides a promising supplement to the phenomenological "jump relaxation model" [1], the numerical "counter-ion model" [18], and geminate recombination [2]. Our results show a surprisingly complicated behaviour in the time domain. The present work also supplies the means to study the dynamics of a one-dimensional random walk in an arbitrary inhomogeneous medium.

2 Inhomogeneous Random Walk

We consider the random walk of a single ion on a one-dimensional periodic regular lattice (c.f. Fig.1). The repeated central segment with length N (note that N must be even because of the central symmetry, and the lattice constant is set to unity) includes a fixed Coulombic center which attracts a mobile ion located at site $l = n + kN$ by the static potential $V_C(l) = -1/|(k + 1/2)N + 1/2 - l|$, illustrated in Fig.1 by the dashed-dotted line. Here, $n = 1, \cdots, N$ labels a site inside the segment and $k = 0, \pm1, \pm2, \cdots$ labels the segments. The charge of the mobile ion and the fixed center is set to unity. As it is seen, we consider the simplified situation (justified below), taking into account only the attractive interaction between the ion and the center located in the same segment.

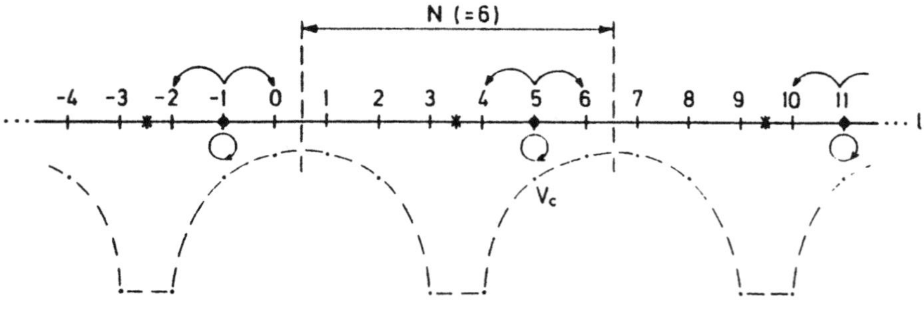

Fig. 1. Schematically illustration of the 1d discrete chain with periodic boundary conditions (for, e.g. $N = 6$). Ions (full circles) are jumping in an attractive Coulomb cutoff potential V_c (indicated by the dashed-doted line) produced by fixed charges (denoted by asterisks). The right and left jumps are indicated by the arrows and sojourns by loops.

The local, discrete-time dynamics is completely defined by the site-dependent jump probabilities according to Metropolis et al. [25]. Note that in one dimension the jump probabilities over the center, towards the center and those for jumps crossing the border between neighboring segments are set to 1/2; other jump probabilities are smaller for finite temperatures. In our model we truncated the range of the Coulomb potential to a given segment. This, however, is not an essential restriction since the periodicity N is a free parameter and can become arbitrarily large. The truncation in the nearest vicinity of the center (c.f. Fig.1) is quite natural since it is caused by the discrete character of the lattice.

The main quantity studied is the time–dependent diffusion coefficient $D(t) = \langle[\Delta l(t)]^2\rangle/2dt$, where d is the dimension of space and $\Delta l(t) = l(t) - l(0)$ the displacement. $l(t)$ is the position of the mobile ion at time t and we assume the initial site $l(0)$ to be nearest to the attractive center i.e., at a mechanically stable site. This is at intermediate and low temperatures also the prevailing configuration at thermal equilibrium. A thermalization procedure is not used here by definition since we study in fact escape or dissociation phenomena.

We also briefly compared our one–dimensional model with a three–dimensional random walk on a regular cubic lattice, where the attractive charge was fixed in the center of the repeated cell. Important for the further considerations is the genaral observation that we are able to reproduce on short, intermediate and long–time scales the substantial aspects of the three–dimensional diffusional behaviour by the one–dimensional model. This is justified by the similarities of Fig.2a and 3a, as well as Fig.2b and 3b, respectively. In Fig.2 and 3 we present the diffusion coefficient as a function of time. The results in Fig.2 were obtained by the standard Monte Carlo simulation [25] for a finite, typical periodicity $N = 10$ (Fig.2a), and for an infinite lattice $N = \infty$ (Fig.2b) for several temperatures. Here the ion was assumed to interact with all attractives centers in the other cells. The jump probabilities were chosen according to Metropolis et al. and the Ewald summation was used to calculate energy differences. In Fig.3 we show the corresponding results for the one–dimensional random walk obtained by a more efficient exact enumeration procedure [26], c.f. Sect.3. Note, that we considered here exponential jump–probabilities [28].

The restriction to the one-dimensional model makes now our further studies more efficient.

3 Elements of Exact Enumeration Procedure

Since the exact enumeration procedure was comprehensively described by Majid et al. [26], we restrict our considerations only to special features and important elements of the procedure used for our one–dimensional case.

In Fig.4 we schematically illustrate how the method allows us to numerically calculate the distribution function $P(l,t|l_0 = 0, t_0 = 0)$ i.e., the conditional probability of finding a random walker (denoted by a full circle) at site l at time t on condition that at time $t_0(= 0)$ it was initially at site $l_0(= 0)$ nearest to the attractive center (denoted by an asterisk). This probability is determined recursively *i)* by the probabilities that the walker is at the nearest neighbour site one step earlier, and *ii)* by site–dependent right–jump $W_R(l)$, left–jump $W_L(l)$, and sojourn $W_0(l)$ probabilities per step. Hence, we can store the elements $P(l,t|l_0,t_0)$ in the form of the matrix $\hat{P}(l_0,t_0)$ consisting of the vectors $\underline{P}(t|l_0,t_0)$.

Since the ion starts its random walk nearest to the center of the attracting Coulomb potential, the distribution function $P(l,t|l_0,t_0)$ is strongly localized

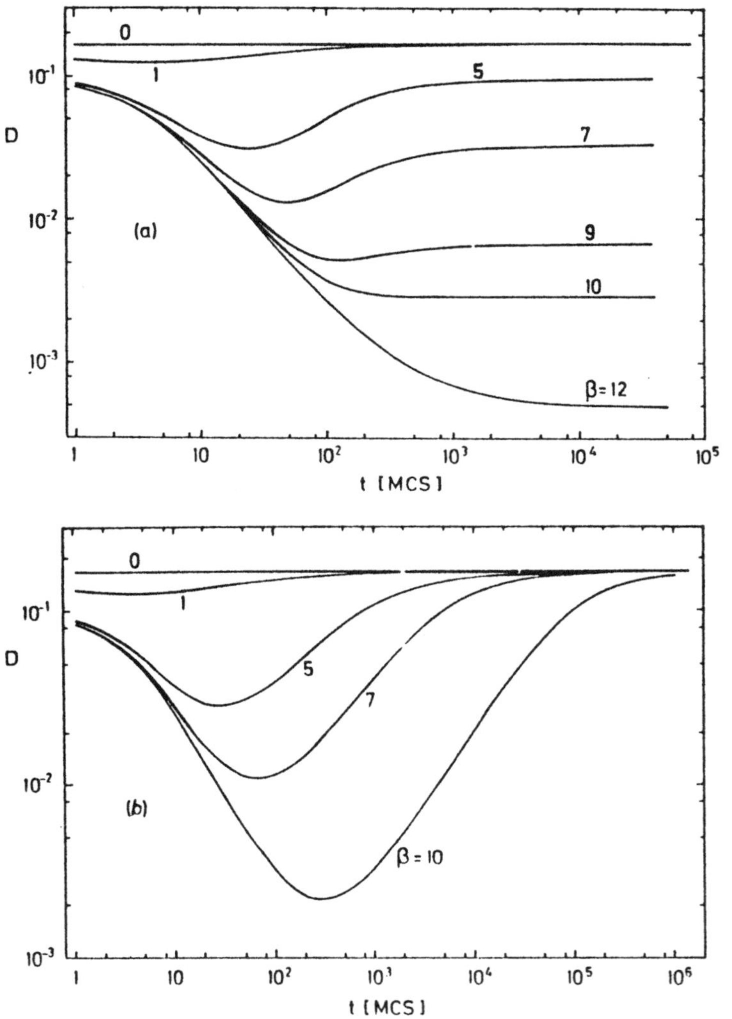

Fig. 2. Diffusion coefficient $D(t)$ for the 3d model as a function of time for several inverse temperatures β for $N = 10$ (a), and for an infinite chain $N = \infty$ (b), obtained by Monte Carlo simulations assuming Metropolis et al. jump probabilities.

around the center, even for long–times, which is clearly visualized in the snap–shot pictures presented in Fig.5. This is the key to improve the efficiency of the exact enumeration procedure in our case since it allows us to cut off ($\approx 90\%$) the long–range tail of the distribution function, even up to 10^6 time steps at intermediate and low temperatures. The errors for the mean– and mean–square displacement are smaller than 0.1%. Of course, when the temperature is low, the localization is better.

The large oscillations of the probability distribution visible for a free discrete random walk (i.e., at $\beta = 0$) confirm its well known feature that the distance from

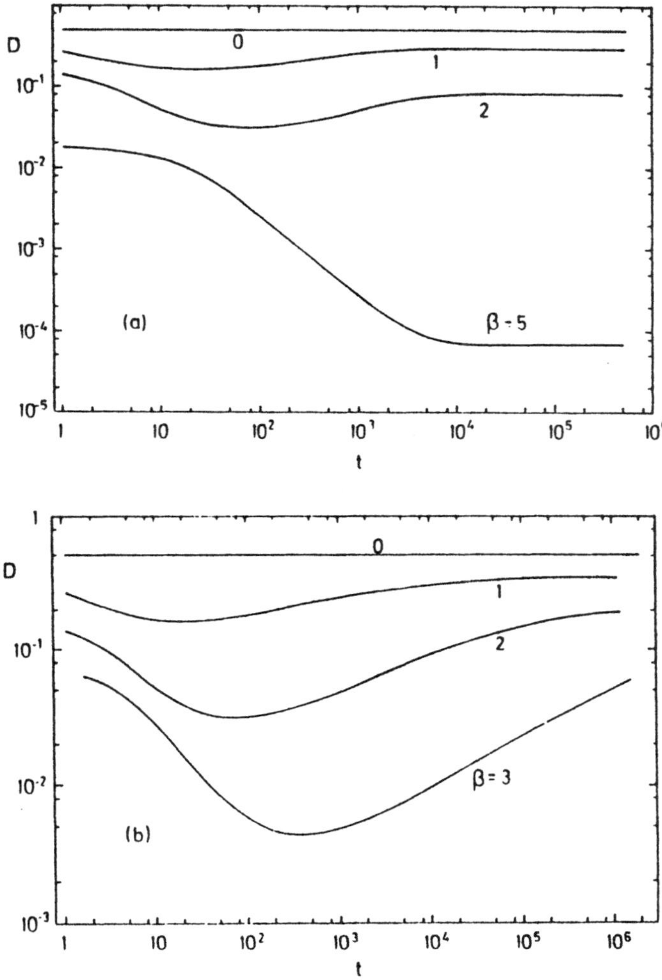

Fig. 3. One-dimensional diffusion coefficient $D(t)$ as a function of time for several inverse temperatures β for $N = 50$ (a), and $N = \infty$ (b), obtained by exact enumeration procedure assuming exponential jump probabilities.

the origin (measured in steps) reached by a walker has the same parity as the total number of steps since its sojourn probability vanishes. Some influence of this effect can also be recognized for finite temperatures. Other type of oscillations are discussed in Sect.4.

Having the matrix $\hat{P}(l_0, t_0)$ enables us to calculate many random walk properties, but in this work we mainly pay attention to the time-dependent diffusion coefficient (discussed in Sect.4). However, to make a complementary analysis we need an analytical formalism (which is in preparation).

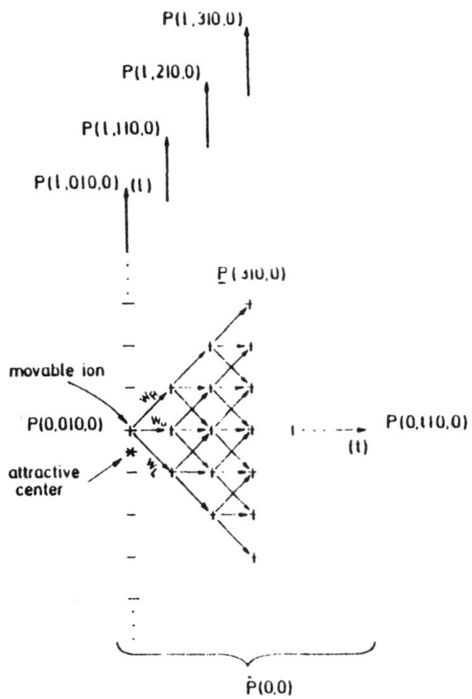

Fig. 4. Schematic illustration of the exact enumeration procedure which recursively constructs path probabilities and hence occupancy probabilities at any site l at any time t. (The notation is described in the text.)

4 Results and Discussion

In this section we report and discuss results on the static and dynamic diffusion coefficient obtained by the standard Monte Carlo simulations and by exact enumeration procedure for our one–dimensional model.

4.1 Static Results

In Fig.6 and 7 we show (by points) the exact asymptotic diffusion coefficient D_∞, given below by (1), as a function of the periodicity N for several fixed inverse temperatures β, and as a function of β for several values of N, respectively. The diffusion coefficient can be easily derived within the linear response theory in the form:

$$D_\infty = D(t \to \infty) = \left[\frac{1}{N^2} \sum_{l=1}^{N} \frac{1}{P_{eq}(l)\, W_R(l)} \right]^{-1} \tag{1}$$

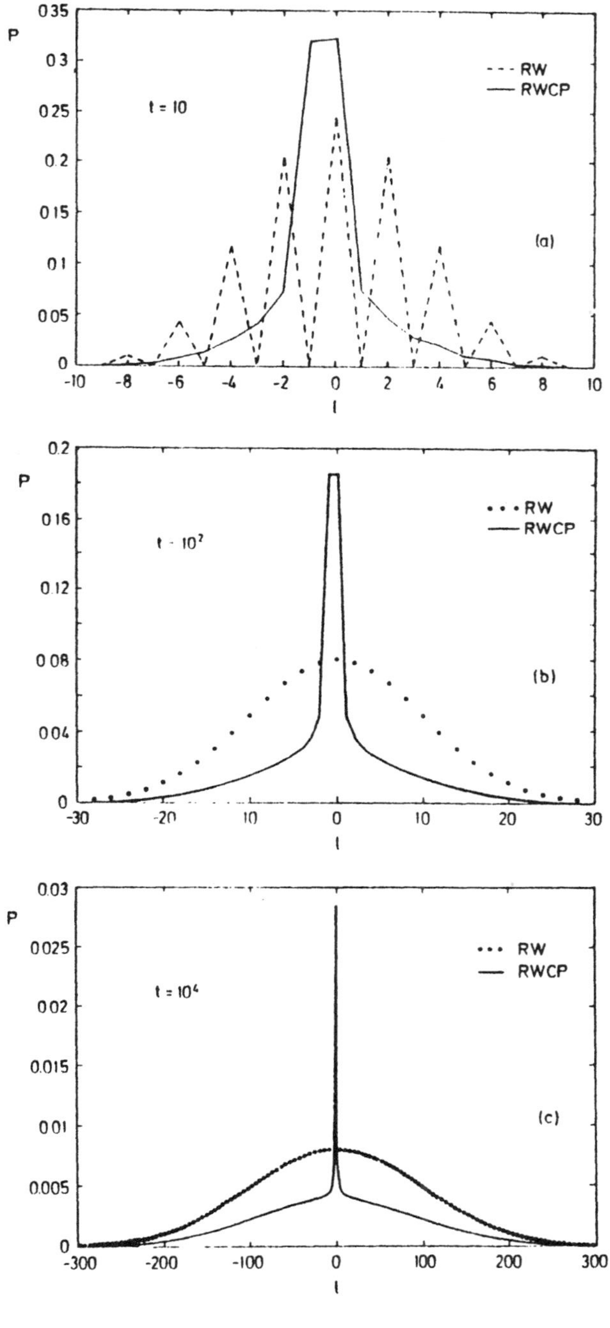

Fig. 5. Evolution of the probability distributions $P(l, t|0, 0)$ presented in the three snap-shot pictures. The solid line represents $\beta = 1$, while $\beta = 0$ is represented by the dashed line (a) and doted lines (b),(c), respectively. (We used dots instead of dashed line to emphasize the main result given by the solid line.)

where $P_{eq}(l)$ is the nonhomogeneous equilibrium probability distribution. The dashed lines in Fig.6 were predicted by the known Smoluchowski formula for the asymptotic diffusion coefficient [29] derived from the Smoluchowski equation.

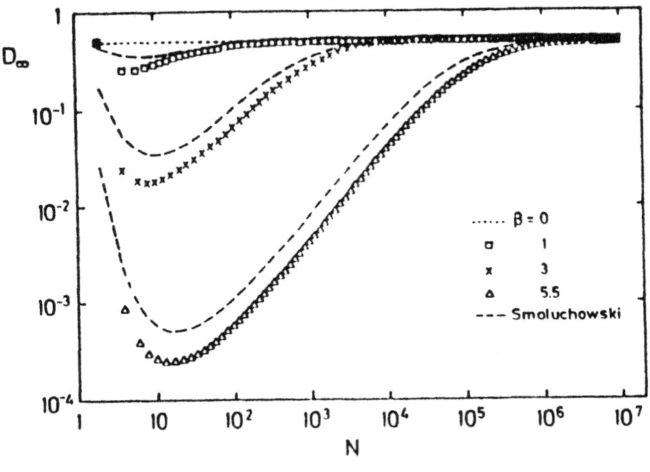

Fig. 6. Exact asymptotic diffusion coefficient D_∞ (points) given by (1), vs. period length N for typical inverse temperatures β. The dashed lines represent the prediction of the formula obtained from the Smoluchowski equation.

The data in Fig.6 exhibit a local minimum vs. N, that is more pronounced at lower temperatures, which can be easily interpreted as follows. When N increases then the height and the width of the Coulomb potential barrier also increases (c.f. Fig.1). At fixed β this leads to a decreasing D_∞. However, when N is quite large another effect arises connected with the known observation [2] that the Coulomb potential is steep near the attractive center, and also has an energetically fairly flat part far from it. Hence, for large N the ion perfoms asymptotically an almost free random walk in this flat, quite extended region. Thus, the mean jump rate of the ion increases and hence D_∞ also increases. Therefore, D_∞ approaches, for every temperature, its limiting value characteristic for free diffusion.

In Fig.7 we present an Arrhenius plot of D_∞ for different N. In the low temperature regime the Van't Hoff–Arrhenius law with activation energy $E_A = V_C(N/2+1) - V_C(N) = 2(1 - \frac{1}{N-1})$ (c.f. Fig.1) well describes the data. This means that the jumping process can be treated asymptotically as a simple overbarrier one at low temperatures.

Hence, our results depend sensitively on the two paramters N and β. We can recognize two limiting regions namely, *i*) the free diffusion (when N is large and β not too high), and *ii*) a strongly temperature dependent one (in the high β regime and N not too large).

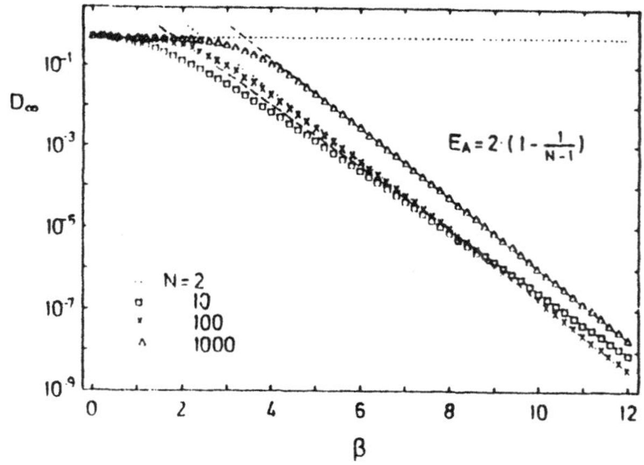

Fig. 7. The dependence of the exact diffusion coefficient D_∞ (points) on inverse temperature β for three typical period lengths (N=2 is a separate case representing free diffusion). The oblique dashed lines represent the predictions of the Van't Hoff–Arrhenius law with an activation enery E_A listed in the figure. (A preexponential factor in the Van't Hoff–Arrhenius law was treated as a free parameter.)

4.2 Dynamic Results

The concept used in the previous section can be extended to the description of our results for the time–dependent diffusion coefficient presented in Fig.8 for several inverse temperatures β at $N = 50$ (Fig.8a) and $N = \infty$ (Fig.8b). The points represent results of Monte Carlo simulations while the full lines result from the exact enumeration procedure. (The jump probabilities were chosen here according to Metropolis et al.. Such a choice of probabilities was made to obtain results which reach the asymptotic values quicker than the exponential [28], although we loose a more complex data structure at initial times.) Very good agreement between both independent methods is observed.

The results from the exact enumeration procedure can be obtained for longer times and with higher precision since the approach is much more efficient (an order of one hundred in this range of time). Moreover, very good agreement with theoretical asymptotic values of the time–dependent diffusion coefficient, given by (1), is obtained.

Note that for intermediate temperatures the crossover regions in Fig.8a are well defined by the two characteristic relaxation times τ_0 and τ_∞. The relaxation time $\tau_\infty = N^2/2D_\infty$ defines in fact the beginning of the asymptotic region, whereas τ_0 estimates the end of the time interval where the particle only oscillates between the two equivalent sites nearest to the attractive center. These temperature dependent, odd–even oscillations are distinctly visible for short times in the plot of the mean–displacement$\langle \Delta l(t) \rangle$ of the walker as a function of time shown in Fig.9. Note that $\langle \Delta l(t) \rangle$ does not vanish here since averaging over symmetrical initial positions is not performed. The relaxation

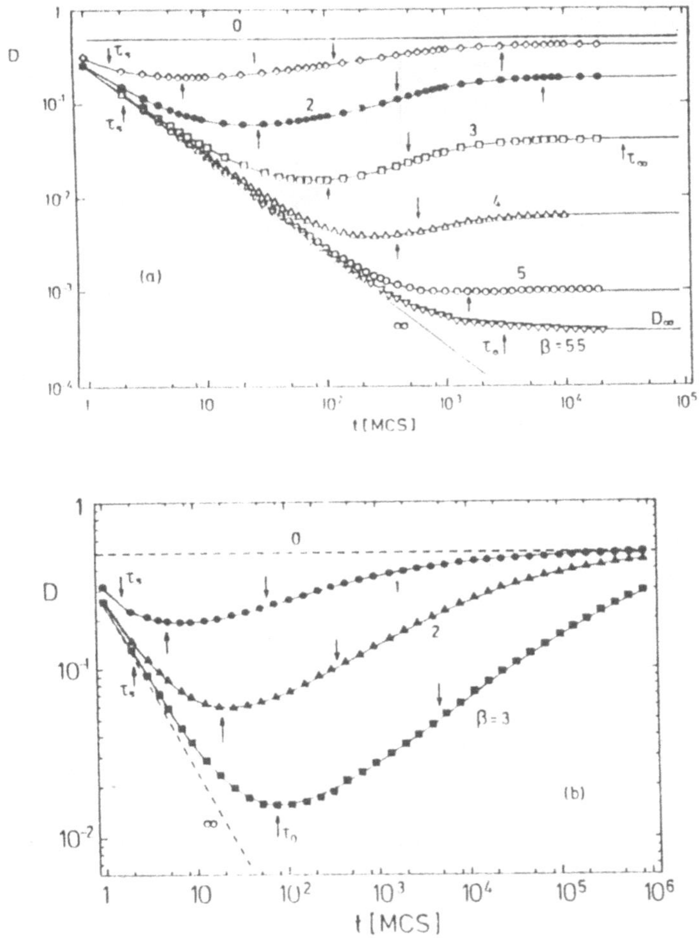

Fig. 8. Diffusion coefficient $D(t)$ vs. time for several inverse temperatures β for $N = 50$ (a), and for $N = \infty$ (b). Downward pointing arrows without labels denote inflection points. The relaxation times τ_s, τ_0, and τ_∞ are defined in the text

time τ_0 is defined as the average residence time of the ion in an isoenergetic trap nearest the center and is easily calculated from the time–dependent survival probability $\Psi(t) = [1 - \frac{1}{2}\exp[-\frac{4}{3}\beta)]^t$ as $\tau_0 = 2\exp(\frac{4}{3}\beta) - 1$. The derivation of $\Psi(t)$ is straightforward because the walker performs in this case a Bernoulli random walk among two equivalent sites. For completeness, we introduced the relaxation time τ_s defined as the average time for an ion to stay at a given site nearest to the center.

It is remarkable that there is a characteristic inverse temperature β_c for a given finite value of N, where the shape of the curves changes (e.g. $\beta_c \approx 5$ for $N =$

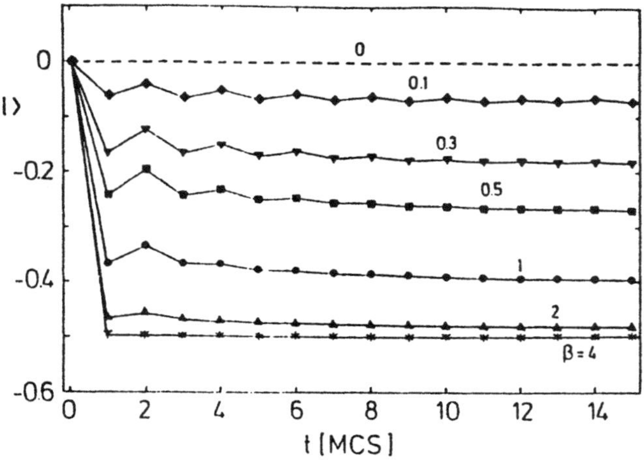

Fig. 9. Time-dependent mean dispacement $\langle \Delta l(t) \rangle$ obtained by Monte Carlo simulation (points) and by exact enumeration procedure (solid lines) for several inverse temperatures β.

50, c.f. Fig.8a). For $\beta < \beta_c$ and $t > \tau_o$ the mean-square displacement grows faster than linear with t ("superdiffusive"), whereas for $\beta > \beta_c$ it grows slower than linear ("subdiffusive"). The above observation may be explained by assuming, in agreement with the previous section, that at lower temperatures the particle cannot obtain enough energy to make an overbarrier jump, and instead performs, during a long time, a restricted random walk in a given segment (valley) mainly around the center. Of course, finally the particle reaches the neighbouring valley. By increasing the temperature the particle also performs a quite intensive local motion around the center inside the segment, but becomes more susceptible to be occasionally absorbed by neighbouring valleys in a reasonably short time.

Fig.8b is a special variation of the Fig.8a and requires an explanation. For infinite periodicity β_c equals infinity (c.f. Fig.10) and $D(t)$ behaves as for $\beta < \beta_c$ with $D_\infty = 1/2$ independent of β. In this case the previous interpretation is in principle valid since the particle reaches a quite extended energetically fairly flat region during a finite time, performing then (an almost) free random walk. This must manifest in $D(t)$ as a superdiffusive approach to D_∞ since $D(\tau_0) < D_\infty$ for $\beta > 0$.

The complementary results for $D(t)$ (at fixed β) were presented in Fig.11 for several values of N. The results extend our earlier observations (c.f. Fig.6) and show that $D(t)$ gets larger with increasing N, and approaches from below its limiting value given by the diffusion coefficient for infinite periodicity.

Having the results shown in Fig.8 (as well as in Fig.11) we introduce initial and intermediate slopes (c.f. Fig.12 and 10, respectively).

In Fig.12 we present the value of the initial slope s as a function of β. (It is easy to calculate this slope analytically as well as numerically from the first two time-steps.) As it is seen $0 \leq s \leq 1$ which is similar to the range of an analogous

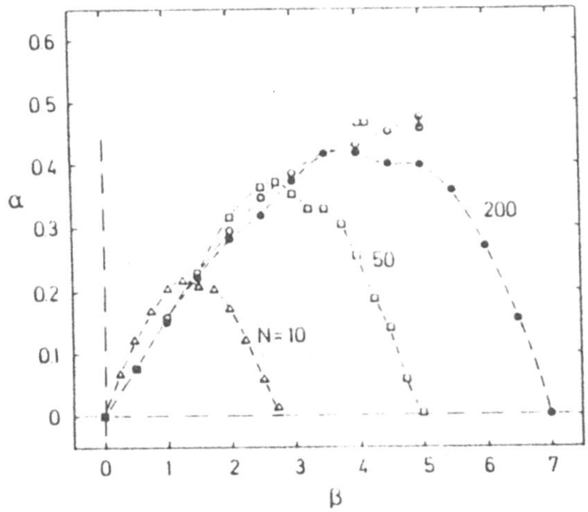

Fig. 10. The 'intermediate' slope α vs. inverse temperature β for several N.

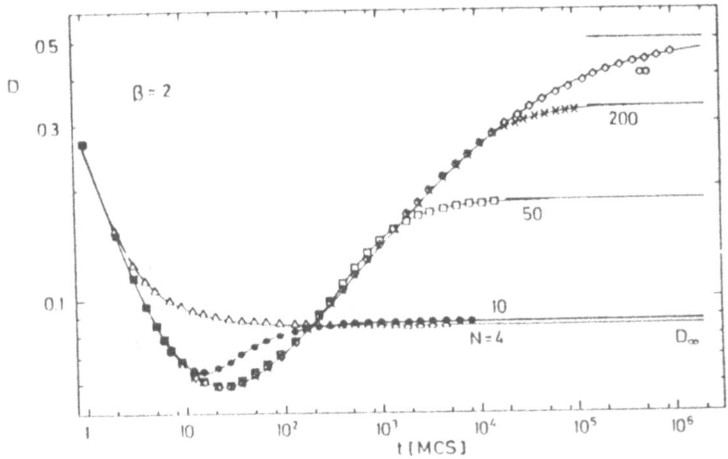

Fig. 11. Diffusion coefficient $D(t)$ vs. time for several periodicities N at a fixed inverse temperature $\beta = 2$. The points represent results of Monte Carlo simulation and the corresponding solid lines result from the exact enumeration procedure

exponent predicted by the "universal dynamic response".

In Fig.10 we present an intermediate slope α determined numerically at the inflection points (denoted by downward pointing arrows in Fig.8) as a function of β, for different N. As it is already suggested by Fig.8a, the slope α has a local maximum vs. β that well defines the temperature at which the ion is most susceptible to reach the energetically fairly flat region and/or to be absorbed by a neighbouring valley. The shift of the position of the maximum of α to higher β and the increase of the maximum with increasing N is consistent with

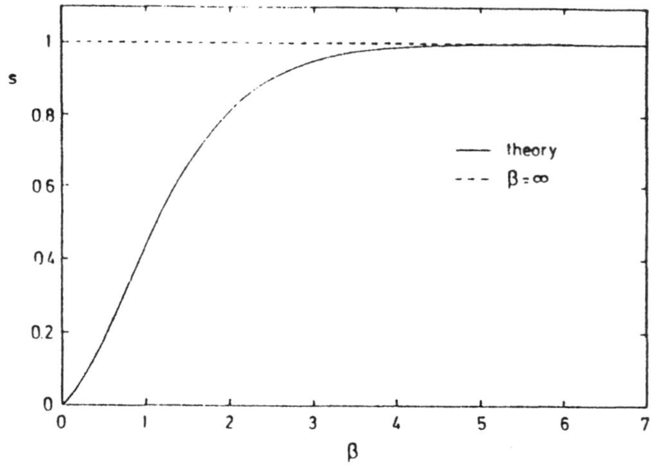

Fig. 12. The initial slope s (solid line) vs. inverse temperature β.

our previous results (c.f. Fig.8). From the presented dependence α vs. β we can easily, numerically find the low-temperature threshold β_c since then $\alpha(\beta_c) = 0$. The slope α seems to be an effective simple description of the ion dynamics at the crossover region. However, it requires further, mainly analytical study.

5 Some Conclusions

The single–ion inhomogeneous random walk approach seems to account for some characteristic features of "universal dynamic response". These features are in fact supplied by Fig.8 being the central result of this work. Namely, there is the initial power–law behaviour of $D(t)$ with an effective exponent s for $t < \tau_0$. For $t > \tau_0$ we observe two different cases: *i*) For $\beta > \beta_c$, $D(t)$ shows a subdiffusive approach to its limit D_∞ at finite N. This is also characteristic for a concentrated Coulomb lattice gas. *ii*) For $\beta < \beta_c$ a superdiffusive approach of $D(t)$ to its limit D_∞ is clearly observed which seems to be characteristic for a single ion inhomogeneous escape random walk, not only in a Coulomb cutoff potential. We suppose that potentials having a steep sufficiently central part and flat sufficiently extended tail are also able to supply results similar to those presented in Fig.8.

For further analysis of the results an analytical formalism is needed. A matrix continued fraction [17, 30], a continuous–time random walk model with different waiting–time distributions for forward and backward jumps [31], or a geminate recombination model [2] seem to be promising.

Acknowledgments

This work was supported in part by Deutsche Forschungsgemeinschaft, SFB 306, and by the Polish KBN Grant No.2 P302 126 04.

References

[1] Funke, K.: Jump relaxation in solid electrolytes. Progr. Solid State Chem. **22** (1993) 251–341

[2] Rackovsky, S., Scher, H.: Theory of geminate recombination on a lattice. IV. Results using large Coulomb radii on the simple cubic and square lattices. J. Chem. Phys. **89** (1988) 7242–7252

[3] Calef, D.F., Deutch, J.M.: Diffusion–controlled reactions. Annu. Rev. Phys. Chem. **34** (1983) 493–524

[4] Rice, S.A.: Diffusion–limited reactions. (Elsevier, Amsterdam, 1985)

[5] Ovchinnikov, A.A., Timashev, S.F., Bely, A.A.: Kinetika Diffusionno-kontroliruemyh Kimicheskih Processov. (Kimija, Moscow, 1986)

[6] Schwarzer, S., Havlin, S., Stanley, E.H.: Multifractal scaling of 3d diffusion–limited aggregation. Physica **A 191** (1992) 117–122

[7] Coniglio, A.: Is diffusion limited aggregation scale invariant? Physica **A 200** (1993) 165–170

[8] Kehr, K.W., Binder, K.: Simulation of diffusion in lattice gases and related kinetic phenomena. Topics in Current Physics. Vol. **36**: Applications of the Monte Carlo Method in Statistical Physics. Ed. Binder, K. (Springer, Berlin, 1984) Chap.6, p.181–221

[9] Taitelbaum, H.: Segregation in reaction–diffusion systems. Physica **A 200** (1993) 155–164

[10] Aslangul, C., Pottier, N., Chvosta, P., Saint–James, D., Skala, L.: Random–random walk on an asymmetric chain with a trapping attractive center. J. Stat. Phys. **69** (1992) 17–34

[11] Onsager, L.: Initial recombination of ions. Phys. Rev. **54** (1938) 554–557

[12] Hong, K.M., Noolandi, J.: Solution of the time–dependent Onsager problem. J. Chem. Phys. **68** (1978a) 5026–5039
Hong, K.M., Noolandi, J.: Solution of the Smoluchowski equation with a Coulomb potential. I. General results. J. Chem. Phys. **68** (1978b) 5163–5171
Hong, K.M., Noolandi, J.: Solution of the Smoluchowski equation with a Coulomb potential. II. Application to fluorescence quenching. J. Chem. Phys. **68** (1978c) 5172–5176

[13] Traytak, S.D.: On the solution of the Debye–Smoluchowski equation with a Coulomb potential. I. The case of a random initial distribution and a perfectly absorbing sink. Chem. Phys. **140** (1990) 281–297
Traytak, S.D.: On the solution of the Debye–Smoluchowski equation with a Coulomb potential. II. An approximation of the time–dependent rate constant. Chem. Phys. **150** (1991a) 1–12
Traytak, S.D.: On the solution of the Debye–Smoluchowski equation with a Coulomb potential. III. The case of a Boltzmann initial distribution and a perfectly absorbing sink. Chem. Phys. **154** (1991b) 263–280

[14] Pedersen, J.B., Lolle, L.I., Jorgensen, J.S.: An optimal numerical solution of diffusional recombination problems. Chem. Phys. **165** (1992) 339–349

[15] Kramers, H.A.: Brownian motion in a field of force and the diffusion model of chemical reactions. Physica **7** (1940) 284–304

[16] Hänggi, P., Talkner, P., Borkovec, M.: Reaction–rate theory: fifty years after Kramers. Rev. Mod. Phys. **62** (1990) 251–341

[17] Risken, H.: The Fokker–Planck Equation. Methods of Solution and Applications. 2.Edition (Springer, 1989)

[18] Knödler, D., Dieterich, W.: Lattice–gas models of dispersive transport in disordered materials. Physica **A 191** (1992) 426–432
[19] Petersen, J., Dieterich, W.: Effects of Coulomb interaction and disordered in a stochastic lattice gas. Phil. Mag. **B 65** (1992) 231–241
[20] Maass, P., Petersen, J., Bunde, A., Dieterich, W., Roman, H.E.: Non–Debye relaxation in structurally disordered ionic conductors: effect of Coulomb interaction. Phys. Rev. Lett., **66** (1991) 52–55
[21] Nitzan, A., Druger, S.D., Ratner, M.A.: Random walk in dynamically disordered systems. Phil. Mag. **B 56** (1987) 853–859
[22] Ratner, M.A., Nitzan, A.: Fast ion conduction: some theoretical issues. Solid State Ionics'87. Proceed. 6th Intern. Conf. Solid State Ionics'87. Eds. Weppner, W., Schutz, H. (Garmisch–Partenkirchen, 1987) p.3–33
[23] Bunde, A.: Anomalous transport in disordered media. Solid State Ionics'87. Proceed. 6th Intern. Conf. Solid State Ionics. Eds. Weppner, W., Schulz, H. (Garmisch–Partenkirchen, 1987) p.34–40
[24] Dieterich, W.: Transport in ionic solids: Theoretical aspects. High conductivity solid ionic conductors – recent trends and applications. (World Scientific, Singapore, 1989) p.17–44
[25] Binder, K.: Introduction: Theory and "technical" aspects of Monte Carlo simulations. Topics in Current Physics. Vol. **7**: Monte Carlo Methods in Statistical Physics. Ed. Binder, K. (Springer, Berlin, 1979) Chap.1, p.1–45
[26] Majid, I., Ben–Avraham, D., Havlin, S., Stanley, H.E.: Exact–enumeration approach to random walks on percolation clusters in two dimensions. Phys. Rev. **B 30** (1984) 1626–1628
[27] Havlin, S., Ben–Avraham, D.: Diffusion in disordered media. Adv. Phys. **36** (1987) 695–798
[28] Bunde, A., Dieterich, W.: Dynamic correlations in charged lattice gas. Phys. Rev. **B 31** (1985) 6012–6021
[29] Dieterich, W., Peschel, I., Schneider, W.R.: Diffusion in periodic potentials. Z. Physik **B 27** (1977) 177–187
[30] Allegrini, M., Arimondo, E., Bambini, A.: Matrix continued–fraction solution for saturation effects in spin–1/2 radio–frequency spectroscopy spectroscopy. Phys. Rev. **A 15** (1977) 718–726
[31] Zwerger, W., Kehr, K.W.: On the frequency dependence of the conductivity in random walk models with internal states. Z. Physik B – Cond. Matt. **40** (1980) 157–166

A Model for the Underpotential Deposition of Metals

Dale A. Huckaby[1] and Lesser Blum[2]

[1] Department of Chemistry, Texas Christian University, Fort Worth, Texas 76129, USA
[2] Department of Physics, University of Puerto Rico, Rio Piedras, Puerto Rico 00931-3343, USA

Abstract. A three-dimensional model consisting of a fluid in contact with a planar wall containing a lattice of adsorption sites is useful for studying phase transitions at the solid-fluid interface. Results have been obtained both for adsorption from a fluid of hard spheres and for deposition of a metal from a dilute solution of metal ions. The latter case is used to model the underpotential deposition of a metal onto a single-crystal electrode. The model explains the qualitative features of the voltammogram of the underpotential deposition of copper on the (111) surface of a gold electrode in the presence of bisulfate.

1 Introduction

For the past five years we have been studying the structured solid-fluid interface [1-9] using a three-dimensional model [10-11] of a fluid in contact with a planar wall which contains an array of sticky adsorption sites. The model, described in detail in Sec. 2, is equivalent to a two-dimensional lattice gas with n-body interactions which are simply related to the n-body correlation functions of the fluid.

For the case of adsorption from a fluid of uncharged hard spheres, we have obtained accurate adsorption isotherms for the model as well as coexistence curves for first-order phase transitions which occur on the surface [1-3]. The effect of triplet correlation functions on these transitions has also been considered [6,7]. These results are described in Sec. 3.

If a solution of metal ions is in contact with an electrode composed of the same metal which is at a potential less than or equal to a characteristic Nernst potential, the ions will be deposited in bulk on the metal electrode. In some cases, however, the metal ions can be deposited at a higher potential on an electrode composed of a different metal. This is called underpotential deposition. Since bulk deposition does not occur at such a high potential, underpotential deposition is useful for obtaining submonolayer deposition of a metal onto a crystalline electrode. Examples include the underpotential deposition of copper on gold or platinum, silver on gold, and lead on silver.

An application of the model to the underpotential deposition of copper from a solution of copper(II) ions onto the (111) surface of a single-crystal gold electrode in the presence of bisulfate is discussed in Sec. 4. In particular, the voltammogram of the deposition process [12] contains two spikes which are well-reproduced by the model and which correspond to first-order phase transitions of submonolayer copper on the gold surface [4,5,8,9]. Both thermodynamic and kinetic aspects of the process have been considered [9,13]. This system is under active experimental and theoretical invetigation by several research groups.

2 The Model

The model [1,3,6,7,10,11] consists of a fluid of N hard spheres, radius σ, in contact with a hard planar wall at $z = -\sigma/2$ which contains a lattice of sticky sites. The partition function is

$$Z = \frac{1}{N!} \int dr^N e^{-\beta H} , \tag{1}$$

where $\beta = 1/kT$, and

$$H = H_0 + \sum_{i=1}^{N} U^s(\mathbf{r}_i) , \tag{2}$$

H_0 being the Hamiltonian for the smooth wall problem. The sticky potential $U^s(\mathbf{r}_i)$ describing the interaction of a sphere at $\mathbf{r}_i$ with the sticky sites is given as

$$e^{-\beta U^s(\mathbf{r}_i)} = 1 + \lambda \sum_{\mathbf{R}_s \in \Lambda} \delta(\mathbf{r}_i - \mathbf{R}_s) , \tag{3}$$

where λ is a stickiness parameter, δ is a Dirac delta function, and Λ is a lattice of sites at the contact plane which are displaced by $\sigma/2$ in the z direction from the sticky sites.

After integrating and rearranging terms we obtain

$$Z / Z_0 = \sum_{n=0}^{N} \frac{[\lambda \rho_1^0(0)]^n}{n!} \sum_{\{\mathbf{R}_i\} \subset \Lambda} e^{-\beta U(\mathbf{R}_1,\ldots,\mathbf{R}_n)} , \tag{4}$$

where Z_0 is the partition function and $\rho_1^0(0)$ is the contact density for the smooth wall problem, and $U(\mathbf{R}_1,\ldots,\mathbf{R}_n)$ is the potential of mean force defined in terms of the n-body contact correlation function as

$$e^{-\beta U(\mathbf{R}_1,\ldots,\mathbf{R}_n)} = g_n^0(\mathbf{R}_1,\ldots,\mathbf{R}_n) . \tag{5}$$

Converting the sum in Eq. (4) from labelled hard sphere positions on Λ to lattice sites of Λ yields

$$\Xi = Z / Z_0 = \sum_{\{t_i\}} [\lambda \rho_1^0(0)]^{\Sigma t_i} e^{-\beta U(\{t_i\})} , \tag{6}$$

where t_i = 1 or 0 depending on whether or not site i is occupied by a sphere in configuration $\{t_i\}$. Equation (6) gives the grand partition function Ξ for an equivalent two-dimensional lattice gas on Λ with a many-body interaction energy $U(\{t_i\})$ and a fugacity $z = \lambda\rho_1^0(0)$. The fraction, θ, of occupied sites of Λ is then

$$\theta = \frac{\lambda}{|\Lambda|}\frac{\partial}{\partial\lambda} \ln \Xi \ . \tag{7}$$

In most of our work we have used the Kirkwood superposition approximation [14]

$$g_n^0(\mathbf{R}_1,\ldots,\mathbf{R}_n) = \prod_{\langle i,j\rangle} g_2^0(|\mathbf{R}_i - \mathbf{R}_j|) \ , \tag{8}$$

which results in an equivalent lattice gas with pair interactions

$$U(\{t_i\}) = \sum_{\langle i,j\rangle} \varepsilon_{i,j}\, t_i t_j \ , \tag{9}$$

where $\varepsilon_{i,j}$ is the pair potential of mean force, which varies with distance.

3 Adsorption from a Fluid of Hard Spheres

If the sticky sites form a triangular lattice with a lattice spacing d satisfying $d<\sigma<\sqrt{3}d$, then the spheres exclude first neighbors. If we use the superposition approximation of Eq. (8), with the pair correlation functions equal to zero for first-neighbor separation and unity for more distant neighbor separation, then the equivalent lattice gas is the hard hexagon model which was solved by Baxter [15].

The hard hexagon model undergoes an order-disorder transition at $\theta_c = (5-\sqrt{5})/10 = 0.2764$ and $[\lambda\rho_1^0(0)]_c = (11+5\sqrt{5})/2 = 11.09$ [14]. Exact expressions have been obtained both for $\lambda\rho_1^0(0)$ as a function of θ [16] and, within the Percus-Yevick approximation, for $\lambda\rho_1^0(0)\sigma^3$ as a function of $\rho\sigma^3$ [17,18], where ρ is the fluid density. Together, these expressions yield isotherms of θ versus $\rho\sigma^3$ for various values of the temperature-dependent parameter λ/σ^3 [1].

If σ is much smaller than the lattice spacing d, we assume all the correlation functions for allowed configurations are unity. This gives an equivalent lattice gas without lateral interactions, which yields the Langmuir isotherm [2,7].

If σ is slightly less than or equal to d, then we assume the superposition approximation, Eq. (8), with the pair correlation function assumed to be unity at second and higher neighbor distances. The model is then equivalent [1,11] to a lattice gas with a first-neighbor attraction ε which is equal to the pair potential of mean force, i.e., $\exp(-\beta\varepsilon) = g_2^0(d)$.

Two-phase coexistence curves in the θ, $\lambda\rho_1^0(0)$ plane have been calculated exactly for a lattice gas with first-neighbor attractions on several two-dimensional lattices. For the triangular lattice the critical parameters are $[g_2^0(d)]_c = 3$ and $[\lambda\rho_1^0(0)]_c = 1/27$ [1]. For the case $\sigma = d$, we used the Percus-Yevick approximation [17,18] to $g_2^0(\sigma)$ as a function of $\rho\sigma^3$ and obtained $\rho_c\sigma^3 = 0.71$ as an estimate of the fluid density at the critical point of the transition [1]. The Carnahan-Starling [19] approximation to $g_2^0(\sigma)$ yields a similar estimate, $\rho_c\sigma^3 = 0.67$ [6].

The inclusion of triplet correlation functions in the superposition approximation [6.7], combined with a closed form approximation of the triplet correlation function for three spheres in contact [6,7] and a Müller-Hartmann-Zittarz [20] approximation of the critical point of the equivalent lattice gas with three-body interactions [21] yielded a somewhat higher estimate of the critical density, $\rho_c\sigma^3 = 0.84$, indicating that triplet correlations are significant at high densities.

Although exact isotherms for the lattice gas with attractions have never been obtained, exact coefficients in series expansions of these isotherms have been calculated for several lattices [22] to high order in terms of the variable $y = [\lambda\rho_1^0(0)\, g_2^0(d)^{q/2}]^{-1}$ at high density, and in terms of y^{-1} at low density, where q is the number of first neighbors to a lattice site.

We constructed high and low density approximants to the isotherms, similar in form to Langmuir's isotherm [2,4,5,7,8,9]

$$\theta_l(y) = \frac{P(y^{-1})}{1 + P(y^{-1})} \tag{10}$$

$$1 - \theta_h(y) = \frac{P(y)}{1 + P(y)} \ , \tag{11}$$

where P is a polynomial with coefficients which are determined by matching the coefficients in the expansion of θ_l and θ_h with the exact coefficients in the known series expansion of θ at low and high density.

Using a switching function such as [7]

$$\eta(y) = \frac{1}{2}\{1 + \mathrm{erf}\,[s(y - y^{-1})]\} \ , \tag{12}$$

an isotherm which is accurate over the entire density range can be constructed as

$$\theta(y) = \theta_l(y)\,\eta(y) + \theta_h(y)\,[1 - \eta(y)] \ . \tag{13}$$

The parameter s in the switching function can be used as a measure of the sharpness of the transition in a realistic experimental situation. Alternatively, a simple approximate form for $P(y^{-1})$ such as [9]

$$P(y^{-1}) = \lambda\rho_1^0(0) + [g_2^0(d) - 1]y^{-n} \tag{14}$$

can be used to form θ_l, where n plays the same role as does s in the switching function. Another possibility is to construct a two-point Padé approximant which matches the exact coefficients in both the high and low density series expansions of θ [2].

4 Underpotential Deposition of Cu on Au (111) in the Presence of Bisulfate

For the case of underpotential deposition, the model hard sphere fluid consists of metal ions of the type which are discharged at the electrode. If these ions have an electrosorption valency ζ, we generalize the fugacity to include the potential ψ at the contact plane of the electrode surface [2,4,5],

$$z = \lambda \rho_1^0(0,\psi) , \tag{15}$$

where

$$\rho_1^0(0,\psi) = \rho_1^0(0,0)\, e^{-\zeta\beta e[\psi - \psi^{ref}]} \tag{16}$$

is the contact density of the metal ions. Here $-e$ is the charge on an electron, and ψ^{ref} is an electrosorption reference potential which depends on the nature of the metal ion and electrode surface.

The isotherm, θ versus ψ, for the deposition is computed using the methods discussed in Sec. 3. The voltammogram, a plot of current intensity I versus voltage ψ, is then derived from the isotherm by assuming the ions discharge as they are deposited, giving the current intensity as

$$I = -\frac{\partial\theta}{\partial t} = -\frac{\partial\theta}{\partial\psi}\frac{d\psi}{dt} . \tag{17}$$

The sweep rate $d\psi/dt$ is kept constant, and the negative sign is in keeping with the convention that the current is positive during the stripping process, when $d\psi/dt$ is positive.

The model with these additions was used in the construction of a model voltammogram for the underpotential deposition of copper on the (111) surface of a gold electrode in the presence of bisulfate ions. The experimental voltammogram contains two spikes which occur during the stripping process at 0.22 V and 0.07 V [12]. The area under the spike at 0.22 V is about twice as large as the area under the spike at 0.07 V.

We supposed the voltammogram spikes corresponded to first-order phase transitions which occurred on the electrode surface, and we proposed the following idealized three-step model for the deposition process [4]. First, at high values of ψ the bisulfate, remaining charged, adsorbs on the gold surface, with three oxygens of each bisulfate ion associating with a triangle of gold atoms, the sulfur atom of the ion being above an adsorption site (see Fig. 1). Packing considerations preclude two oxygens from being associated with the same gold atom. As is apparent from Fig. 1, this is equivalent to excluding sulfur atoms of neighboring bisulfate ions from being above neighboring adsorption sites. This adsorption process is thus equivalent to the hard hexagon model described in Sec. 3. The bisulfate ions form a $\sqrt{3} \times \sqrt{3}$ structure, occupying a triangular sublattice of the adsorption sites.

Second, as the potential is made less positive, the copper ions are deposited and discharged on the honeycomb lattice of free adsorption sites, undergoing a first-order phase transition. Finally, as the potential is made less positive, a first-order phase

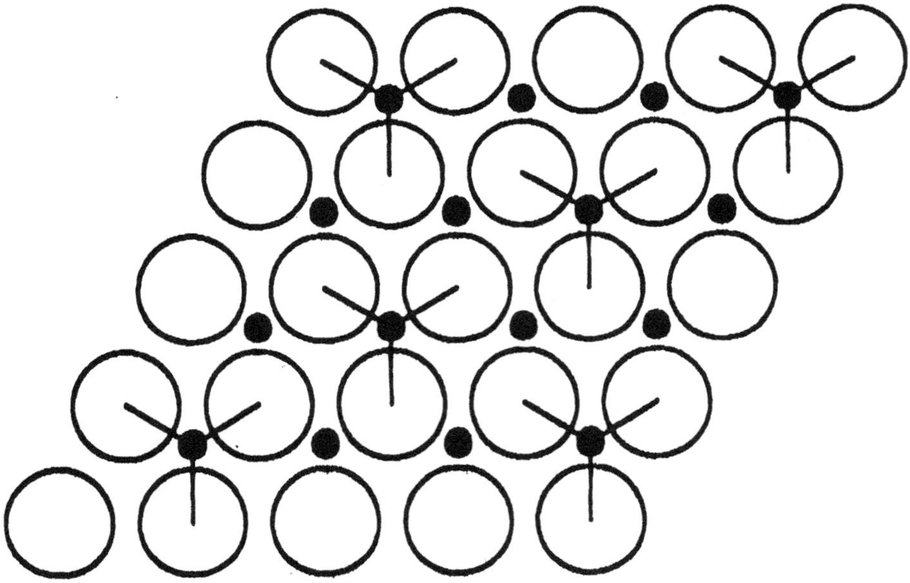

Fig. 1. The open circles represent gold atoms, the dark dots represent adsorption sites, and adsorbed bisulfate ions are each represented by three lines. This figure first appeared in [4].

transition occurs as copper ions displace the bisulfate ions and are deposited and discharged to complete a copper monolayer.

The overall adsorption isotherm for the copper deposition, θ_{Cu}, was approximated as

$$\theta_{Cu} = \frac{2}{3}\theta_H + \frac{1}{3}\theta_T , \tag{18}$$

where θ_H is an isotherm constructed for the deposition of copper on a honeycomb lattice, forming two thirds of a monolayer, and θ_T is an isotherm constructed for the deposition of copper on a triangular lattice, forming the final third of the monolayer.

For reasonable values of the parameters, a model voltammogram was constructed which qualitatively matched the two spikes of the experimental voltammogram [4]. Recent experiments have shown that our model for the underpotential deposition of copper on the (111) surface of gold in the presence of bisulfate is essentially correct [23-25].

Our original treatment of the model, in which the molecular processes were treated as independent events is, however, an idealization. In order to explain the presence of a large "foot" on the high potential side of the experimental voltammogram peak at 0.22 V, we assumed the bisulfate begins to desorb and becomes disordered before the copper begins to deposit, the copper bringing the bisulfate back to the surface to reform the $\sqrt{3} \times \sqrt{3}$ bisulfate structure at a potential near that at which the copper forms the honeycomb array [8]. This copper and bisulfate coadsorption is in agreement with recent experiments [25].

We provided a tractable model for the copper and bisulfate coadsorption process by treating the bisulfate adsorption as a hard hexagon lattice gas, the coadsorption of copper being treated as effectively causing a change in the reference potential for bisulfate, which we approximated as [8,9]

$$\psi_S^{ref} = \theta_{Cu}\psi_S^{Cu} + (1-\theta_{Cu})\psi_S^{Au} \quad . \tag{19}$$

The reference potentials for bisulfate on gold and on copper were determined from a consideration of the potential, ψ_{crit}, at which the bisulfate fugacity reaches the critical value of the hard hexagon phase transition, $z_{crit} = 11.09$. Setting $\lambda\rho_1^0(0,0) = 1$, at 25°C Eq. (15) and Eq. (16) yield

$$\psi_S^{ref} = \psi_{crit} - 0.06 \text{ V} \ . \tag{20}$$

If no copper is present in the system, we assumed the bisulfate layer becomes disordered at about $\psi_{crit} = 0.45$ V, yielding

$$\psi_S^{Au} = 0.39 \text{ V} \ . \tag{21}$$

This value is near the experimental potential of zero charge of gold (111) in the presence of sulfate [24].

If both copper and bisulfate are present, we assumed the bisulfate becomes ordered again at about the potential of the phase transition of the copper on the honeycomb lattice, $\psi_{crit} = 0.22$ V. Because of the hole-particle symmetry of the copper lattice gas, $\theta_{Cu} = 1/3$ at this transition, and Eqs. (19) - (21) yield [9]

$$\psi_S^{Cu} = -0.30 \text{ V} \ . \tag{22}$$

This value agrees with experiment [26,27].

By approximating the copper deposition for $\psi > 0.22$ V using a Langmuir isotherm, a model voltammogram peak was obtained which contained a foot [8,9] similar to the one seen in experiment [12]. In addition, kinetic effects [9,13] and the effects of diffusion into the interface [13] have also been treated, the inclusion of such effects being sufficient to produce a fitted model voltammogram which agrees quite well with experiment [13].

We plan to use the model to study other examples of underpotential deposition, and we plan to extend the model to treat the effects of lateral diffusion and the properties of incommensurate phases.

Acknowledgements

L. B. was supported by the Office of Naval Research and by the EPSCoR program EHR-91-08775, and D. H. was supported by the Robert A. Welch Foundation grant P-0446 and by the TCU Research Fund.

References

[1] D. A. Huckaby and L. Blum, J. Chem. Phys. **92**, 2646 (1990).
[2] L. Blum and D. A. Huckaby, J. Chem. Phys. **94**, 6887 (1991).
[3] L. Blum and D. A. Huckaby, in Liquids, Freezing, and Glass Transition, edited by J. P. Hansen, D. Levesque, and J. Zinn-Justin, (Elsevier, Amsterdam, 1991), p. 983.
[4] D. A. Huckaby and L. Blum, J. Electroanal. Chem. **315**, 255 (1991).
[5] D. A. Huckaby and L. Blum, in X-Ray Methods in Corrosion and Interfacial Electrochemistry, edited by A. Davenport and J. G. Gordon II, (Electrochemical Society, Pennington, New Jersey, 1992), p. 139.
[6] D. A. Huckaby and L. Blum, J. Chem. Phys. **97**, 5773 (1992).
[7] D. A. Huckaby and L. Blum, in Condensed Matter Theories, vol. 8, edited by L. Blum and F. B. Malik, (Plenum, New York, 1993) p. 637.
[8] L. Blum and D. A. Huckaby, in Microscopic Models of Electrode-Electrolyte Interfaces, edited by J. W. Halley and L. Blum, (Electrochemical Society, Pennington, New Jersey, 1993), p. 232.
[9] L. Blum and D. A. Huckaby, J. Electroanal. Chem., in press.
[10] M. L. Rosinberg, J. L. Lebowitz, and L. Blum, J. Stat. Phys. **44**, 153 (1986).
[11] J. P. Badiali, L. Blum, and M. L. Rosinberg, Chem. Phys. Lett. **129**, 149 (1986).
[12] D. M. Kolb, K. Al Jaaf-Golze, and M. S. Zei, Dechema Monographien, (VCH, Weinheim, 1986), vol. 102, p. 53.
[13] L. Blum, M. Legault, and P. Turq, J. Electroanal. Chem., in press.
[14] J. G. Kirkwood, J. Chem. Phys. **3**, 300 (1935).
[15] R. J. Baxter, J. Phys. A **13**, L61 (1980).
[16] G. S. Joyce, J. Phys. A **21**, L983 (1988).
[17] M. Wertheim, J. Math. Phys. **5**, 643 (1964).
[18] J. A. Barker and D. Henderson, Rev. Mod. Phys. **48**, 587 (1976).
[19] N. F. Carnahan and K. E. Starling, J. Chem. Phys. **51**, 635 (1969).
[20] E. Müller-Hartmann and J. Zittarz, Z. Physik **B27**, 261 (1977).
[21] J. Dóczi-Réger and P. C. Hemmer, Physica **109A**, 541 (1981).
[22] C. Domb, in "Phase Transitions and Critical Phenomena," edited by C. Domb and M. S. Green (Academic, New York, 1974), vol. 3, pp. 1 and 375.
[23] T. Hachiya, H. Honbo, and K. Itaya, J. Electroanal. Chem. **315**, 275 (1991).
[24] G. L. Borges, K. K. Kanazawa, J. G. Gordon II, K. Ashley, and J. Richer, J. Electroanal. Chem. **364**, 281 (1994).
[25] Z. Shi and J. Lipkowski, J. Electroanal. Chem. **365**, 303 (1994).
[26] P. Zelenay, L. M. Rice, and A. Wieckowski, Surf. Sci. **256**, 253 (1991).
[27] L. M. Rice-Jackson, Ph.D. Thesis, Univ. of Illinois, Urbana (1990).

Oxygen Diffusion in 123-YBCO

Marcel Ausloos[1] *and Andrzej Pękalski*[2]

[1] SUPRAS, Institute of Physics B5, University of Liège
B-4000 Liège, Belgium
[2] Institute of Theoretical Physics, University of Wrocław
pl. M. Borna 9, PL-50-205 Wrocław, Poland

Abstract: The problem of oxygen diffusion in 123–YBCO is discussed. We briefly recall the type of experiments, their results and theoretical interpretations. We observe the large spread of data values. The tracer diffusion coefficient interpretation is based on the ASYmmetric Next Nearest Neighbor Ising model and its phase diagram. Inclusion of oxygen-vacancy next nearest neighbor (NNN) hopping in the basal planes of YBCO leads to reconciling experimental data and theory. We also observe that the newly found V–Λ transition in the tracer diffusion coefficient at low temperature can be a source of interesting effects including the mark of superconductivity onset.

1 Introduction

There are several important problems related to oxygen off–stoichiometry in recently (1988) discovered high critical temperature superconductors (HTS) [1]. The main questions are whether oxygen off-stoichiometry has an effect on the superconductivity origin, and next what role the off–stoichiometry plays on the normal state and superconducting properties. Therefore the first goal has been to obtain the phase diagram of the oxygen content induced crystallographic structure(s).

The phase diagram calculations showing various oxygen vacancy ordered phases were based on essentially one model but used different techniques and various values of the parameters. It is believed that the normal state properties are influenced by the ordering phase transitions and also that the superconducting properties are related to certain ordering of the system [2, 3].

Finally, and this is not the least of the interests of studying the oxygen-vacancy distribution related properties, is the use of vacancies as pinning centers for magnetic flux lines [4].

One early discovery was that in 123–YBCO (= $YBa_2Cu_3O_{6+2c}$) the value of the off–stoichiometry parameter c was to be small in order to obtain a high critical temperature T_c. Accordingly, the oxygenation process and the stabilization of the oxygenation level are important conditions for the use of such materials under technological and scientific constraints. Thus, there is much work on oxygen in– and out–diffusion. A review has been recently presented on such topics [5]. Some discussion and much literature can be found in the Ph.D. thesis of Poulsen [6] and LaGraff [7]. In Sect. 2, we will recall experimental investigations on such processes, but rather briefly since much work has been covered in ref. [5].

We shall briefly recall the theoretical models and techniques in Sect. 3. Two interesting Ph. D. theses summarizing the literature, and discussing the models are those of Schleger [8] and Fiig [9].

In– and out– diffusion are concerned with kinetic effects which are due to the non equilibrium nature of the system, i.e. when the oxygen concentration does not correspond to the equilibrium one at the given temperature, or/and when a (rather important) oxygen chemical potential gradient exists. A grand canonical ensemble can (and must) of course be used when the atomic concentration (i.e. a concentration gradient) varies in the sample. Furthermore, in– and out– diffusion are to be seen as surface or bulk processes. The question is raised whether one can distinguish between such processes.

In Sect. 4 we will be concerned with a fixed oxygen vacancy concentration in the system. We will measure the so-called tracer diffusion coefficient D^* (in absence of chemical potential gradient, but allowing for a concentration gradient of course) in contrast to the chemical diffusion coefficient D, which is the one appropriate for the in– and out– diffusion. In Sect. 5 we will present and discuss relevant theoretical results on the tracer diffusion coefficient mainly. We will emphasize findings at low and high temperatures. In so doing we can better (briefly) comment upon the fundamental question whether superconductivity and normal state properties are governed by anomalies in the diffusion coefficient(s).

2 Experimental Investigations of In– and Out– Diffusion

The unit cell of 123-YBCO is shown in Fig. 1. Due to the oxygen off-stoichiometry, vacancies occur at so-called O(5) sites in deficient CuO planes. The O(1) sites on copper chains are occupied in the ground state of $YBa_2Cu_3O_7$, but are vacant in $YBa_2Cu_3O_6$. Apical O(4) oxygens are thought to be relevant for the superconductivity mechanism. The crystallographic structure depends on the temperature and oxygen content (Fig. 2). Three major phases are observed: tetragonal (T), orthorhombic I (OI) and orthorhombic II (OII). The crystallographic structure can thus be modified by the oxygenation process.

Experimental investigations of in– and out– oxygen diffusion have been reviewed by Baikov et al. [5] for 123–YBCO (ca. 112 refs.). Let us recall that Baikov [5] distinguishes between direct and indirect methods. The direct methods are gravimetry, volumetry, electronic and ionic conductivity, calorimetry, X-ray radiography, neutron diffraction, Raman spectroscopy, and nuclear reaction. Indirect methods relate the physical or chemical characteristics to the mobility of oxygen through a theoretical concept: internal friction, structural and

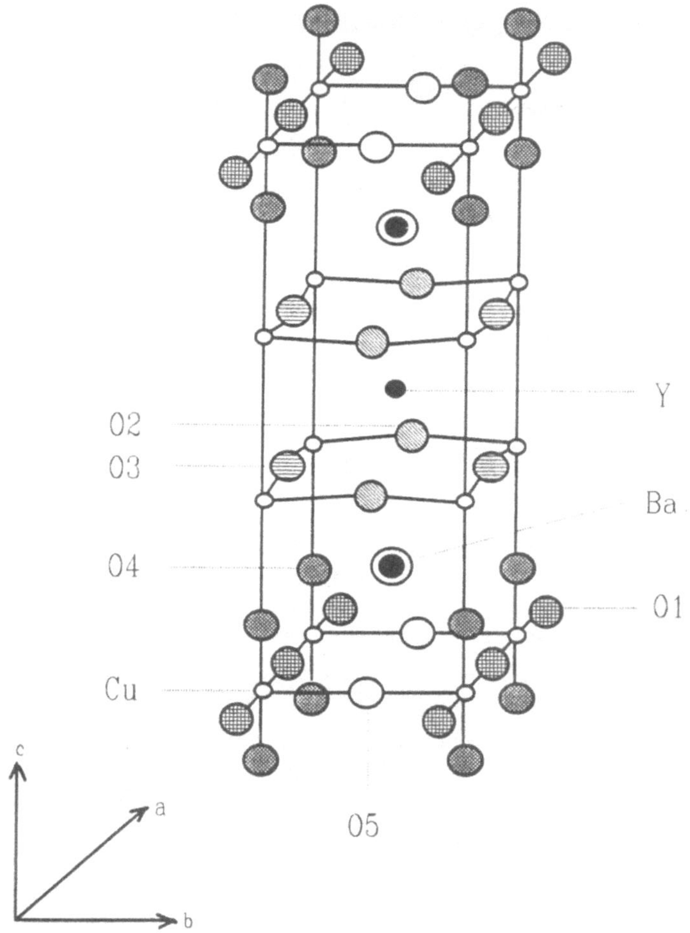

Fig. 1. Crystallographic structure of 123–YBCO.

spectroscopic studies, degradation and aging processes, positron annihilation, Rutherford back-scattering, acoustic wave propagation, NMR, NQR,....

One should distinguish between the techniques which leave the structure unchanged and are thus steady state-like studies, and those which strongly modify the oxygen-vacancy distribution, and thus pertain to structural phase transition studies, in particular $O-T$ and $OI-OII$ transitions. One should also distinguish between single crystals and (porous) polycrystalline materials. It is obvious that due to the presence of grain boundaries the oxygenation process will be easier, but not necessarily more uniform in the latter materials. Such studies have also shown the marked anisotropy of oxygen intake. It can be easily demonstrated (Fig. 3) through e.g. high resolution polarized light microscopy

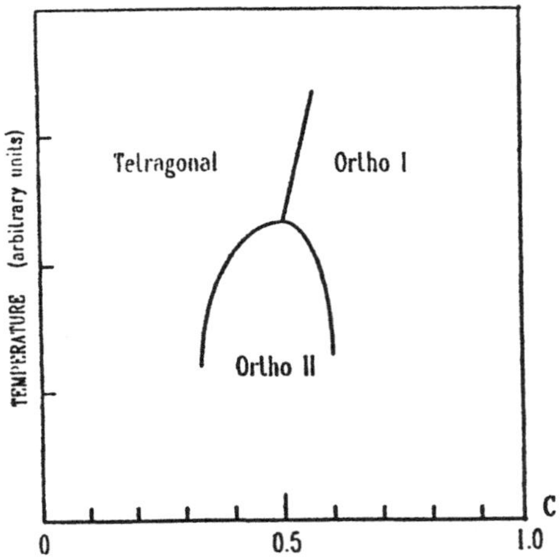

Fig. 2. Schematic phase diagram of 123–YBCO at intermediate temperatures.

that oxygen density is higher around intrinsic or extrinsic crack regions.

Detailed analysis should most of all distinguish between bulk and surface (where some chemical change of oxygen is inevitable) diffusion controlled kinetics: see ref. [5] for a discussion. The relevance of the starting material (tetragonal or orthorhombic or multiphase, twinned or untwinned) should be emphasized. Atomic substitution can also influence the diffusion coefficient.

From the examined experimental data [5], it can be noticed that there is an enormous spread of values for the diffusion coefficient ($\log D$ varies between -4 and -14, in the same temperature range), the activation energy (from tenths of eV up to 1.7 eV), and the Arrhenius pre-exponent factor between 10^{-10} (!) to 170 cm^2/s). This testifies of a great variety of migration paths and jump phenomena depending on the experimental conditions. In our opinion the great difficulty is to distinguish between bulk and surface diffusion controlled phenomena.

3 Models for Diffusion, but First for Ordering

The models for the description of in– and out– diffusion must necessarily take into account the above mentioned experimental and sample conditions. Interpretation of data taken by indirect methods, and resulting models, will not be considered here since they pertain to a different range of interest. Furthermore the surface barrier problem has been well discussed in ref. [5]. Thus we concentrate here on theoretical work concerned by bulk diffusion. In fact, some warning must be made here concerning a too quick identification of the most common used model(s) for bulk diffusion with experimental measurements which often are (we still emphasize the fact) really controlled by surface diffusion mechanisms. We consider that such a distinction is not well made in ref. [5] alas.

Microscopic diffusion models find their roots generally in the theory of random walks on a lattice, be it regular or disordered. The particles (in the system

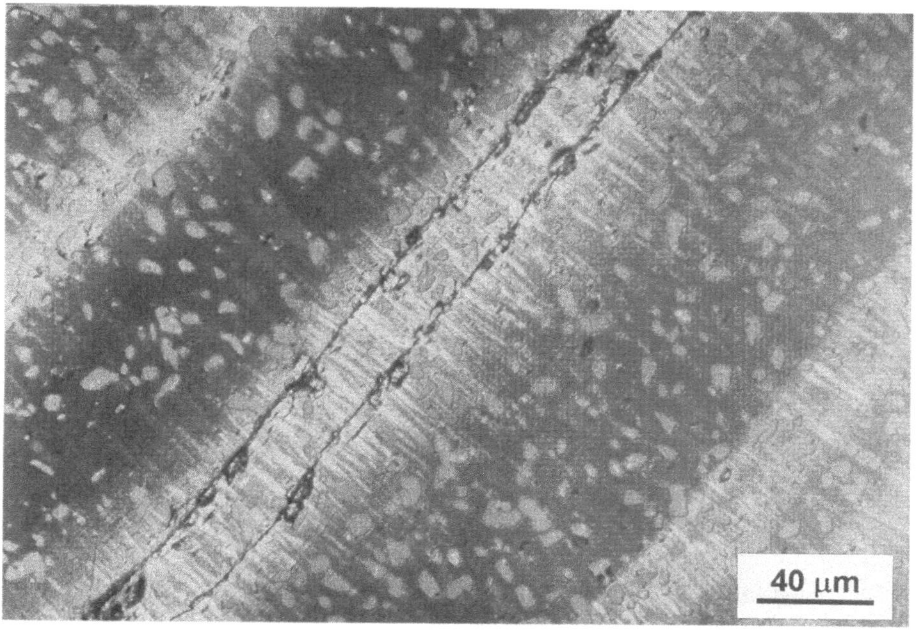

Fig. 3. High resolution polarized light microscopy showing that oxygen density is higher around intrinsic or extrinsic cracks.

discussed here, oxygen ions) interact through some potentials which can be defined on sites or on bonds. A lattice site can be either occupied by an oxygen ion or empty. The most simple model to cover such a situation is the Ising model or the equivalent lattice gas model. The exact values (or even the sign and the range) of the interaction energies, and the number of neighbors to be included, is still a matter of controversy (see [7]).

Most of the studies have dealt with the square 2-dimensional ASYmmetric Next Nearest Neighbor Ising (ASYNNNI) model [10]. Three types of effective pair interactions for the central O atom are defined (see Fig.4 for the site definitions): the interaction with NN at sites (1, 2, 3, 4) is defined by a repulsive $V_1 > 0$, while with NNN (5, 7 - indirect, via the Cu atom bridge) it is through an attractive $V_2 < 0$, and with NNN (6, 8 - direct) by a repulsive $V_3 > 0$. For further reference the x and y axes have (0, 4) and (0, 1) site directions (Fig. 4) respectively. Each lattice site can be vacant or occupied by an O ion. It is found in agreement with experimental data [11] that (Fig. 2) there are essentially three major phases tetragonal (T) at high temperature and near $c = 0$, ortho-I (OI) near $c = 0.5$ which contains copper-oxygen chains parallel to the b axis, and an ortho-II (OII) phase with a double cell (near $c = 0.25$) with every second chain removed. The OII phase is observed in small domains only and samples are usually then defected.

Spatially modulated superstructures can also exist at zero temperature just like for a devils staircase [12]. In fact diffraction patterns for superstructures

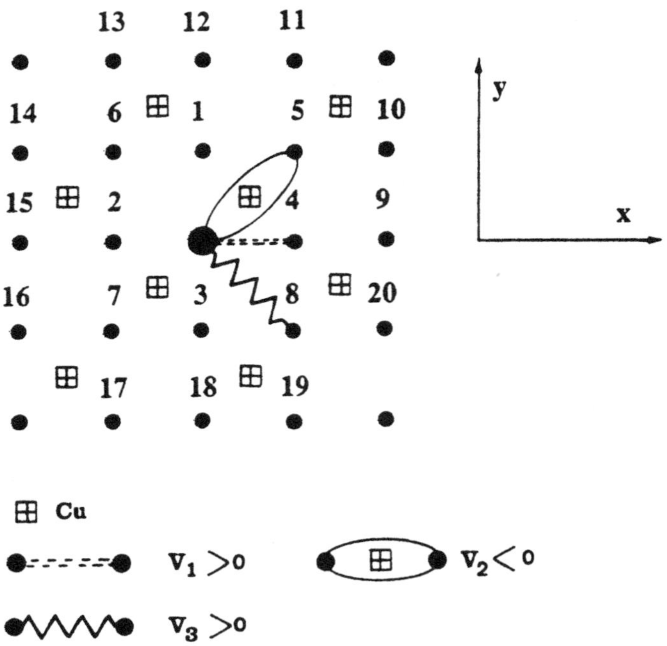

Fig. 4. Sketch of the plaquette for studying oxygen- vacancy interchange in 2-dimensional ASYNNNI model [10]. Three types of effective pair interactions for the central O atom are defined: the interaction with NN at sites (1, 2, 3, 4) is defined by a repulsive $V_1 > 0$, while with NNN (5, 7 - indirect, via the Cu atom bridge) it is through an attractive $V_2 < 0$, and with NNN (6, 8 - direct) by a repulsive $V_3 > 0$. The x and y axes have (0, 4) and (0, 1) site directions respectively.

have been seen at $x = 1/8, 1/2, 3/8, 4/7, 3/5, \ldots$ [13].

Beside such a discussion, one important point is to warn about studies restricted to two dimensions for describing a three dimensional system. Fiig [9] has shown the relevance of the c–axis parallel bond between O ions for determining the predicted phase diagram.

Last but not least, Schleger [8] has pointed out that inconsistencies arise in the lattice gas models as studied in most of the previous references. He points out to more chemically oriented defect models. Schleger [8] claims that for the ordering (mainly the phase diagram) behavior one must include the electronic structure with some detail, since the electronic properties dominate the thermodynamics.

4 Experimental Investigations on Diffusion for Fixed Vacancy Concentration

In this section, we distinguish between the so-called tracer diffusion coefficient D^* and the chemical diffusion coefficient D. Fick's first law reads

$$J = -D^* \left(\frac{dc}{dx}\right)$$

in an isotropic system. When the concentration gradient varies with position one writes rather

$$\frac{dc}{dt} = -D^* \frac{d^2c}{dx^2}.$$

It is clear that due to the anisotropy of the crystallographic structure in YBCO (see Fig.1), a vector formalism should be used and the coefficient D^*, called the tracer diffusion coefficient, or the diffusivity, should be a tensor. Each of its components can be written as

$$D^* = \alpha \cdot f \cdot G \cdot a^2,$$

in terms of a mean jump frequency G, a jump distance a, a geometric factor α, and a correlation factor f which accounts for the fact that ions do not necessarily move through random walk. When $f = 1$ diffusion is a random walk and $f < 1$ indicates that there is a greater chance that the diffusing ion will return to the site it has just left. The diffusion coefficient is often analyzed as a function of temperature T through an Arrhenius law in terms of an activation energy E, i.e.

$$D^* = D_0 \exp(-E/kT).$$

The jump frequency is also often written in a form of an activation law as

$$G = G_0 \exp(-U/kT).$$

If the correlation factor is temperature dependent, E and U are related by [14]

$$E = U - k\left(\frac{d\,(ln f)}{d\,(ln(1/T))}\right)$$

Hence a departure from an Arrhenius law is a sign of correlation effects.

We have mentioned that most investigations of oxygen diffusion in HTS have been carried out in presence of an oxygen chemical potential gradient. Therefore the measurements gave the value of the chemical diffusion coefficient D, related to D^* by

$$D = D^* \left(\frac{1 + d\,(ln\gamma)}{d\,(lnc)]}\right)$$

where γ is the activity coefficient. The term in parentheses is called the thermodynamic factor. The latter can be different from unity in nonideal substances like 123–YBCO.

Clearly the methods of measurements must be distinguished, i.e. one must check whether the pressure of oxygen around the sample varies or not. This goes true for impedance measurements as well. Most of the data scattering recalled here above can probably be traced to this condition.

A type of local probe experiment which avoids the in– and out– diffusion and can be used for fixed oxygen concentration is the SIMS technique [7]. Another, but inherently averaging the processes, is the internal friction based on relaxation technique [7]. Except ref. [15] to our knowledge, all works pertain to temperatures above 300 K. References to other techniques mentioned here can be found in ref. [5].

Major findings [7] to be recorded include:

1. diffusivity is a very weak function of partial oxygen pressure, and of oxygen concentration dependence
2. Arrhenius plots suggest a single type of diffusion mechanism in the range 300-850 C, with an activation energy of the order of 1 eV, irrespective of the O or T phase,
3. diffusivity is highly anisotropic with $D^*_{ab}/D^*_c = 10^{-4}$ and 10^{-2} at 400 and 800 C respectively. The time indicative of relaxation in stress-stain cycles is of the order of 10^{-15} sec considered to be characteristic of O(1) - O(5) jumps. The relaxation time follows an Arrhenius law with an activation energy of the order of 1 eV also.

5 Theoretical Investigations on Diffusion for Fixed Vacancy Concentration

Theoretical work must touch upon the following:

- the anisotropy of oxygen diffusion
- the concentration and temperature dependence of the diffusivity
- the activation energy value
- the effect of correlations
- the influence of defects, intrinsic like twins or extrinsic like substitutional atoms, cracks, et.

The first rather detailed analysis of self-diffusion is by Kishio et al. [16]. Bakker et al. [17] have led others into focussing the attention on diffusion as mainly a CuO plane problem, eventhough they warn that this is only a part of the general problem of oxygen-vacancy mobility in YBCO.

Ronay and Nordlander [18] have drawn attention to the fact that activation-free displacement is possible in $[1/2, b, 0]$ channels of vacancies in the basal plane for $c = 0.5$ and $c = 0$. Other low values of the activation energy can likely be found for specific oxygen concentrations when ordered phases [12] corresponding to Fibonacci or rational values of the concentration exist. In these cases ordered domains and free row patterns exist. They are seen in Monte Carlo simulations

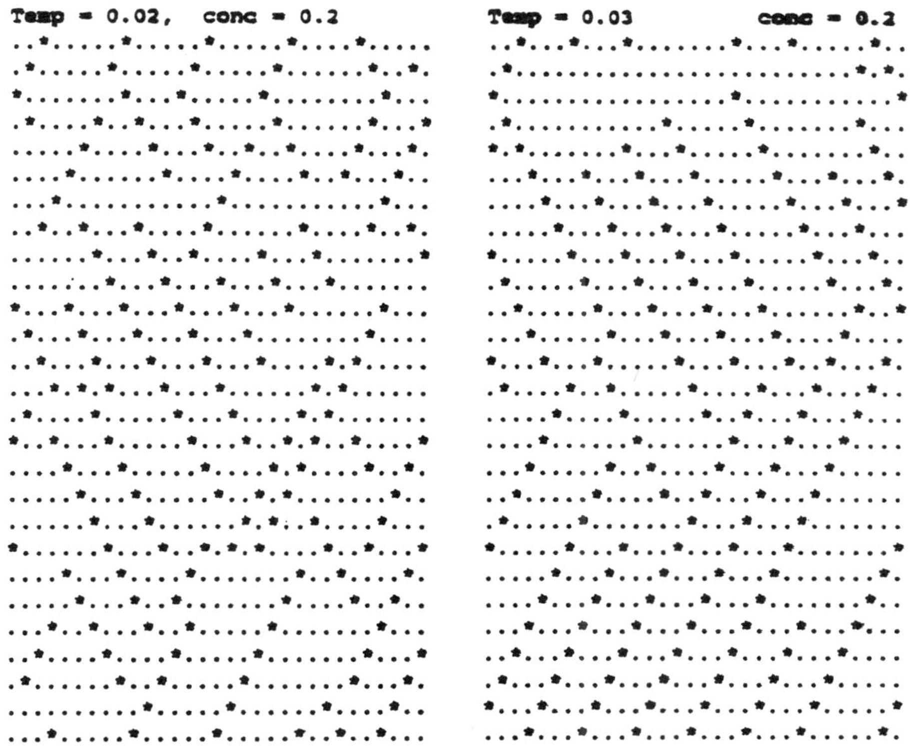

Fig. 5. Ordered chevron domain and free row patterns allowing for the O diffusivity in the *ab* plane of 123–YBCO at concentration $c = 0.2$ and temperatures $T = 0.02$ and 0.03 eV [21].

(Fig. 5). In the same spirit, Islam et al. [19] have used atomistic calculations for the O diffusivity.

Let us also quote the molecular dynamics calculation of Zhang and Catlow [20] at 1400 K.

The ASYNNNI model has been introduced in [10] and used in a Monte Carlo study to calculate the tracer and chemical diffusion of oxygen ions in HTS. It was assumed that the diffusion takes place only in the *ab*-plane - hence in the simulation the oxygen ions move along the edges of the square grid. The potential barrier, Q, surrounding an oxygen ion was put equal to zero. An ion could jump only to one of its nearest neighboring sites, provided it was empty. The Kawasaki dynamics and the standard Metropolis algorithm have been used. The tracer diffusion coefficient shows a strong dependence on the oxygen content and on the temperature. Within the plane the diffusion was anisotropic at low temperatures and the activation energy showed a maximum.

An extension [21] of the work [10] consisted in choosing a different algorithm. In [10] the site to which a given particle may jump was chosen at random. In [21] the site selection is weighted through a probability depending on the energy difference between the initial and the possible final configurations. Also the limitation $Q = 0$ has been lifted. The algorithm turned out to yield the same results for the same data. However when looking at lower temperatures than those examined in [10] an interesting change of the diffusion coefficient dependence on the oxygen concentration has been found. At low temperatures the diffusivity has a minimum at concentration $c = 0.5$ (corresponding to every second chain of oxygen sites being occupied), whereas above a certain temperature the diffusivity has a maximum (Fig. 6). We called the phenomenon the V–Λ transition. The tracer diffusion coefficient was also concentration dependent and the activation energy had a maximum.

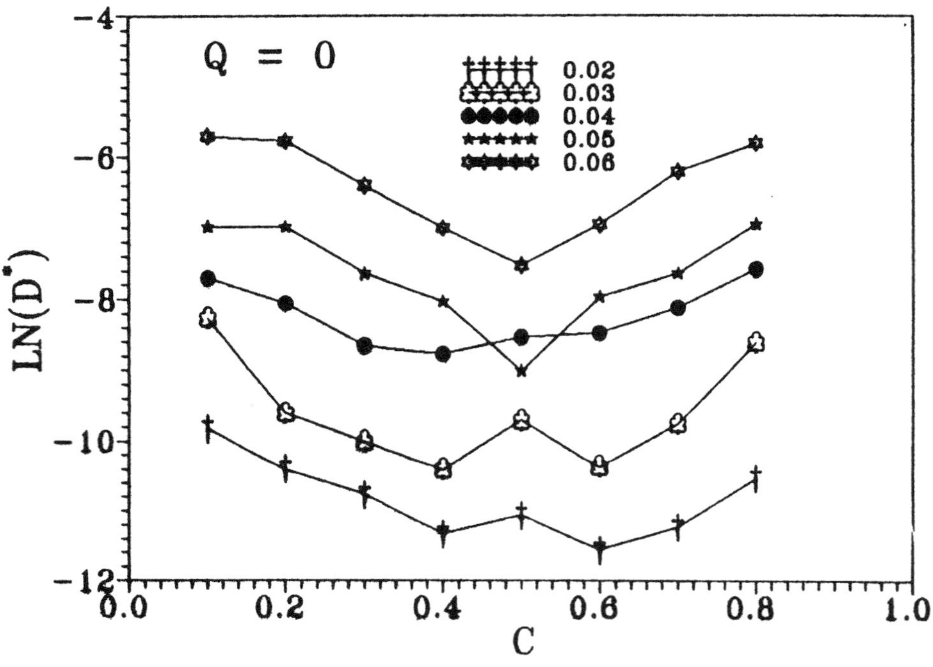

Fig. 6. Appearance of a $V - \Lambda$ transition as a function of oxygen concentration c at low temperature. Symbols correspond to various temperatures in eV.

Another refinement of the Salomons and deFontain paper [10] was presented recently [22], where the oxygen ions were allowed to jump also to their next-nearest neighboring sites, but not across the Cu atoms.

The major findings are:

- the tracer diffusion depends only weakly on the concentration of oxygen ions,
- the anisotropy in the ab plane only occurs for untwinned crystals and in ordered domains: the ratio D^*_+/D^*_- varies from 6.75 to 2.5 at temperatures between 0.02 eV and 0.12 eV,

- the Arrhenius plots give nearly straight lines within the examined temperature range of 0.02 to 0.12 eV,
- the correlation factor depends significantly on c only for low T, i.e. below 0.04 eV,
- the diffusion is taking place along the copper oxygen chains, i.e. O(5)-O(5) or O(1)-O(1), but at low concentrations of oxygen ions the vacancy diffusion, i.e. O(1) to O(5) jumps, is also very important,
- the activation energy is very low due to channels which exist in untwinned systems. It has a maximum at $c = 0.25$ when the basal plane can be the most disordered.

We conclude that the variation in activation energy values which is reported, e.g. in [5] should be correlated with the degree of twins, beside taking into account specific surface features like cracks and chemical inhomogeneities. This implies a deep understanding of the diffusion mechanism as a function of concentration, i.e. a description of the microscopic features like the number of chain fragments which form twins, tweeds and chevron patterns.

One should note that the 2-dimensional NNN Ising model has been studied much before the field of HTS. Indeed the model applies to surface coverage by adatoms [23]. Murch has investigated the NN repulsive and NNN attractive interaction in a lattice gas at 50% coverage [24], while Sadiq and Binder have studied the NN repulsive and NNN repulsive interaction for various temperatures and coverages [25]. In fact a recent series of papers by Uebing and Gomer studies various possibilities among the interactions [26]. See the review article by Gomer for a more complete presentation of various quantities of interest, and related models [27]. See also the recent work of Bulnes et al. [28] on the simulation of tracer diffusion on heterogeneous energy surfaces.

6 Effect of Oxygen Diffusion on Normal State Properties

The above discussion applied to effects in the CuO planes. Normal state properties are much governed by electronic properties in the CuO_2 planes. We can speculate that anomalous diffusion properties in the CuO planes influence the behavior of charge carriers in the CuO_2 planes by Josephson-like or other couplings. Among the normal state properties let us mention that a change in proton mobility occurs in parallel to oxygen diffusion changes [5].

It has been also observed that the superconductivity fluctuations are marked in the so-called excess electrical conductivity up to ca. $2T_c$ [29], i.e. near to the V–Λ transition temperature. Secondly, the Hall number also presents a break in the vicinity of this temperature [30]. In the latter case, the effect markedly depends on the oxygen stoichiometry. The thermoelectric power near room temperature also presents some anomalous behavior [31] which could be related to a change in diffusivity.

The above points do not rule out the role of the apical oxygen in controlling the origin of superconductivity in the HTS. On the contrary, two dimensional and three dimensional effects are likely intermixed in HTS.

Acknowledgements

This work has been partially financed by the Impulse Program on High Temperature Superconductivity of the Belgian Prime Minister Policy Services (SPPS) under contract SU/02/13. AP thanks CEGRI and FNRS (Brussels) for financial support.

References

[1] J. G. Bednorz and K. A. Müller, Z. Phys. **B 64**, 189 (1986)
[2] J. D. Jorgensen et al., Physica **C 153–155**, 578 (1988); J. D. Jorgensen et al., Physica **C 185-189**, 184 (1991)
[3] R. K. Kremer, V. Hizhnyakov, E. Sigmund, A. Simon and K. A. Müller, Z. Phys. **B 91**, 169 (1993)
[4] R. Feenstra, D. K. Christen, C. E. Klabunde and J. D. Budai, Phys. Rev. **B 45**, 7555 (1992)
[5] Yu. M. Baikov, E. K. Shalkova and T. A. Ushakova, Superconductivty: Phys. Chem. Technol. **6**, 349 (1993)
[6] H. F. Poulsen, Ph. D. thesis, Univ. Copenhagen (1991)
[7] J. R. LaGraff, Ph. D. thesis, Univ. Illinois-Urbana (1992)
[8] P. R. Schleger, Ph. D. thesis, Univ. British Columbia (1992)
[9] T. Füg, Ph. D. thesis, Techn. Univ. Denmark (1994)
[10] D. de Fontaine, L. T. Wille and S. C. Moss, Phys. Rev. **B 36**, 5709 (1987)
[11] N. H. Andersen, B. Lebech and H. F. Poulsen, Physica **C 172**, 31 (1990)
[12] D. Adelman, C. P. Burmester, L. T. Wille, P. A. Sterne and R. Grondsky, J. Phys.: Condens. Matter **4**, L585 (1992)
[13] A. A. Aligia and J. Garces, Phys. Rev. **B 49**, 524 (1994); Solid State Comm. **87**, 363 (1993)
[14] J. R. Manning, Phys. Rev. Letters, **1**, 365, (1958)
[15] G. Cannelli, M. Canali, R. Cantelli, F. Cordero, S. Ferraro, M. Ferretti and F. Trequattrini, Phys. Rev. **B 45**, 931 (1992)
[16] K. Kishio, K. Suzuki, T. Hasageawa, T. Yamamoto, K. Kitazawa and K. Fweki, J. Solid State Chem. **82**, 192 (1989)
[17] H. Bakker, J. P. A. Westerfeld, D. M. R. Lo Cascio and D. O. Welch, Physica **C 157**, 25 (1989)
[18] M. Ronay and P. Nordlander, Physica **C 153-155**, 834 (1988)
[19] M. S. Islam and R. C. Baetzold, J. Mater. Chem. **4**, 299 (1994)
[20] X. Zhang and C. R. A. Catlow, Phys. Rev. **B 46**, 457 (1992)
[21] A. Pękalski and M. Ausloos, J. Chem. Phys. **100**, 3175 (1994), Physica **C 226**, 188 (1994)
[22] M. Ausloos and A. Pękalski, unpublished
[23] M. Bowker and D. A. King, Surf. Sci. **71**, 583 (1978)
[24] G. E. Murch, Phil. Mag. **A 43**, 871 (1981)
[25] A. Sadiq and K. Binder, Surf. Sci. **128**, 350 (1983)
[26] C. Uebing and R. Gomer, J. Chem. Phys. **95**, 7626, 7636, 7641, 7648 (1991)
[27] R. Gomer, Rep. Prog. Phys. **53**, 917 (1990)
[28] F. Bulnes, K. Sapag, J. L. Riccardo V. Pereyra and G. Zgrablich, J. Phys.: Condens. Matter **5**, A223 (1993)

[29] M. Ausloos, F. Gillet, Ch. Laurent and P. Clippe, Z. Phys. **B 84**, 13 (1991)
[30] S. Sergeenkov, M. Ausloos and M. Mehbod, J. Phys.: Condens. Matter (in press, 1994)
[31] J. Takeda, T. Nishikawa and M. Sato, J. Phys. Soc. Jpn. **6**, 2571 (1993)

Gating of Voltage-Dependent Sodium Channels in Excitable Membranes - a continuous Process?

Michał Bartoszkiewicz

Department of Biophysics, Medical School, Wrocław, Poland

1 Introduction

1.1 Excitability and voltage-dependent ionic conductance

The most fundamental property of nerve cells (neurons) is their ability to generate and conduct electric signals in response to various stimuli [1].

In order to be "readable" for a neuron, every stimulus must be "translated" into a change in voltage across the membrane of the cell body (Fig. 1). This change is then propagated (towards another cell) along a cell process called axon which may be regarded as a cylinder. In the case of short-distance communication between neurons the physical picture of propagating the signal is not complicated. Namely, the spread of a voltage change applied to the cell body of a short neuron may be satisfactory described by the same linear cable equation which was used by Kelvin to theoretically treat signal transmission along an undersea telegraph cable [2].

However, this so-called passive signalling is not sufficient in the case of long axons. The space constant, that is the distance over which the amplitude of the voltage change falls off by a factor of $1/e$, is for an axon of order 10^{-3} m, whereas the same axon may be more than one meter long! To overcome the problem of this relatively strong signal attentuation, larger neurons have developed an ability to generate and conduct special electric signals called action potentials or nerve impulses. From the physical point of view the action potential is again a local, temporary change of voltage across the axon membrane, but this time it will travel along the membrane with constant amplitude. A cell able to generate and conduct an action potential is called excitable (not only a neuron, but also a muscle cell may be exictable).

Not every stimulus, that is, not every local change of voltage across the excitable membrane produces an action potential. Firstly, the electric potential difference between the interior of the cell body and its surrounding (this voltage is called by electrophysiologists the membrane potential) must increase. Secondly, an action potential will be fired only when the membrane potential of the cell body exceeds some threshold value - if this condition is not met, the excitable cell will respond in the passive way (like a telegraph cable or a passive neuron).

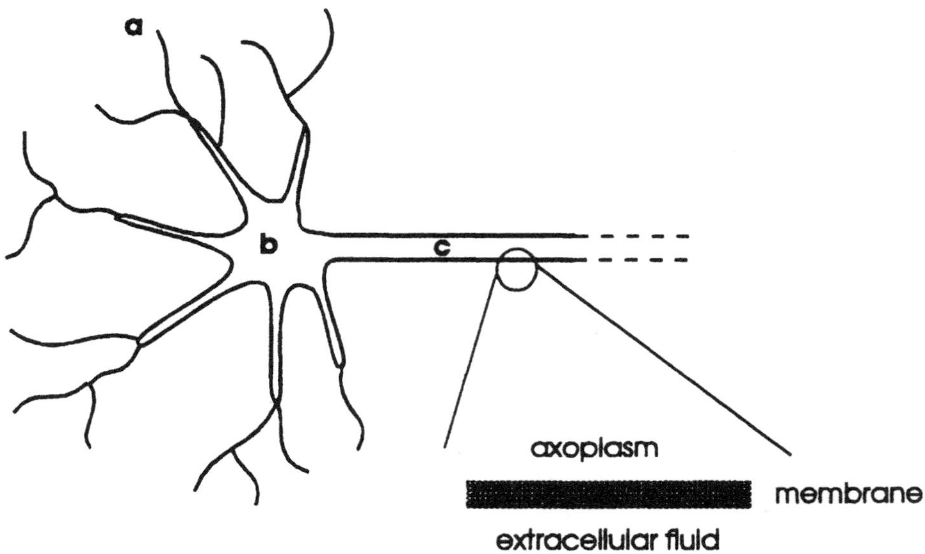

Fig. 1. A neuron. Dendrites (a) receive information from other cells. The cell body (b) is the metabolic center of the cell. The long tubular process (c) whose initial part was drawn is called axon - its length varies from milimeters to meters, depending on the function of the cell. At rest, there exists a high electric potential difference between the interior of the neuron and the extracellular fluid (about -70 mV), called the resting potential of the cell.

If, however, the threshold value of membrane potential is achieved, a burst of electrical activity is observed - there is further, spontaneous rise in membrane potential (depolarization) in the stimulated region (Fig. 2). The extensive work of Cole, Curtis, Hodgkin, Huxley and Katz has shown this triggering of the initial depolarization to be due to a transient increase in membrane conductance for sodium ions [1] (because the electrochemical potential of sodium is higher in the extracellular fluid than inside the axon, increasing the sodium conductance leads to an increase in sodium influx in the axon and thus to a depolarization). The triggered local depolarization (action potential) propagates along the axon by means of local, longitudinal electric currents which flow between the excited and the adjacent unexcited regions. Since the currents are large enough to bring the latter to the threshold, the action potential "jumps" from the excited region to the neighboring resting one - the electric signal may thus travel with constant amplitude [1].

1.2 Gating currents, voltage–dependent sodium channels and gating particles

One may say that the phenomenon of excitability is based on the voltage dependence of sodium conductance of the excitable membrane. What is the molecular

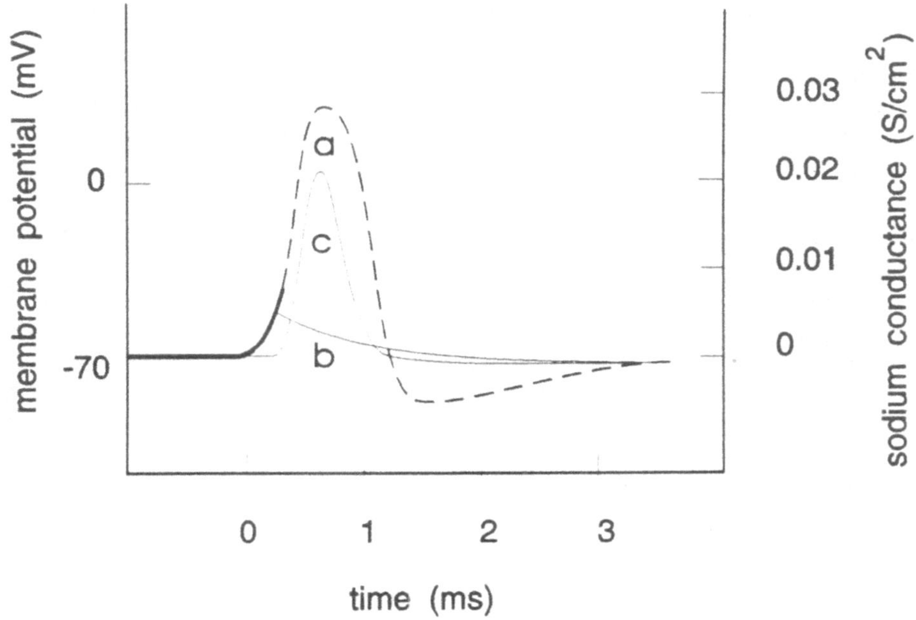

Fig. 2. Action potential (local picture). When the membrane potential is locally brought to the threshold value, there is a further, spontaneous depolarization of the stimulated fragment of the membrane (a). If the initial depolarization does not reach the threshold, the axon will respond like a telegraph cable (b). The action potential is due to a transient increase in sodium conductance (c). The recovery to the resting potential takes place by means of an outward, voltage-dependent current carried by potassium ions.

mechanism of this voltage dependence? Hodgkin and Huxley wrote in 1952 [3]: "Details of the mechanism will probably not be settled for some time, but it seems difficult to escape the conclusion that the changes in ionic permeability depend on the movement of some component of the membrane which behaves as though it had a large charge or dipole moment. If such components exist it is necessary to suppose that their density is relatively low and that a number of sodium ions cross the membrane at a single active patch. Unless this were true one would expect the increase in sodium permeability to be accompanied by an outward current comparable in magnitude to the current carried by sodium ions" and added :"...there is no evidence from our experiments of any current associated with the change in sodium permeability".

The current accompanying the change in sodium conductance that Hodgkin and Huxley were not able to detect is in newer terminology called "gating current". More than twenty years later it has been shown that the current exists [4]. However, the amplitude of the gating current is so small (as related to the sodium current) that the "structural" prediction of Hodgkin and Huxley cited above still had a solide physical basis. What is more, the prediction turned out to be completely correct.

The "single active patch" supposed by Hodgkin and Huxley to be used by a number of sodium ions during excitation appeared to be a small, selective "aqueous pore" formed by a large glycoprotein called sodium channel [1] (because of the high Born energy [5], any hydrophylic ions practically cannot penetrate the lipid "core" of a cell membrane and thus need special transport mechanisms). The channel, like many other ionic channels of biological membranes, may be "closed" (not permeable to ions) or "open" (permeable), as directly proven by single-channel recording [1] (there is a third, non-conducting state called "open inactivated" which will be mentioned below). Since the sodium channel plays a central role during excitation, it is of course expected that its opening probability depends on voltage. This is indeed the case - the probability dramatically increases above threshold (Fig. 3). Note that the sodium influx which leads to a further depolarization during an action potential causes in turn a further increase in sodium conductance (see Fig. 2) - a good example of positive feedback in biological systems.

After establishing the existence of voltage-dependent sodium channels it became clear that the component of the excitable membrane whose spatial position controls the membrane conductance for sodium ions should be a part of the sodium channel. The component, called nowadays "gating particle", "gating charge" or "voltage sensor", is expected to have two properties. Firstly, it should be able to "measure" the local electric field strength within the membrane. Secondly, it should have the property of "switching" the channel protein between the closed and the open state (and vice versa), the probability of the process depending strongly on the result of the electric field strength "measurement". It is very likely that the major components of the gating particle (voltage sensor) are placed in a highly charged part of the sodium channel protein called the S4 segment [7].

2 Gating currents - exact definition, properties and modelling

2.1 Definition and basic properties of gating current

The gating current was introduced above as an electric current due to a non-equilibrium displacement of charged molecules within the membrane (accompanying a change in sodium conductance). Since many currents will flow across the excitable membrane when a voltage step is applied, the experimental determination of the current is not simple. Usually, ionic currents are eliminated by using channel blockers (toxins). The membrane is held under constant membrane potential (holding potential) and then suddenly depolarized to a new command voltage (voltage-clamp conditions). The gating current, $i_g(t)$, is now calculated according to the following definition :

$$i_g(t) = I(t) - I_c(t) - I_l(t)$$

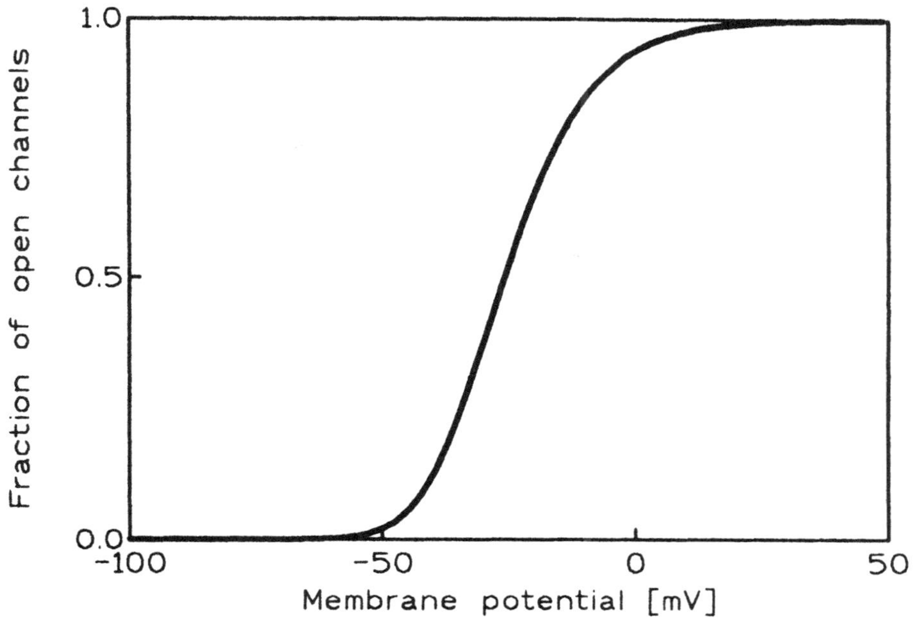

Fig. 3. Membrane potential dependence of fraction of sodium channels open (rat-brain-II sodium channels expressed in *Xenopus* oocytes), calculated from Eq. (8) of Ref. 6. The dependence was experimentally determined by measuring the macroscopic sodium conductance under voltage-clamp conditions (membrane potential held constant after depolarization) . Since the activation of the macroscopic conductance is a kinetic process (in a voltage-clamp its relaxation time is of order 10^{-4}s , the exact value depending on membrane potential), it should be added that the value of fraction of channels open refers to the final "steady-state" (the quotation-marks reflect the fact that even under voltage-clamp conditions the sodium conductance eventually falls to zero due to another kinetic process called "inactivation" of sodium channels which is responsible for the decay of sodium conductance in Fig.2).

where $I(t)$ is the total current response, $I_c(t)$ the capacitance current and $I_l(t)$ the "leakage" current of undetermined basis. Both the (linear) capacitance current and the leakage current must be determined by independent procedures (using membrane potentials at which the gating current is negligible) .

Fig. 4 shows the time course of a typical gating current. The rising phase of the current is most probably artefactual. Its shape depends strongly on the voltage risetime which varies from a few to tens of microseconds.

The gating current is approximately two orders of magnitude smaller than the sodium current (in the case of the squid axon the amplitude of gating current is of order 10 $\mu A/cm^2$). Also, the "relaxation time" of the current (hundreds of microseconds) is smaller than that of the corresponding ionic current. At the beginning of gating current investigations the current was successfully fitted by one exponential. Recent measurements have found the kinetics of gating current to be much more complex [8, 9]. To present the data, the gating current is

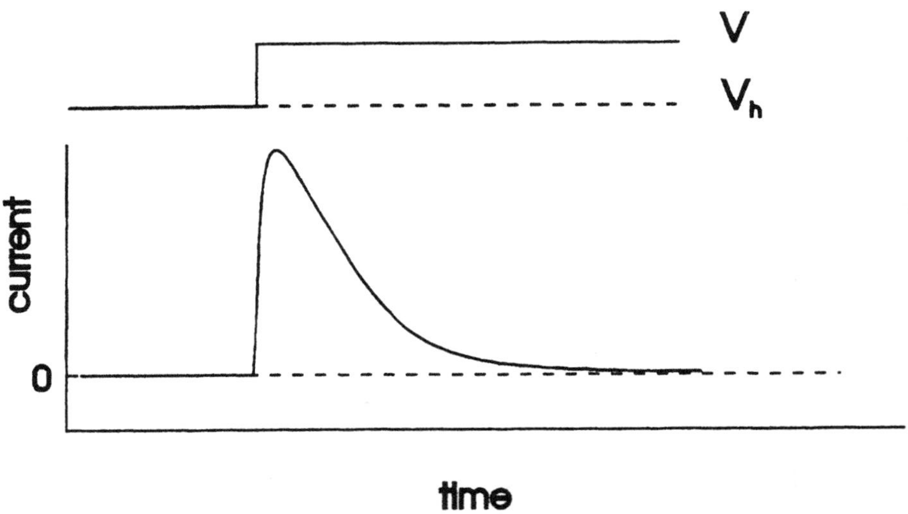

Fig. 4. The shape of a typical gating current measured in response to a depolarization (the ON response). The time constants of the decay s well as the amplitude of the current depend on both membrane potential V and the holding potential V_h. Further explanation in the text.

usually approximated by a sum of decaying exponentials. In the language of this approximation the gating current has at least four components, including a fast decay of about 10 μs [8]. Additionally, for depolarizations above 0 mV the initial spike is followed by a characteristic shoulder which has been proven not to be an artifact [8].

Another important variable characterizing the gating process is the so-called (displaced) gating charge Q defined as the time integral of gating current. Fig. 5 shows its voltage dependence. The Q(V) curve is shifted to the left as compared to the fraction of channels open in Fig. 3 (this immediately excludes simple models in which there is only one closed state of the channel). As expected, the Q-V relationship depends on the holding potential (that is, on the initial distribution of gating particles).

Though the kinetics of a single channel may be observed in single-channel recording [1], this kind of measurements does not provide any information concerning the complex physical events which precede the final transition between the closed and the open state (the transition itself is seen in single-channel recording as an instantaneous rise in electric current). The measurements of macroscopic gating currents are thus still a powerful tool leading to better understanding of the actual physical nature of voltage gating.

2.2 Models of gating

The amino acid sequence of the basic unit of the voltage-dependent sodium channel is known [1]. Also, some functional elements of the protein, like the

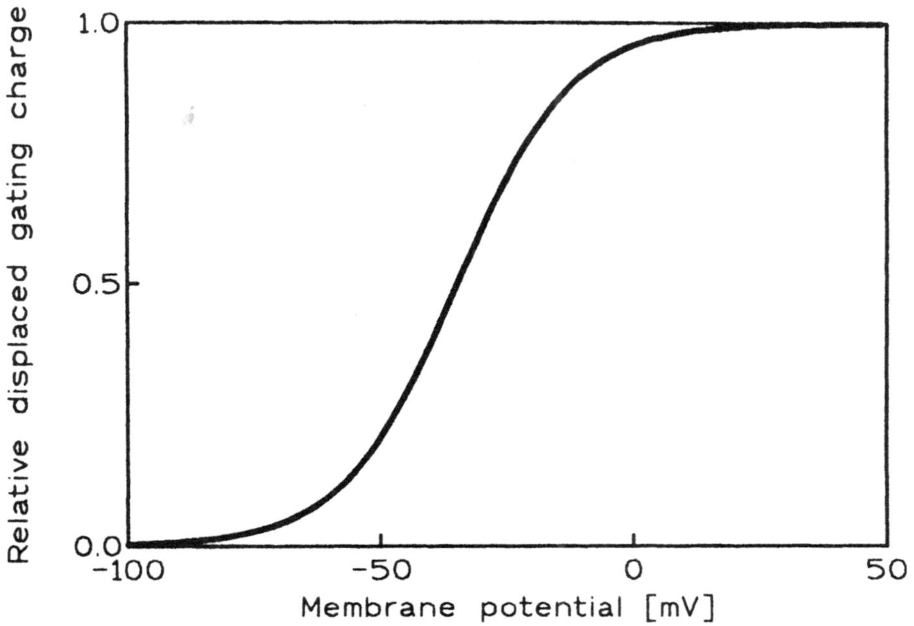

Fig. 5. Voltage dependence of relative gating charge, calculated from Eq. (7) of Ref. 6. As in the case of data in Fig. 3, the measurements were done on rat-brain-II sodium channels expressed in *Xenopus* oocytes. Holding potential -100 mV.

voltage sensor or the permeation path, have most likely been recognized, at least in part (see Ref. 10 for a discussion). However, both the large molecular mass and the complex structure of the channel protein make a detailed physical analysis of the dynamics of the channel impossible. In particular, the molecular mechanism of gating still remains obscure. Mathematical models of gating are therefore forced to start with very general physical ideas.

The models usually assume that the whole gating current is due to the displacement of gating particles. Though there are, of course, other charged components of an excitable membrane (e.g. some amino acid residues), there is some evidence that the major part of the current indeed reflects the movement of those charges which control the state of the sodium channel [11].

Kinetic models of gating The first kinetic model of gating was created by Hodgkin and Huxley [3]. Though their equations originally served to describe purely phenomenologically the kinetics of sodium conductance activation measured at various membrane potentials in a voltage-clamp, then they were ascribed a tempting physical interpretation.

As mentioned above, Hodgkin and Huxley found that the number of gating particles per channel should be low (as compared to the number of ions permeating a sodium channel during a single opening). Let the number be n. Let us next assume that the gating particles are identical and independent, and that they may occupy two states, A (at one side of the membrane) and B (at another side of the membrane), the transitions between them being instantaneous. Let us finally postulate that the probability m of finding a gating particle in state A at time t after a change in membrane potential is governed by the simple kinetic equation:

$$\frac{dm(t,V)}{dt} = \alpha(V)(1-m) + \beta(V)m \tag{1}$$

where α and β are the voltage-dependent (forward and backward) rate constants. Now, if a channel is open only when all respective gating particles occupy the state A, then the fraction of open channels $F(t,V)$ and the gating current $i_g(t,V)$ will read:

$$F(t,V) = m^n(t,V) \tag{2}$$

$$i_g(t,V) \sim \frac{dm(t,V)}{dt} \tag{3}$$

Putting:

$$n = 3 \tag{4}$$

solving Eq. (1) for V = const and introducing the solution into relation (2), Hodgkin and Huxley very nicely fitted the family of time courses of sodium conductance obtained in a voltage-clamp for a broad range of membrane potentials. Since sodium channels are also subject to another kinetic process called inactivation, to calculate the fraction of conducting channels $G(t,V)$ Hodgkin and Huxley additionally assumed that:

$$G(t,V) = F(t,V)\,h(t,V) \tag{5}$$

where h is the probability that a channel is not inactivated (statistical independence assumption). An "inactivating particle" was introduced and a kinetic equation was postulated for the probability $h(t,V)$ identical in shape to that describing the behavior of the gating particle. However, since only the process of activation (closed-open not inactivated transitions) and the reverse one (open not inactived-closed transitions) are of interest here, I am omitting all the complex problems connected with the inactivation of channels. Though activation and inactivation appeared not to be entirely independent, they may be regarded as separate processes. This means in particular that Eqs. (2) and (4) describe quite correctly the whole time course of sodium conductance when inactivation is removed by means of some chemical agents (that is, when transitions to the inactivated state are impossible).

Though the sodium conductance is "spatially discrete", supplementing the classical cable equation by the Hodgkin-Huxley equations for ionic conductances gives a very good description of action potential propagation (all the equations are nonlinear this time!). The physical reason for this convergence is the fact that the space constant of a typical axon is much greater than the distance between channels [12].

Can the HH model describe quite correctly not only the family of sodium currents and the propagation of action potential, but also the gating currents?

As far as the complex kinetics of gating current mentioned above is concerned, the answer is negative. Indeed, when the membrane potential is constant, then a respective solution of Eq. (1) reads:

$$m(t) = a + b \exp(-t/c) \tag{6}$$

where a, b and c are appropriate constants. Inserting now Eq. (6) into relation (3) we see that the gating current predicted by the HH model has only one exponential component. Besides, the formalism fails to properly describe some experiments in which the membrane is repolarized before gating current falls to zero (see Ref. 13 for a discussion).

It is obvious that the HH model is equivalent to the following sequential model:

$$C_1 \underset{\beta}{\overset{3\alpha}{\rightleftarrows}} C_2 \underset{2\beta}{\overset{2\alpha}{\rightleftarrows}} C_3 \underset{3\beta}{\overset{\alpha}{\rightleftarrows}} 0 \tag{7}$$

The HH scheme may thus be regarded as the first sequential kinetic model of gating. Also the later kinetic models of gating are often sequential, but they include a higher number of closed states. Moreover, the gating particles are not considered to be identical [14, 15], that is, no strong constraints like those in Eq. (7) are put on the rate constants. The last assumption has recently been supported by the structure-function investigations. The sodium channel protein includes four internal repeats (homology units), each of them containing six transmembrane helice stretches (segments S1-S6) [1]. As mentioned before, the major components of gating particles are most likely placed in the S4 segment. However, identical mutations in the S4 segment of different homology units do not lead to identical changes in gating current [7].

A higher number of closed states allows to account for a variety of experimental facts like the gating current dependence on voltage prepulses or the complex kinetics of the current. Information obtained from single-channel measurements may be employed when constructing more complex models [15]. The models are flexible enough to take into account new experimentals facts by introducing new transitions. For instance, the fast component of gating current mentioned above may be described within an existing multistate kinetic model of gating when a single transition carrying one elementary charge is added [8].

In kinetic models all transitions are instantaneous, carry various electric charges and depend on voltage. Since the number of states often exceeds 10, the question arises if such a big number of conformational changes is possible in the channel protein during transition to the open state. Of course, when considering a big number of closed states the kinetic models approach a continuum description.

Continuous models of gating and the electrodiffusion model The continuous models of gating assume that there is a slow movement of gating particles [16] (or, a slow conformational change of the channel protein [17]) followed by a fast transition to the open state. The relaxation time of the slow process is assumed to be comparable to that of the macroscopic gating current (hundreds of microseconds). Let us ask the fundamental question if this assumption is justified in view of what is known about the internal dynamics of proteins. It has recently been pointed out that a protein may, within the native state, undergo numerous conformational changes whose relaxation time varies from 10^{-11} to 10^5 s [18], so that the answer seems to be positive. One of the possible molecular realisations of the process is that proposed by Catterall within the so-called sliding helix model [19]. The main idea of the model is that a change in membrane potential causes a charged transmembrane helix (a part of the channel protein) to undergo a screwlike motion. Since electric charges are displaced, their movement gives rise to gating current. As far as I know, the relaxation time of the process has not been determined yet, but from the physical point of view it really might be of order 10^{-3} s.

The electrodiffusion model of Neumcke, Nonner and Stämpfli [16] assumes that gating particles are point charges which may move in a dielectric "slab" of the membrane. The movement is perpendicular to the surface of the membrane. The gating particles are identical and do not interact with each other. The state of a channel (open or closed) depends on the position of the gating particles in such a way that the channel is open only if all respective gating particles are in a specified region of the slab ($x_{\mathrm{th}} \leq x \leq d$, where d is the thickness of the slab). The electrodiffusion model may thus be considered as a continuum generalization of the Hodgkin-Huxley model described above.

The probability density $P(x,t)$ of finding a gating particle between x and $x + dx$ at time t after applying a voltage step is postulated to be governed by the following Smoluchowski equation (constant force approximation):

$$\frac{\partial P(x,t)}{\partial t} = \frac{1}{f}\left(-\frac{U_0 + q\,V_m}{d}\,\frac{\partial P(x,t)}{\partial x} + k\,T\,\frac{\partial^2 P(x,t)}{\partial x^2}\right) \tag{8}$$

where f is the phenomenological friction coefficient, U_0 that part of the gating particle potential energy which does not depend on the external field, q the charge of the gating particle, V_m the membrane potential and $k\,T$ the Boltzmann factor. When $V_m = \text{const}$, Eq. (8) is linear. The boundary conditions reflect the fact that a gating particle cannot leave the slab, whereas the probability densities $P(x,0)$ and $P(x,\infty)$ are assumed to be given by the Boltzmann distributions corresponding to the holding and the target potential, respectively.

The number of channels open, $f_{\mathrm{op}}(t)$, and the gating current, $i_{\mathrm{g}}(t)$, now compute [20]:

$$f_{\mathrm{op}} = \int_{x_{\mathrm{th}}}^{d} P^n(x,t)\,dx \tag{9}$$

$$i_{\mathrm{g}} = \frac{Nq}{d}\,\frac{d}{dt}\int_0^d x\,P(x,t)\,dx \tag{10}$$

where n is the number of gating particles per channel and N the number of channels.

A diffusion type equation may be considered as a continuum limit to a set of equations describing conformational transitions along a generalized coordinate [17, 18]. Here I ascribe to Eq. (8) the original physical meaning, literally treating the gating particle as some charged part of the protein which moves in a potential field.

The extended electrodiffusion model The electrodiffusion model was created to describe the gating of voltage-dependent sodium channels in the Ranvier node of the frog [16]. However, the model is not able to correctly describe, within a broad interval of membrane potentials, the sodium conductance activation data obtained on squid giant axons. Namely, when the membrane potential is changed, the fraction of channels open obtained from Eq.(9) fails to properly follow the respective changes in both the slope of sodium conductance and the lag in the turn-on of the conductance [22].

On the other hand, the electrodiffusion model yields a reasonable description of the "steady-state" data. The question arises whether the basic physical idea of the model is not adequate to describe the kinetics of the investigated process or the sodium conductance activation is more complex.

In its original formulation, the model contains only one variable "responsible" for the kinetics of sodium conductance - the friction coefficient f (whereas the remaining variables describe equilibrium properties of the system, like the F-V or the Q-V relationship). However, there is some evidence that the step between the displacement of the gating particle and the final opening of the respective channel is not instantaneous [21]. To allow for this hypothesis within the model, I have replaced Eq. (9) by:

$$f_{\mathrm{op}} = \int_{x_{\mathrm{th}}}^{d} P(x, t-\tau)\, dx \tag{11}$$

where τ is the mean duration of the step.

Though the step between gating charge displacement and channel opening is considered to be "electrically silent" [21], I assumed that the variable τ may depend on voltage. This assumption is justified if the (hypothetical) electric current due to the step is considerably smaller than the gating current. Such a possibility cannot be excluded [10].

Finally, I allowed also the friction coefficient f to depend on voltage. Since at membrane potential of 10 mV the electric field strength within the membrane is of order 10^6 V/m, such a behavior of the coefficient seems to be not only possible, but also expected.

With the above modifications, the electrodiffusion model was able to quite successfully describe the whole family of macroscopic sodium conductances measured on squid giant axons in response to depolarizations ranging from -40 to 40 mV [22]. Moreover, it predicted many important features of gating current. Fig. 6 presents the time course of the predicted gating current obtained for three different depolarizations. In very good agreement with the measurements of Forster and Greeff mentioned above, there is a spike at the beginning of the curve. Again

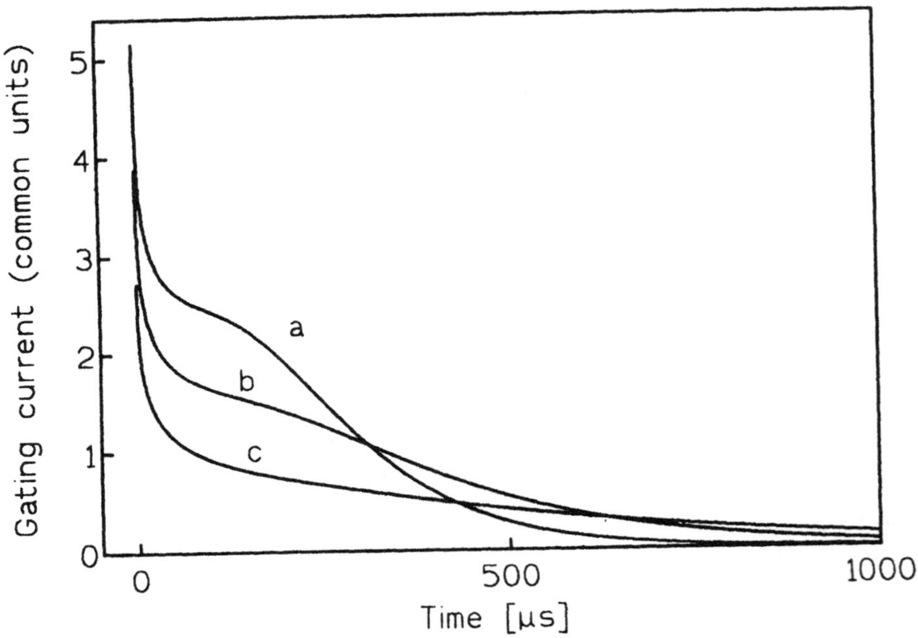

Fig. 6. Gating currents evaluated from Eq.(10) for : 40 mV (a), 20 mV (b) and -10 mV (c). Holding potential -100 mV.

in accordance with the measurements, for depolarizations above 0 mV the spike is followed by a shoulder whose "amplitude" decreases with decreasing depolarization. I therefore dare to say that the time course of gating current measured on squid giant axons is of (electro)diffusional character. The agreement with the experiment is not only qualitative. Though, when presented as a sum of decaying exponentials, the predicted gating current has two fast decaying components, their sum approximates very good the single component found experimentally [22] (because of the shoulder, the predicted gating current may be approximated in this way only for low membrane potentials). The extended model also predicts quite reasonably the Q-V relationship.

References

[1] B. Hille: *Ionic channels of excitable membranes* Sinauer Associates, Sunderland, Mass. (1992)

[2] H. C. Tuckwell: *Introduction to theoretical neurobiology* Cambridge University Press (1988).

[3] A. L. Hodgkin and A. F. Huxley: J. Physiol. **117**, 500 (1952).

[4] C. M. Armstrong, F. Bezanilla: Nature **242**, 459 (1973).

[5] J. O. M. Bockris, A. K. N. Reddy: *Modern electrochemistry.* Macdonald, London (1970).

[6] F. Conti, W. Stühmer: Eur. Biophys. J. **17**, 53 (1989).

[7] W. Stühmer, F. Conti, H. Suzuki, X. Wang, M. Noda, N. Yahagi, H. Kubo, S. Numa: Nature **339**, 597 (1989).
[8] I. C. Forster, N. G. Greeff: Eur. Biophys. J. **21**, 99 (1992).
[9] R. D. Keynes, N. G. Greeff, I. C. Forster: Proc. R. Soc. **B 240**, 411 (1990).
[10] F. J. Sigworth: Quart. Rev. Biophys. **27**, 1 (1994).
[11] C. M. Armstrong: Physiol. Rev. **61**, 644 (1981).
[12] J. W. Mozrzymas, M. Bartoszkiewicz: J. Theor. Biol. **162**, 371 (1993).
[13] F. Bezanilla: J. Membrane Biol. **88**, 97 (1985).
[14] J. Patlak: Physiol. Rev. **71**, 1047 (1991).
[15] C. A. Vandenberg, F. Bezanilla: Biophys. J. **60**, 1511 (1991).
[16] B. Neumcke, W. Nonner, R. Stämpfli: Int. Rev. Biochem. **19**, 129 (1978).
[17] D. G. Levitt: Biophys. J. **55**, 489 (1989).
[18] M. Kurzyński: FEBS Lett. **328**, 221 (1993).
[19] W. A. Catterall: Trends Neurosci. **9**, 7 (1986).
[20] B. Neumcke: Thesis (1978).
[21] F. Conti, I. Inoue, F. Kukita, W. Stühmer: Eur. Biophys. J. **11**, 137 (1984).
[22] M. Bartoszkiewicz: (1994) Paper in preparation for publication.

The Schrödinger Problem

Piotr Garbaczewski and Robert Olkiewicz

Institute of Theoretical Physics, University of Wrocław,
PL-50204 Wrocław,Poland

1 The Schrödinger problem: microscopic dynamics from the input-output statistics

We invert the well developed strategy of studying dynamics in terms of probability densities and investigate the problem of the most likely microscopic propagation scenario, which is consistent with the given *a priori* (possibly phenomenological) input-output statistics data for the process taking place in a finite time interval. A subclass of solutions includes the familiar Smoluchowski diffusions.

It is clear that a stochastic process is any conceivable evolution which we can analyze in terms of probability. We shall be particularly interested in situations when the involved probability measures give rise to densities (probability distributions which are absolutely continuous with respect to the Lebesgue measure). In many branches of physics ranging from deterministic (the folk lore phrase "studying chaos with densities" pertains to the currently fashionable topic) to quantum problems, densities of probability measures do naturally arise. The quantum issue should receive a particular attention in connection with the Born statistical postulate, which implies that quantum theory deals with probability densities. However, quite generally the stochastic analysis is disregarded against the pragmatic viewpoint of deducing as many experimentally verifiable or rather falsifiable data as possible, even at the price of manipulating with the ill defined or not defined at all (safe bypassing of rather fundamental difficulties) probabilistic framework.

The main idea behind what we call the *Schrödinger problem* is an attempt to get an insight (in fact through modeling) into an unknown in detail physical process with a finite time of duration, in terms of random motions consistent with the prescribed input - output statistics i.e. the boundary distributions for repeatable single particle (sample) procedures. In less specific lore, we can simply look for a stochastic evolution (diffusion of probabilities) which interpolates between the boundary probability measures, in particular for the (invariant) measure preserving dynamics.

Given a dynamical law of motion (for a particle as example), in many cases one can associate with it (compute or approximate the observed frequency data) a probability distribution and various mean values. In fact, it is well known that inequivalent finite difference random motion problems may give rise to the same continuous approximant (e.g. the diffusion equation representation of discrete processes). The *inverse operation* of deducing the detailed (possibly individual, microscopic) dynamics, which either implies or is consistent with the given probability distribution (and eventually with its own time evolution) cannot thus have a unique solution.

For clarity of discussion, we shall confine our attention to continuous Markov processes, whose random variable $X(t), t \geq 0$ takes values on the real line R^1, and in particular can be restricted (constrained) to remain within the interval $\Lambda \subset R^1$, which may be finite or (semi-) infinite but basically an open set.

In the above input-output statistics context, let us invoke a probabilistic problem, originally due to Schrödinger : *given two strictly positive (on an open interval) boundary probability distributions $\rho_0(x), \rho_T(x)$ for a process with the time of duration $T \geq 0$. Can we uniquely identify the stochastic process interpolating between them?*

Perhaps unexpectedly in the light of our previous comments, the answer is known to be affirmative, if we assume the interpolating process to be Markovian. In fact, we get here a unique Markovian diffusion, which is specified by the joint probability distribution

$$m(A,B) = \int_A dx \int_B dy\, m(x,y) \tag{1}$$

$$\int dy\, m(x,y) = \rho_0(x)$$

$$\int dx\, m(x,y) = \rho_T(y)$$

where

$$m(x,y) = \Theta_*(x,0)\, k(x,0,y,T)\, \Theta(y,T) \tag{2}$$

and the two unknown (not necessarily Lebesgue integrable) functions $\Theta_*(x,0)$, $\Theta(y,T)$ come out as solutions of *the same sign* of the integral identities (1). Provided, we have at our disposal a bounded strictly positive integral kernel $k(x,s,y,t)$,
$0 \leq s < t \leq T$. Then:

$$\Theta_*(x,t) = \int k(0,y,x,t)\Theta_*(y,0)dy \tag{3}$$

$$\Theta(x,s) = \int k(s,x,y,T)\Theta(y,T)dy$$

and the sought for interpolation has a probability distribution $\rho(x,t) = (\Theta_*\Theta)$ $(x,t), t \in [0,T]$.

2 Markov diffusions solving the Schrödinger problem: the role of natural boundaries

To have a definite Markov solution in hands, we must decide what is the most appropriate choice for the dynamical semigroup kernel in the above. Apparently it is the crucial step in the construction of any explicit random propagation consistent with the boundary measure data. The reader should be warned that the whole family of Levy processes and their perturbations (in the sense of Kato) is here allowed. The conventional Brownian diffusion and the equally conventional Poisson jump process are rather specialized examples in this context.

We wish to discuss diffusive solutions only, and take for granted that the traditional Fokker-Planck equation sets rules of the game for the interpolating probability density. Then we look for the corresponding fundamental law of random displacements and choose the transition probability density for the Markovian diffusion process in the form (the so called h-transformation, invented long ago by Doob and Hille is here involved)

$$p(y,s,x,t) = k(y,s,x,t)\,\frac{\Theta(x,t)}{\Theta(y,s)} \tag{4}$$

with $s \leq t$. This transition density is required to come out from the forward Kolmogorov equation (e.g. the Fokker-Planck equation) as its fundamental solution ($p \rightarrow \delta(x-y)$ as $t \downarrow s$). For convenience we simplify the whole problem by utilising a diffusion constant $D > 0$ (this choice narrows slightly the allowed framework):

$$\partial_t p = D\triangle_x p - \nabla_x(bp) \tag{5}$$

$$\rho(x,t) = \int p(y,s,x,t)\,\rho(y,s)\,dy$$

with $\rho_0(x) = \rho(x,0)$ and the drift $b(x,t)$ given by :

$$b(x,t) = 2D\,\frac{\nabla\Theta}{\Theta}(x,t) \tag{6}$$

In addition we demand that the backward diffusion equation is solved by the same transition density (with respect to another pair of variables)

$$\partial_s p = -D\triangle_y p - b\nabla_y p \tag{7}$$

$$p = p(y,s,x,t)\,,\; s \leq t\,,\; b = b(y,s)$$

It implies that we deal here with a *unique* diffusion process, whose transition density is a *common* fundamental solution for both the backward and forward Kolmogorov equations.

To understand the rôle of the integral kernel $k(y,s,x,t)$ in (1)-(7) let us assume that $\Theta(x,t)$ is given in the form (drifts are gradient fields as a consequence):

$$\Theta(x,t) = \pm\exp\Phi(x,t) \Rightarrow b(x,t) = 2D\nabla\Phi(x,t) \tag{9}$$

$$x \in (r_1, r_2)$$

and insert (4) to the Fokker-Planck equation (5). Then, if $p(y,s,x,t)$ is to solve (5), the kernel $k(y,s,x,t)$ must be a fundamental solution of the generalised diffusion equation:

$$\partial_t k = D\triangle_x k - \frac{1}{2D}\Omega(x,t)\, k \tag{10}$$

$$k(y,s,x,t) \to \delta(x-y) \, \text{as}\, t \downarrow s$$

$$\Omega(x,t) = 2D[\partial_t \Phi + \frac{1}{2}(\frac{b^2}{2D} + \nabla b)]$$

and to guarrantee (3), it must display the semigroup composition properties.

Notice that (4), (9) imply that the backward diffusion equation (7) takes the form of the adjoint to (10):

$$\partial_s k = -D\triangle_y k + \frac{1}{2D}\Omega(y,s)\, k \tag{11}$$

$$k = k(y,s,x,t)$$

If the process takes place in-between boundaries at infinity $r_1 = -\infty, r_2 = +\infty$, the standard restrictions on the auxiliary potential Ω (Rellich class) and hence on the drift potential $\Phi(x,t)$, yield the familiar Feynman-Kac representation of the fundamental solution $k(y,s,x,t)$ *common* for (10) and (11):

$$k(y,s,x,t) = \int \exp[-\frac{1}{2D}\int_s^t \Omega(X(u),u)du]\, d\mu[s,y \mid t,x] \tag{12}$$

which integrates $\exp[-(1/2D)\int_s^t \Omega(X(u),u)du]$ weighting factors with respect to the conditional Wiener measure i.e. along all sample paths of the Wiener process which connect y with x in time $t-s$. More elaborate discussion is necessary, if at least one of the boundary points is *not* at infinity.

Let us notice that the time independence of Ω is granted if either Φ is independent of time, or depends on time at most linearly. Then the standard expression $\exp[-H(t-s)](y,x)$ for the kernel k clearly reveals the involved semigroup properties, with $H = -D\triangle + (1/2D)\Omega(x)$ being the essentially self adjoint operator on its (Hilbert space) domain.

We shall make one more step narrowing the scope of our discussion by admitting diffusions (1)-(7) whose drift fields are time-independent:$\partial_t b(x,t) = 0$ for all x. We know that both the free Brownian motion and the Brownian motion in a field of force in the Smoluchowski approximation, belong to this class of processes. We know also that the boundary value problems for the Smoluchowski equation have a profound physical significance, albeit the attention paid to various cases is definitely unbalanced in the literature. It is then interesting to observe that the situation we encounter in connection with (1)-(7) is very specific from the point of view of Feller's classification of one-dimensional diffusions encompassing effects of the boundary data. Our case is precisely the Feller diffusion respecting (confined between) the *natural boundaries*. An equivalent statement is that boundary points r_1, r_2 are *inaccessible* barriers for the process i.e. there is *no* positive probability that any of them can be reached from the interior of (r_1, r_2) within a *finite* time for all $X(0) = x \in (r_1, r_2)$.

In the mathematical literature a clear distinction is made between the backward and forward Kolmogorov equations. The backward one defines the *transition density* of the process, while the forward (Fokker-Planck) one determines the *probability distribution* (density) of diffusion. With a given backward equation one can usually associate the whole family of forward (Fokker-Planck) equations, whose explicit form reflects the particular choice of boundary data. This fundamental distinction seemingly evaporates in our previous discussion (1)-(11), but it is by no means incidental. In fact, according to Feller: in order that there exists one and only one (homogeneous ; $p(y, s, x, t) = p(t - s; y, x)$) process satisfying $-\partial_t u = D\Delta u + b\nabla u$ in a finite or infinite interval $r_1 < x < r_2$ it is necessary and sufficient that both boundaries are inaccessible (the probability to reach either of them within a finite time interval must be zero). A general feature of the inaccessible boundary problems is that the density of diffusion vanishes at the boundaries:$\rho(r_1) = 0 = \rho(r_2)$.

The standard (unrestricted) Brownian motion on R^1 is the most obvious example of diffusion with natural boundaries. It is not quite trival to construct explicit examples, if one of the boundaries is not at infinity. The classic example of diffusion on the half-line with natural boundaries at 0 and $+\infty$ is provided by the so called Bessel process. As mentioned before, diffusions with inaccessible barriers might have drifts which are unbounded on (r_1, r_2). Hence, our discussion definitely falls into the framework of diffusion processes with singular drift fields, which is *not* covered by standard monographs on stochastic processes.

We skip the standard details concerning the probability space, filtration and the process adapted to this filtration and notice that a continuous random process $X(t), t \in [0, T]$ with a probability measure P is called a process of the diffusion type if its drift $b(x)$ obeys:

$$P[\int_0^T \mid b(X(t)) \mid dt < \infty] = 1 \tag{13}$$

and, given the standard Wiener process (Brownian motion) $W(t)$, the integral identity (D constant and positive)

$$X(t) = \int_0^t b(X(s))\, ds + \sqrt{2D}W(t) \tag{14}$$

holds true P-almost surely (except possibly on sets of P-measure zero). It means that $W(t) = (1/\sqrt{2D}[X(t) - \int_0^t b(X(s))ds]$ is a standard Wiener process with respect to the probability measure P of the process $X(t)$.

For diffusions with natural boundaries, we remain within the regularity interval of $b(X(t))$ for all (finite) times, and (13) apparently is valid. Therefore, the standard rules of the stochastic Itô calculus can be adopted to relate the Fokker-Planck equation (7) with the natural boundaries to the diffusion process $X(t)$, which "admits the stochastic differential"

$$dX(t) = b\big(X(t)\big)dt + \sqrt{2D}dW(t) \tag{15}$$

$$X(0) = x_0 \, , \, t \in [0,T]$$

for all (finite) times. The weak (in view of assigning the density $\rho_0(x)$ to the random variable $X(0)$) solution of (15) is thus well defined.

For stochastic differential equations of the form (15), the explicit Wiener noise input, because of (9) implies that irrespective of whether natural boundaries are at infinity or not, the Cameron-Martin-Girsanov method of measure substitutions which parallel transformations of drifts, is applicable. Even though the drifts are generally unbounded on (r_1, r_2) and the original theory is essentially based on the boundedness demand. It is basically due to the fact that the probabilistic Cameron-Martin formula relating the probability measure P_X of $X(t)$ with the Wiener measure P_W (strictly speaking it is the Radon-Nikodym derivative of one measure with respect to another) reduces to the familiar Feynman-Kac formula (with the multiplicative normalisation). The problem of the existence of the Radon-Nikodym derivative (and this of the absolute continuity of P_X with respect to P_W, which implies that sets of P_W-measure zero are of P_X-measure zero as well) is then replaced by the standard functional analytic problem of representing the semigroup operator kernel via the Feynman-Kac integral with respect to the conditional Wiener measure.

The Feynman-Kac formula is casually viewed to encompass the unrestricted (the whole of R^n) motions, however it is known to be *localizable*, and its validity extends also to finite and semi-infinite subsets of R^1 (R^nmore generally) as demonstrated in the context of the statistical mechanics of continuous quantum systems. More specifically, it refers to the Dirichlet boundary conditions for self-adjoint Hamiltonians, which ensure their essential self-adjointness (to yield the Trotter formula).

Let us emphasize the importance of (15), and of the Itô differential formula induced by (15) for smooth functions of the random variable $X(t)$. Its *first* consequence is that given $p(y,s,x,t)$, for any smooth function of the random variable the forward time derivative in the conditonal mean can be introduced (we bypass in this way the inherent non-differentiability of sample paths of the process)

$$lim_{\triangle t\downarrow 0}\frac{1}{\triangle t}\Big[\int p(x,t,y,t+\triangle t)f(y,t+\triangle t)dy - f(x,t)\Big] = (D_+ f(X(t),t) \quad (16)$$

$$= (\partial_t + b\nabla + D\triangle)f(X(t),t)$$

$$(D_+(X)(t) = b(x,t)\ X(t) = x$$

so that the second forward derivative associates with our diffusion the local field of accelerations:

$$(D_+^2 X)(t) = (D_+ b)(X(t),t) = (\partial_t b + b\nabla b + D\triangle b)(X(t),t) = \nabla\Omega(X(t),t) \quad (17)$$

with the (auxiliary potential $\Omega(x,t)$ introduced before in the formula (10). Since we have given $\rho(x,t)$ for all $t \in [0,T]$, the notion of the backward transition density $p_*(y,s,x,t)$ can be introduced as well

$$\rho(x,t)p_*(y,s,x,t) = p(y,s,x,t)\rho(y,s) \quad (18)$$

which allows to define the backward derivative of the process in the conditional mean (it is quite illuminating to appply this discussion in case of the most traditional Brownian motion)

$$lim_{\triangle t\downarrow 0}\frac{1}{\triangle t}[x-\int p_*(y,t-\triangle t,x,t)ydy]=(D_-X)(t)=b_*(X(t),t)$$

$$=[b-2D\nabla ln\rho](X(t),t) \tag{19}$$

$$(D_-f)(X(t),t)=(\partial_t+b_*\nabla-D\triangle)f(X(t),t)$$

Apparently, the validity of (17) extends to $(D_-^2X)(t)$ as well, and there holds

$$(D_+^2X)(t)=(D_-^2X)(t)=\partial_t v+v\nabla v+\nabla Q=\nabla\Omega \tag{20}$$

$$v(x,t)=\frac{1}{2}(b+b_*)(x,t)\,,\;u(x,t)=\frac{1}{2}(b-b_*)(x,t)=D\nabla ln\rho(x,t)$$

$$Q(x,t)=2D^2\,\frac{\triangle\rho^{1/2}}{\rho^{1/2}}$$

Clearly, if b and ρ are time-independent, then (20) reduces to the identity

$$v\nabla v=\nabla(\Omega-Q) \tag{21}$$

while in case of constant (or vanishing) current velocity v, the acceleration formula (21) reduces to

$$0=\nabla(\Omega-Q) \tag{22}$$

which establishes a very restrictive relationship between the auxiliary potential $\Omega(x)$ (and hence the drift $b(x)$) and the probability distribution $\rho(x)$ of the stationary diffusion. The pertinent random motions have their place in the mathematically oriented literature.

Let us notice that (20) allows to transform the Fokker-Planck equation (7) into the familiar continuity equation, so that the diffusion process $X(t)$ admits a recasting in terms of the manifestly *hydrodynamical* local conservation laws (we adopt here the kinetic theory lore)

$$\partial_t\rho=-\nabla(\rho v) \tag{23}$$

$$\partial_t v+v\nabla v=\nabla(\Omega-Q)$$

$$\rho_0(x)=\rho(x,0)\,,\;v_0(x)=v(x,0)$$

which form a closed (in fact, Cauchy) nonlinearly coupled system of differential equations, strictly equivalent to the previous (7), (17).

In view of the natural boundaries (where the density $\rho(x,t)$ vanishes), the diffusion respects a specific ("Euclidean looking") version of the Ehrenfest theorem:

$$E[\nabla Q]=0\Rightarrow \tag{24}$$

$$\frac{d^2}{dt^2}E[X(t)]=\frac{d}{dt}E[v(X(t),t)]=E[(\partial_t v+v\nabla v)(X(t,t)]=E[\nabla\Omega(X(t),t)]$$

Notice that the auxiliary potential of the form $\Omega = 2Q - V$ where V is any Rellich class representative, defines drifts of Nelson's diffusions for which $E[\nabla Q] = 0 \Rightarrow E[\nabla \Omega] = -E[\nabla V]$ and the "standard looking" form of the second Newton law in the mean arises.

At this point it seems instructive to comment on the essentially hydrodynamical features (compressible fluid/gas case)of the problem (23), where the "pressure" term ∇Q might look annoying from the traditional kinetic theory perspective. Although (23) has a conspicuous Euler form, one should notice that if the starting point of our discussion would be a typical Smoluchowski diffusion (7), (15) whose drift is given by the Stokes formula (i.e. is proportional to the external force $F = -\nabla V$ acting on diffusing molecules), then its external force factor is precisely the one retained from the original Kramers phase-space formulation of the high friction affected random motion. In the Euler description of fluids and gases, the very same force which is present in the Kramers (or Boltzmann in the traditional discussion) equation, should reappear on the right-hand-side of the local conservation law (momentum balance formula) (23). Except for the harmonic oscillator example, in view of (10) it is generally *not* the case in application to diffusion processes. As it appears, the validity of the stochastic differential representation (15) of the diffusion (5) implies the validity of the hydrodynamical representation (23) of the process. It in turn gives a distinguished status to the auxiliary potential $\Omega(x,t)$ of (10)-(12). We encounter here a fundamental problem of what is to be interpreted by a physicist (observer) as the *external force field manifestation* in the diffusion process. Should it be dictated by the drift form following Smoluchowski and Kramers, or rather by $\nabla \Omega$ entering the evident (albeit "Euclidean looking") second Newton law, respected by the diffusion ?

In the standard derivations of the Smoluchowski equation, the deterministic part (force and friction terms) of the Langevin equation is postulated. What however, if the experimental data pertain to the local conservation laws like (23), and there is no direct (experimental) access to the microscopic dynamics ?

If the field of accelerations $\nabla \Omega$ is taken as the primary defining characteristics of diffusion we deal with, then we face the problem of deducing all drifts, and hence diffusions, which give rise to the same acceleration field, and thus form a class of *dynamically equivalent diffusions.*

Let us analyze the *second consequence* of the unattainability of the boundaries, which via (13) gives rise to (15). On the same footing as in case of (13), we have satisfied another probabilistic identity:

$$P[\int_0^T b^2(X(t))dt < \infty] = 1 \tag{25}$$

For a diffusion $X(t)$ with the differential (15), we know that (25) is a sufficient and necessary condition for the absolute continuity of the measure $P = P_X$ with respect to the Wiener measure P_W. Since, for any (Borel) set A, $P_W(A) = 0$ implies $P_X(A) = 0$, the Radon-Nikodym theorem applies and densities of these measures can be related. It is worthwhile to mention the demonstration due to

Fukushima that the mutual absolute continuity (the previous implication can be reversed) holds true for most measures we are interested in.

In the notation (12), the conditional Wiener measure $d\mu[s,y \mid t,x]$ gives rise to the familiar heat kernel, if we set $\Omega = 0$ identically. It in turn induces the Wiener measure P_W of the set of all sample paths, which originate from y at time s and terminate (can be located) in the Borel set A after time $t-s$:

$$P_W[A] = \int_A dx \int d\mu[s,y \mid t,x] = \int_A d\mu \tag{26}$$

where, for simplicity of notations, the $(y, t-s)$ labels are omitted and $\int d\mu[s,y \mid t,x]$stands for the standard path integral expression for the heat kernel.

Having defined an Itô diffusion $X(t)$, (5), (15) with the natural boundaries, we are interested in the analogous (with respect to (26)) path measure P_X

$$P_X[A] = \int_A dx \int d\mu_X[s,y \mid t,x] = \int_A d\mu_X \tag{27}$$

The absolute continuity $P_X \ll P_W$ implies the existence of the strictly positive Radon-Nikodym density, which we give in the Cameron-Martin-Girsanov form

$$\frac{d\mu_X}{d\mu}[s,y \mid t,x] = \exp\Big[\int_s^t \frac{1}{2D} b(X(u))dX(u) - \frac{1}{2}\int_s^t \frac{1}{2D}[b(X(u))]^2 du\Big] \tag{28}$$

Notice that the standard normalisation appears, if we set $D = 1/2$ which implies $D\triangle \rightarrow \frac{1}{2}\triangle$ in the Fokker-Planck equation.

On account of our demand (9) and the Itô formula for $\Theta(X(t),t)$ we have

$$\frac{1}{2D}\int_s^t b(X(t))dX(t) = \Phi(X(t),t) - \Phi(X(s),s) - \int_s^t du\,[\partial_t\Phi + \frac{1}{2}\nabla b](X(u),u) \tag{29}$$

so that, apparently

$$\frac{d\mu_X}{d\mu}[s,y \mid t,x] = \exp[\Phi(X(t),t) - \Phi(X(s),s)]\,\exp\,[-\frac{1}{2D}\int_s^t \Omega(X(u),u)du] \tag{30}$$

with $\Omega = 2D\partial_t\Phi + D\nabla b + (1/2)b^2$ introduced before in (10), by means of the substitution of (4) in the Fokker-Planck equation.

In case of natural boundaries at infinity, the connection with the Feynman-Kac formula (12) is obvious, and we have

$$P_X[A] = \int_A \frac{d\mu_X}{d\mu} d\mu = \int_A dx \int \frac{d\mu_X}{d\mu}[s,y \mid t,x] d\mu[s,y \mid t,x] \tag{31}$$

where the second integral refers to the path integration of the Radon-Nikodym density with respect to the conditional Wiener measure.

In the context of (31) and (12) we can safely assert that the pertinent processes ($X(t)$ and $W(t)$) have *coinciding sets of sample paths*. The stochastic process "realizes" them merely (via sampling) with a probability distribution (frequency) different from this for the Wiener process $W(t)$.

The situation drastically changes, if we wish to exploit the "likelihood ratio" formulas (28), (30) for diffusions confined between the unattainable (natural) boundaries, at least one of which is *not* at infinity. In view of the absolute continuity of P_X with respect to P_W, we must be able to select a subset of Wiener paths which coincide with these admitted by the process $X(t)$, except on sets of measure zero (both with respect to P_X and P_W).

3 Brownian motion and Smoluchowski diffusions

A mathematical idealisation of the individual Brownian particle dynamics, in case of the free evolution in the high friction regime, is provided by the configuration space (Wiener) projection of the phase space (Ornstein-Uhlenbeck) process. One deals then with the stochastic differential equation

$$dX(t) = \sqrt{2D}\, dW(t) \tag{32}$$

$$X(0) = x_0 \in R^3,\ t \in [0,T],\ D > 0$$

which is a symbolic expression representing an ensemble of possible instantaneous values (sample locations in space), generated by the random noise W(t) according to a definite statistical law. Eq. (32) is known (via the stochastic Itô calculus) to imply the Kolmogorov equation for the transition probability density (heat kernel here) i.e. a fundamental law of random displacements of the process, which gives rise to the Fokker-Planck (heat) equation for the time developement of the probability distribution of diffusing particles

$$\partial_t \rho = D\triangle\rho \tag{33}$$

$$\rho(x,0) = \rho_0(x)$$

Then, $\rho(x,t)$ is the probability distribution of the random variable $X(t)$, given the distribution $\rho_0(x)$ of its initial values $X(0)$ in R^3.

By introducing the (irrotational, $rot v = 0$) local velocity field

$$v = -D\frac{\nabla\rho}{\rho} \Rightarrow \partial_t\rho = -\nabla(\rho v) \tag{34}$$

for all conceivable choices of the smooth function $\rho_0(x)$ the heat equation, if combined with the assumption (34), inevitably gives rise to the local conservation law (the momentum balance equation in the kinetic theory lore)

$$\partial_t v + (v\nabla)v = -\frac{1}{m}\nabla Q \tag{35}$$

$$Q = 2mD^2\,\frac{\triangle\rho^{1/2}}{\rho^{1/2}}$$

$$v_0 = -D\frac{\nabla\rho_0}{\rho_0}$$

where m stands for the hitherto absent (albeit included in the definition of the diffusion constant D via the fluctuation-dissipation theorem) mass parameter of diffusing particles, while the potential Q is recognized to have the standard functional form of the familiar de Broglie-Bohm "quantum potential", except for the opposite sign.

In case of an arbitrary non-symmetric distribution $\rho_0(x)$ we have fulfilled the following property, which is maintained in the course of the diffusion process $(X(t) \in R^3)$:

$$\frac{1}{m}\partial_i Q = \frac{1}{\rho}\sum_{j=1}^{3}\partial_j P_{ij} \tag{36}$$

$$P_{ij} = D^2\,\rho\partial_i\partial_j\, ln\,\rho$$

where $\nabla \equiv (\partial_1, \partial_2, \partial_3)$ and $i, j, = 1, 2, 3$. Apparently, $P_{ij} = \delta_{ij}D^2\rho\Delta ln\rho$ in the totally isotropic case. The unconventional "pressure" term $(-\frac{1}{m}\nabla Q)$ in (35) is a distinctive characteristic of all diffusions derivable (via conditioning as example) from the Brownian motion proper and is a collective, statistical ensemble measure of momentum transfer per unit of time and per unit of volume: *away* ($-\nabla Q$ corresponds to the conventional Brownian propagation with the obvious tendency of a particle to leave the area of the higher concentration) or *towards* ($+\nabla Q$) the infinitesimal surrounding of the given spatial location $x \in R^3$ at time t, in the very same rate.

The conventional Brownian dynamics is a very special solution of the general Cauchy problem composed of the mass conservation law (33) and the momentum balance equation (35) with the initial data $\rho_0(x), v_0(x)$ in principle unrelated, in contrast to the assumption (34). Then, we arrive at the rich family of Markovian diffusions, all of which are the *descendants* of the Brownian motion, the Brownian motion itself included.

To be more specific, let us consider the boundary probability distributions $\rho_0(x) = \rho(x, 0), \rho_T(x) = \rho(x, T)$ for a stochastic diffusion process in R^3, confined to the time interval $[0, T] \ni t$. We realise that the dynamical semigroup operator $\exp(tD\Delta)$ provides us with the *probabilistic semigroup transition mechanism*, in the sense that the strictly positive semigroup (heat in our case) kernel is given:

$$h(y, 0, x, t) = (4\pi Dt)^{-1/2}\exp\left[-\frac{(x-y)^2}{4Dt}\right] = [\exp(tD\Delta)](y, x) \tag{37}$$

Following Schrödinger, we ask for the joint probability distribution $m(x, y) = \theta_*(x, 0)\, h(x, 0, y, T)\, \theta(y, T)$ whose marginals $\int dx\, m(x, y) = \rho_T(y)$, $\int dy\, m(x, y) = \rho_0(x)$ coincide with the previously prescribed boundary data for the random propagation in the interval $[0, T]$. It is clear, that for arbitrarily chosen (not necessarily disjoint) areas A and B in R^3, the probability to find in B a particle which originated from A at time 0 and was subject to the random (Brownian, e.g. Wiener) perturbations in the whole run of duration T, reads $m(A, B) = \int_A dx \int_B dy\, m(x, y)$.

With the data $\theta_*(x)$ and $\theta_T(x)$ we can construct respectively the forward and backward diffusive propagation by means of the kernel $h(y,0,x,t)$:

$$\partial_t \theta_* = D\,\Delta\theta_* \tag{38}$$

$$\partial_t \theta = -D\,\Delta\theta$$

$$\theta_*(x,0) = \theta_{*0}(x)\,,\ \theta(x,T) = \theta_T(x)\,,\ t\in[0,T]$$

where

$$\theta_*(x,t) = \int h(y,0,x,t)\,\theta_{*0}(y)\,dy \tag{39}$$

$$\theta(x,t) = \int h(x,t,y,T)\,\theta_T(y)\,dy$$

$$0 \le t \le T$$

The local conservation laws (33), and (35) are satisfied by:

$$\rho(x,t) = (\theta\theta_*)(x,t) \tag{40}$$

$$v(x,t) = D\nabla\, ln\,\frac{\theta}{\theta_*}(x,t)$$

$$x \in R^3\,,\ t\in[0,T]$$

A complete statistical information about the *most likely* way the individual particles propagate, is provided by the transition density $p(y,s,x,t) = h(y,s,x,t)\frac{\theta(x,t)}{\theta(y,s)}$ which solves the Kolmogorov (Fokker-Planck) equation associated with the (individual particle motion recipe) stochastic differential equation

$$dX(t) = b(X(t),t)dt + \sqrt{2D}dW(t) \tag{41}$$

$$b(x,t) = (u+v)(x,t)$$

$$u(x,t) = D\frac{\nabla\rho}{\rho}$$

Notice that the standard Brownian motion comes here in a trivial way by substituting $\theta_* = \rho(x,t), \theta = 1$ for all times $t\in[0,T]$.

We can still have a more detailed insight into the standard Brownian dynamics. Let us consider the initial probability distribution of the random variable $X(0)$ of the Wiener (Brownian in the high friction regime) process in the form

$$\rho_0(x) = (\pi\alpha^2)^{-1/2}\exp[-\frac{x^2}{\alpha^2}] \tag{42}$$

Then its statistical evolution is given by the familiar heat kernel

$$p(y,s,x,t) = [4\pi D(t-s)]^{-1/2}\exp[-\frac{x^2}{4D(t-s)}] \tag{43}$$

$$\rho(x,t) = [\pi(\alpha^2+4Dt)]^{-1/2}\exp[-\frac{x^2}{\alpha^2+4Dt}]$$

where $s \le t$.

Let us notice that since the density distribution is now defined for all times $t > s$ we can introduce a convenient device allowing to reproduce a statistical past of the (irreversible on physical grounds, but admitting this specific inversion mathematically)

$$p_*(y,s,x,t) = p(y,s,x,t)\frac{\rho(y,s)}{\rho(x,t)} \tag{44}$$

with the properties (set $s = t - \Delta t$)

$$\int p_*(y,s,x,t)\rho(x,t)dx = \rho(y,s) \quad s \leq t \tag{45}$$

$$\int y\, p_*(y,s,x,t)dy = x\,\frac{\alpha^2+4Ds}{\alpha^2+4Dt} = x \;-\; \frac{4Dx}{\alpha^2+4Dt}\Delta t = x \;-\; b_*(x,t)\Delta t$$

where $b_*(x,t) = -2D\nabla\rho(x,t)/\rho(x,t)$ and quite trivially $b(x,t) = 0$. Notice furthermore that by defining $v(x,t) = \frac{1}{2}b_*(x,t)$, as a consequence of the heat equation we have satisfied $(\rho v)(x,t) = \int p(y,s,x,t)\rho_0(y)v_0(y)dy$ and equations (38).

Our previous discussion was entirely devoted to the free evolution, and it is quite natural to address the issue of the effects of external force fields on the random propagation. If to accept the high friction regime, like in case of (33), we should consider the Brownian motion in a field of force, in the Smoluchowski approximation.

The Fokker-Planck equation governing the time developement of the spatial probability distribution in case of the phase space noise with high friction, in the Smoluchowski form reads

$$\partial_t\rho \;=\; D\,\Delta\rho \;-\; \nabla(b\rho) \tag{46}$$

$$b(x,t) \;=\; \frac{1}{\beta}F(x), \;\;,\;\; \rho_0(x) = \rho(x,0)$$

where β is the friction constant and the external force we assume to be conservative

$$F(x) \;=\; -\nabla\Phi(x) \tag{47}$$

It is well known that the substitution

$$\rho(x,t) \;=\; \theta_*(x,t)\,\exp[-\frac{\Phi(x)}{2D\beta}] \tag{48}$$

converts the Fokker-Planck equation into the generalised diffusion equation for $\theta_*(\dot{x},t)$

$$\partial_t\theta_* \;=\; D\,\Delta\theta_* \;-\; \frac{V(x)}{2mD}\theta_* \tag{49}$$

where (the mass m was here introduced per force, but with a very concrete purpose of embedding our discussion in the formalism of the "Euclidean quantum mechanics", the name coined by J. C. Zambrini for a natural extension of the standard nonequilibrium statistical physics)

$$V(x) = \frac{m}{\beta}(\frac{F^2}{2\beta} + D\nabla F) \tag{50}$$

Since F^2, D, β are positive, a sufficient condition for the auxiliary potential $V(x)$ to be bounded from below (its continuity is taken for granted) is that the source term $g(x)$ in the familiar Poisson equation

$$\nabla F = -\Delta\Phi = g \tag{51}$$

is bounded from below: $g(x) > -c, c > 0, c$ is finite. Under this boundedness condition, we know that the equation (49) defines the fundamental *semigroup transition mechanism* underlying the Smoluchowski diffusion. Indeed, by (49) we have in hands the well defined semigroup operator $\exp[-t(-D\Delta + V/2mD)]$, whose integral kernel is a strictly positive solution of (49) with the initial condition $\lim_{t\to 0} h(y,0,x,t) = \delta(y-x)$.

The kernel is defined by the Feynman-Kac formula (in terms of the conditional Wiener measure, which sets an obvious link with the Brownian propagation). It is immediate that

$$\theta_{*0}(x) = \rho_0(x)\exp[\frac{\Phi(x)}{2D\beta}] \longrightarrow \tag{52}$$

$$\theta_*(x,t) = \int h(y,0,x,t)\theta_*(y,0)dy$$

while, apparently

$$\theta(x,t) = \exp[-\frac{\Phi(x)}{2D\beta}] = \int h(x,t,y,T)\theta_T(y)dy = \theta_T(x) \tag{53}$$

for all $t \in [0,T]$. Indeed $\theta(x,t)$, (53) solves

$$\partial_t\theta = -D\Delta\theta + \frac{V}{2mD}\theta \tag{54}$$

where $\partial_t\theta = 0$ and

$$D\Delta\theta = [\frac{(\nabla\Phi)^2}{4D\beta^2} - \frac{\Delta\Phi}{2\beta}]\theta = \frac{V}{2mD}\theta \tag{55}$$

as should be. Since the deterministic evolution governed by the Smoluchowski equation gives rise to a definite terminal (in the interval $[0,T]$) outcome $\rho_T(x)$ given $\rho_0(x)$, a straightforward inspection demonstrates that the Schrödinger system is solved by $\theta_{*0}(x)$ and $\theta_T(x)$ with the kernel $h(V;y,s,x,t)$. As a consequence, we have completely specified the unique Markov-Bernstein diffusion interpolating between $\rho_0(x)$ and $\rho_T(x)$, which is identical with the Smoluchowski diffusion itself. We know here the transition probability density (e.g. the law of random displacements modified by the presence of external force fields)

$$p(y,s,x,t) = h(y,s,x,t)\,\frac{\theta(x,t)}{\theta(y,s)} \tag{56}$$

which is responsible for the *most likely* particle propagation scenario. We have also automatically satisfied the local conservation laws

$$\partial_t v = -\nabla(\rho v) \tag{57}$$

$$\partial_t v + (v\nabla)v = \frac{1}{m}\nabla(V - Q)$$

$$\rho(x,0) = \rho_0(x)\,,\; v(x,0) = v_0(x)$$

where $\rho(x,t), v(x,t)$ are defined by the formula (40). Notice that in the detailed derivation, the above momentum balance equation does not appear directly, but in the indirect way by taking the gradient of the much weaker (Hamilton-Jacobi) identity

$$V - Q = 2mD[\partial_t S + D(\nabla S)^2] \tag{58}$$

$$S(x,t) = \frac{1}{2}\,ln\,\frac{\theta}{\theta_*}$$

In our case, apparently

$$v(x,t) = D\nabla(-\frac{\Phi}{D\beta} - ln\rho) = -\frac{1}{\beta}\nabla\Phi - D\frac{\nabla\rho}{\rho} \longrightarrow \tag{59}$$

$$\partial_t\rho = \nabla[\frac{1}{\beta}(\nabla\Phi)\rho] + D\triangle\rho$$

to be compared with the Smoluchowski equation.

The above discussion admits various generalisations. As example, by choosing a definite (reference) Smoluchowski force potential and then the auxiliary (induced) one V, we have fixed the strictly positive kernel $h(V;y,s,x,t)$. By playing with different choices of the boundary data ρ_0, ρ_T (unrelated to the initially considered) and seeking solution of the Schrödinger system, we can generate a rich class of the (conditional) random motions, all of which are governed by the local conservation laws with the potential V. However, their forward drifts $b(x,t)$ would have the functional form completely divorced from the simple Smoluchowski expression.

We can as well start from the general Cauchy problem with the completely arbitrary V (except for being continuous and bounded from below). Then, the corresponding Smoluchowski diffusion can be reproduced only if the potential allows to decouple from the defining identity, the force field F.

4 Quantum mechanical games : schrödinger wave mechanics as the theory of markov diffusion processes

Now, we shall analyze some consequences of the primordial for quantum theory, albeit frequently underestimated statistical postulate due to Max Born: the identification of the squared modulus of the Schrödinger wave function with the probability density ("of something if anything", but undoubtedly of a certain probability measure) is what makes quantum mechanics a part of the theory of stochastic processes, and in particular of Markov diffusions.

4.1 A specific example of the invariant probability measure: measure preserving stochastic dynamics

We indicate at this point certain amusing features of the harmonic attraction. Let us consider the Sturm-Liouville problem on $L^2(R^1)$

$$-D\Delta\psi + \frac{\omega^2 x^2}{4D}\psi = \epsilon\psi \tag{60}$$

The substitutions:$\alpha^4 = \omega^2/4D^2, \lambda = \epsilon/\omega, x = \xi/\alpha$ give rise to the equivalent eigenvalue problem

$$(-\frac{1}{2}\Delta_\xi + \frac{\xi^2}{2})\phi = -\lambda\phi \tag{61}$$

$$\phi(\xi) = \psi(\frac{\xi}{\alpha}) = \psi(x)$$

with the well known solution (normalised relative to x)

$$\lambda_n = n + \frac{1}{2} \leftrightarrow \epsilon_n = (n + \frac{1}{2})\omega \,,\; n = 0, 1, 2, ...$$

$$\psi_n(x) = \phi_n(\xi) = (\frac{\alpha}{2^n n! \sqrt{\pi}})^{1/2} \exp[-\frac{\xi^2}{2}]\, H_n(\xi) \tag{62}$$

$$H_0 = 1 \,,\; H_1 = 2\xi \,,\; H_2 = 2(2\xi^2 - 1) \,,\; H_3 = 4\xi(2\xi^2 - 3) \,, ...$$

Except for $n = 0$ the solutions $\phi_n(\xi)$ are not positive definite and change sign at nodes. We have

$$n = 0 \,,\; \psi_0(x) > 0 \,,\; x \in (-\infty, +\infty)$$

$$n = 1 \,,\; \psi_1(x) > 0 \,,\; x \in (0, +\infty)$$

$$\psi_1(x) < 0 \,,\; x \in (-\infty, 0)$$

$$n = 2 \,,\; \psi_2(x) > 0 \,,\; x \in (-\infty, -1/\sqrt{2}) \cup (1/\sqrt{2}, +\infty)$$

$$\psi_2(x) < 0 \,,\; x \in (-1/\sqrt{2}, +1/\sqrt{2})$$

and so on. It is convenient to continue further considerations with respect to the rescaled $\xi = \alpha x$ variables, in view of the form $-\frac{1}{2}\Delta_\xi + \frac{\xi^2}{2} = H$ of the Hamiltonian predominantly used in the mathematical physics literature. To proceed in this notational convention it is enough to set $x \to \xi$ and $D \to \frac{1}{2}$ and thus utilize $b = \nabla\Theta/\Theta, \Omega = \frac{1}{2}(b^2 + \nabla b), \nabla\Omega = b\nabla b + \frac{1}{2}\Delta b$.

Although we need Θ, Θ_* of the same sign, and $\rho(x)$ to be strictly positive, we can first make a formal identification $\Theta = \Theta_* = \phi_n, n = 0, 1, 2, ...$ and notice that

$$n = 0\,,\, b_0 = -\xi \to \Omega_0 = \frac{\xi^2}{2} - \frac{1}{2}$$

$$n = 1\,,\, b_1 = \frac{1}{\xi} - \xi \to \Omega_1 = \frac{\xi^2}{2} - \frac{3}{2}$$

$$n = 2\,,\, b_2 = \frac{4\xi}{2\xi^2 - 1} - \xi \to \Omega_2 = \frac{\xi^2}{2} - \frac{5}{2}$$

Obviously $\nabla\Omega_n = \xi$ for all n. Irrespective of the fact that each of $b_n, n > 0$ shows singularities, the auxiliary potentials are well defined for all x, and for different values of n they acquire an additive renormalisation $-\lambda_n = -(n + \frac{1}{2})$.

The case of $n = 0$ is a canonical example of the Feynman-Kac integration, and the classic Mehler formula involves the Cameron-Martin-Girsanov density as well.

Indeed, the integral kernel $[\exp(-Ht)](y, x) = k(y, 0, x, t)$ for $H = -\frac{1}{2}\Delta + (\frac{1}{2}x^2 - \frac{1}{2})$ is known to be given by the formula:

$$k(y, 0, x, t) = \pi^{-1/2}(1 - e^{-2t})^{-1/2} \exp[-\frac{x^2 - y^2}{2} - \frac{(e^{-t}y - x)^2}{2}] \tag{63}$$

$$(e^{-Ht}\Theta)(x) = \int k(y, 0, x, t)\Theta(y)dy$$

where the integrability property

$$\int k(y, 0, x, t) \exp[\frac{x^2 - y^2}{2}]dy = 1 \tag{64}$$

is simply a statement pertaining to the transition density of the homogeneous diffusion, which preserves the Gaussian distribution $\rho(x) = (\Theta\Theta_*)(x) = \frac{\alpha}{\sqrt{\pi}} \exp(-\xi^2)$.

4.2 The imaginary time substitution as a mapping between two families of diffusion processes

Let us invoke the analytic continuation in time concept, which is a notorious technical tool to pass to the so called Euclidean framework whenever any problems with the mathematically rigorous processing appear in the context of quantum theory. In fact it is also well known that the easiest way to generate explicit examples of Markov (actually Markov-Bernstein) diffusions is by analytic continuation of solutions of the Schrödinger equation. For V continuous and bounded from below, the generator $H = -2mD^2\Delta + V$ is essentially selfadjoint, and then the kernel $h(x, s, y, t) = [\exp[-(t - s)H]](x, y)$ of the related dynamical semigroup is strictly positive, so the previous Markov-Bernstein process considerations do follow immediately for time-independent potentials V. On the other hand it is quite traditional to relate this dynamical semigroup evolution to the quantum mechanical unitary evolution operator $\exp(iHt)$ by the imaginary time substitution $t \rightarrow it$. In the most pedestrian and naive interpretation of this fact, one might be tempted to invent the concept of "diffusion process in the imaginary time". Actually nothing like that is here allowed, and if taken seriously, becomes self-contradictory.

The routine illustration for the imaginary time transformation is provided by considering the force-free propagation, where apparently the formal recipe gives rise to (one should be aware that to execute a mapping for concrete solutions, the proper adjustment of the time interval boundaries is indispensable):

$$i\partial_t \psi = -D\triangle\psi \longrightarrow \partial_t \overline{\theta_*} = D\triangle\overline{\theta_*}$$
$$i\partial_t \overline{\psi} = D\triangle\overline{\psi} \longrightarrow \partial_t \overline{\theta} = -D\triangle\overline{\theta}$$
$$it \rightarrow t \tag{65}$$

Then

$$\psi(x,t) = [\rho^{1/2}\exp(iS)](x,t) = \int dx' G(x-x',t)\psi(x',0)$$
$$G(x-x',t) = (4\pi iDt)^{-1/2}\exp[-\frac{(x-x')^2}{4iDt}] \tag{66}$$
$$\overline{\theta_*}(x,t) = \int dx' h(x-x',t)\overline{\theta_*}(x',0)$$
$$h(x-x',t) = (4\pi Dt)^{1/2}\exp[-\frac{(x-x')^2}{4Dt}]$$

where the imaginary time substitution recipe

$$h(x-x',it) = G(x-x',t)\ ,\ h(x-x',t) = G(x-x',-it) \tag{67}$$

seems to persuasively suggest the previously mentioned "evolution in imaginary time" notion, except that one *must decide in advance*, which of the two considered evolutions:the heat or Schrödinger transport, would deserve the status of the "real time diffusion".

At this point let us recall that given the initial data

$$\psi(x,0) = (\pi\alpha^2)^{-1/4}\exp(-\frac{x^2}{2\alpha^2}) \tag{68}$$

the free Schrödinger evolution $\partial_t\psi = -D\triangle\psi$ implies

$$\psi(x,t) = (\frac{\alpha^2}{\pi})(\alpha^2 + 2iDt)^{-1/2}\exp[-\frac{x^2}{2(\alpha^2+2iDt)}] \tag{69}$$

with $(\frac{m}{2}(D_+D_- + D_-D_+)X(t) = 0$ applies):

$$\rho(x,t) = |\psi(x,t)|^2 = \frac{\alpha}{[\pi(\alpha^4+4D^2t^2)]^{1/2}}\exp(-\frac{x^2\alpha^2}{\alpha^4+4D^2t^2})$$

$$= \int p(y,0,x,t)\rho(y,0)dy \tag{70}$$

$$p(y,0,x,t) = (4\pi Dt)^{-1/2} \exp[-\frac{(x-y-\frac{2D}{\alpha^2}yt)^2}{4Dt}]$$

where $p(y,0,x,t)$ is the (distorted Brownian) transition probability density for Nelson's diffusion derivable from $\psi(x,t)$. On the other hand we can straightforwardly pass to

$$\psi(x,-it) = \overline{\theta_*}(x,t) = (\frac{\alpha^2}{\pi})^{1/4}(\alpha^2+2Dt)^{-1/2} \exp[-\frac{x^2}{2(\alpha^2+2Dt)}] \tag{71}$$

Let us confine t to the time interval $[-T/2, T/2]$ with $DT < \alpha^2$. Then we arrive at

$$\begin{aligned} \partial_t \overline{\theta_*} &= D\triangle\overline{\theta_*} \\ \partial_t \overline{\theta} &= -D\triangle\overline{\theta} \\ -\frac{T}{2} &\le t \le \frac{T}{2} \end{aligned} \tag{72}$$

$$\overline{\theta} = (\frac{\alpha^2}{\pi})^{1/4}(\alpha^2-2Dt)^{-1/2} \exp[-\frac{x^2}{2(\alpha^2-2Dt)}]$$

where

$$\overline{\rho}(x,t) = (\overline{\theta\theta_*})(x,t) = [\frac{\alpha^2}{\pi(\alpha^4-4D^2t^2)}]^{1/2} \exp[-\frac{\alpha^2x^2}{\alpha^4-4D^2t^2}] \tag{73}$$

with the interesting, and certainly unpredictable if to follow the traditional Brownian intuitions, outcome:

$$\overline{\rho}(x,-T/2) = \overline{\rho}(x,T/2) \tag{74}$$

However strange the probabilistic evolution appropriate for (74) would seem, it does not need an imagination effort, to realize that it refers to a conditional Brownian motion (in fact the Brownian bridge with smooth ends) for which the acceleration formula $D_+^2 X = D_-^2 X = 0$ holds true. Here the intermediate probability density (73) can be represented as the conditional transition probability density formula (identifiable as the Bernstein transition density)

$$\overline{\rho}(x,t) = P(x_1,t_1;x,t;x_2,t_2) = \frac{h(0,-\alpha,x,t)h(x,t,0,\alpha)}{h(0,-\alpha,0,\alpha)} \tag{75}$$

$$h(0,-\alpha,x,t) = [4\pi D(t+\alpha)]^{-1/2} \exp(-\frac{x^2}{4D(t+\alpha)})$$

Clearly nothing like the "imaginary time diffusion" is here involved. We have rather executed a mapping from one real time diffusion to another, with the *incompatible* dynamical principles (previously introduced microscopic conservation laws) at work. Since the Schrödinger equation plays here the role of the linear problem associated (linearisation) with the nonlinear diffusion equations, there are *not* the diffusions themselves which are related directly by the Wick rotation. The link can be established on the auxiliary (for the Nelson diffusion) level of description:the corresponding linear problem (Schrödinger equation which itself generates nonlinear diffusions) can be mapped into the linear diffusion problem, with all the reservations concerning the proper choice of the time interval boundaries.

4.3 Free Schrödinger dynamics as the diffusion process

By defining

$$p(y,0,x,t) = (4\pi Dt)^{-1/2} \exp[-\frac{(x-y+2Dty/\alpha^2)^2}{4Dt}] \tag{76}$$

we realise that

$$\int p(y,0,x,t)(\pi\alpha^2)^{-1/2}\exp(-y^2/\alpha^2)dy =$$

$$\frac{\alpha}{[\pi(\alpha^4+4D^2t^2)]^{1/2}}\exp[-\frac{x^2\alpha^2}{\alpha^4+4D^2t^2}] = \rho(x,t) \tag{77}$$

and

$$\int p(y,0,x,t)[\frac{2Dy}{\alpha^2}(\pi\alpha^2)^{-1/2}]\exp[-\frac{y^2}{\alpha^2}]dy =$$

$$\frac{2D(\alpha^2-2Dt)x}{\alpha^4+4D^2t^2}\rho(x,t) = -b(x,t)\rho(x,t) \tag{78}$$

where evidently

$$v(x,t) = b(x,t) - D\nabla\rho(x,t)/\rho(x,t) \tag{79}$$

solves local conservation identities (laws) with $V = 0$ and via the familiar Madelung transcription of the free Schrödinger dynamics $i\partial_t\psi(x,t) = -D\Delta\psi(x,t)$ with $\psi = \exp(R+iS), \rho = \exp(2R), v = 2D\nabla S$ the link between the Brownian type diffusion and the quantum mechanical evolution is established.

However, it seems instructive to have a detailed demonstration that the pertinent dynamics is a well defined solution of the Schrödinger problem as formulated in Section 1. To simplify considerations we shall rescale the variables so that effectively $D = 1$ appears everywhere. Certainly, we deal with the evolution associated with the continuous mapping:

$$\rho_0(x) = (2\pi)^{-1/2}\exp[-\frac{x^2}{2}] \longrightarrow \rho(x,t) = [2\pi(1+t^2)]^{-1/2}\exp[-\frac{x^2}{2(1+t^2)}] \tag{80}$$

We have defined the transition probability density effecting the "quantum job" from the initial time instant 0 till any finite time t. However, our diffusion process is definitely *not* homogeneous in time, hence the fundamental trasport mechanism for arbitrary times is very different from what the previously utilized formula might suggest. Indeed, let us consider :

$$p(y,s,x,t) = [4\pi(t-s)]^{-1/2}\exp[-\frac{(x-cy)^2}{4(t-s)}] \tag{81}$$

$$c = c(s,t) = [\frac{(1-t)^2 + 2s}{1+s^2}]^{1/2}$$

which by setting $c(0,t) = 1-t$ reduces to the previously considered $p(y,0,x,t) = (4\pi t)^{-1/2}\exp[-\frac{(x-y-yt)^2}{4t}]$. One can easily calculate the drift $b(x,t) = -\frac{(1-t)}{1+t^2}x$ following the standard stochastic methods and check the validity of both the continuity and momentum balance equations. It is however more interesting to realize that by taking $\psi(x,t) = \exp(R+iS)$ where$R(x,t), S(x,t)$ are real functions, we can as well introduce the new real functions $\theta = \exp(R+S), \theta_* = \exp(R-S)$ such that :

$$R(x,t) = -\frac{1}{4}ln2\pi(1+t^2) - \frac{x^2}{4(1+t^2)} \tag{82}$$

$$S(x,t) = \frac{x^2}{4}\frac{t}{1+t^2} - \frac{1}{2}arctan\ t$$

implies

$$\theta(x,t) = [2\pi(1+t^2)]^{-1/4}\exp(-\frac{x^2}{4}\frac{1-t}{1+t^2})\exp(-\frac{1}{2}arctan\ t) \tag{83}$$

$$\theta_*(x,t) = [2\pi(1+t^2)]^{-1/4}\exp(-\frac{x^2}{4}\frac{1+t}{1+t^2})\exp(\frac{1}{2}arctan\ t)$$

where, strikingly there holds:

$$\partial\theta = -\Delta\theta + \Omega\theta \tag{84}$$

$$\partial\theta_* = \Delta\theta_* - \Omega\theta_*$$

$$\Omega(x,t) = \frac{x^2}{2(1+t^2)^2} - \frac{1}{1+t^2} = 2\frac{\Delta\rho^{1/2}}{\rho^{1/2}}$$

Moreover, the function $h(y,s,x,t) = p(y,s,x,t)\frac{\theta(y,s)}{\theta(x,t)}$ is a fundamental solution of the above equations:

$$\theta_*(x,t) = \int h(y,s,x,t)\theta_*(y,s)dy \tag{85}$$

$$h(y,s,x,t) = [4\pi(t-s)]^{-1/2}(\frac{1+t^2}{1+s^2})^{1/4}\exp\frac{1}{2}(arctan\ t - arctan\ s)$$

$$\exp[-\frac{(x-cy)^2}{4(t-s)} - \frac{y^2}{4}\frac{1-s}{1+s^2} + \frac{x^2}{4}\frac{1-t}{1+t^2}]$$

Although the form of the strictly positive semigroup kernel $h(y,s,x,t)$ does not look that promising,it is possible to check through a direct (albeit a little bit involved) computation that the dynamical semigroup implemented identity

$$\lim_{\Delta s\to 0}\frac{1}{\Delta s}[1 - \int h(y,s,x,s+\Delta s)dx] = \Omega(y,s) \tag{86}$$

is valid, as expected from the fundamental solution of the generalized diffusion equation. All probabilistic features characteristic for solution of the Schrödinger problem were there-by recovered.

Ackonwledgement

The talk was delivered by the first named author with support coming from the KBN research grant No 2 P302 057 07

References

[1] E. Schrödinger, Ann. Inst. Henri Poincare, **2**, 269 (1932)
[2] B. Jamison, Z. Wahrsch. verw. Geb. **30**, 65 (1974)
[3] J. C. Zambrini, J. Math. Phys. **27**, 3207 (1986)
[4] R. Carmona, in:Taniguchi Symp. PMMP, Katata 1985, Academic Press, Boston, 1987
[5] M. Nagasawa, Prob. Theory Relat. Fields, **82**, 109 (1989)
[6] P. Garbaczewski and J. P. Vigier, Phys. Rev. **A 46**, 4634 (1992)
[7] P.Garbaczewski, Phys.Lett. **A 172**, 208 (1993)
[8] P.Garbaczewski, Phys.Lett. **A 178**, 7 (1993)
[9] Ph. Blanchard and P. Garbaczewski, Natural boundaries for the Smoluchowski equation and affiliated diffusion processes, Phys. Rev. E, (1994), in press

Fractal Dimension and Roughness of Profiles and Surfaces*

J. G. Moreira[1], *J. Kamphorst Leal da Silva*[1] *and S. Oliffson Kamphorst*[2]

[1] Departamento de Física, Instituto de Ciências Exatas,
Universidade Federal de Minas Gerais,
C.P. 702, 30161-970 Belo Horizonte, Brazil
[2] Departamento de Matemática, Instituto de Ciências Exatas,
Universidade Federal de Minas Gerais,
C.P. 702, 30161-970 Belo Horizonte, Brazil

1 Introduction

The complex forms of nature are too irregular to fit into a traditional geometrical setting. On the other hand, they are nicely described in terms of the concepts of fractal geometry [1]. These forms often present statistical scale invariance and can be characterized by few parameters like fractal dimensions and scaling exponents. A self-similar fractal is invariant under an isotropic length scale transformation – all the different directions scale in the same way. We can cite as examples coastlines [1], percolation clusters [1, 2], colloidal aggregates [3] etc. When an object is invariant under a transformation with different length scales in different directions it is a self-affine fractal [4]. Recent works have demonstrated that a wide class of processes lead to objects with self-affine properties: plots of random walks [1, 2], interfaces in far-from-equilibrium systems [5] and interfaces resulting from growth processes [5, 6, 7].

In many circumstances the interface $f(\mathbf{x})$ generated from a smooth surface by some stochastic process becomes rough. These interfaces can be characterized by its roughness $W(f,L)$. It is defined as the fluctuation around the average of the height f at scale L, namely

$$W^2(f,L) = \langle f^2(\mathbf{x})\rangle_L - \langle f(\mathbf{x})\rangle_L^2 \ . \qquad (1)$$

where $\langle ...\rangle_L$ means average over x in the scale L. For self-affine interfaces, the roughness $W(f,L)$ scales with the linear size L of the surface by an exponent H ($W \sim L^H$), called the roughness or Hurst exponent. This exponent has been received a lot of attention in the last years and has been used to characterize different phenomena in many distinct subjects as for example, biology [8] and fractures [9].

In this work we consider the connection between generalized roughness and fractal dimension for interfaces in dimension $d = 2$ (profiles) and $d = 3$ (surfaces).

* This research was supported in part by Conselho Nacional de Desenvolvimento Científico e Tecnológico (CNPq) and in part by Fundação de Amparo à Pesquisa do Estado de Minas Gerais (Fapemig), Brazilian agencies.

The profiles and surfaces $f(\mathbf{x})$ are characterized by a fractal dimension D and by a roughness or Hurst exponent H. By fractal dimension we mean the Minkowski-Bouligand dimension, box dimension [1] or other equivalent one. Essentially the fractal dimension is related to the scaling properties of d-dimensional volumes of a cover for the interface and the Hurst exponent measures the scaling of the average of a local roughness $w(\mathbf{x}, \epsilon)$, evaluated in a ϵ−neighborhood of $\mathbf{x}$. The rate of convergence to zero of $W(f, \epsilon) = \int w(\mathbf{x}, \epsilon)\, d\mathbf{x}$ as ϵ goes to zero defines the roughness exponent H. We show that H is related to the fractal dimension D through $H = d - D$.

We propose a new numerical algorithm for the evaluation of D, based on a roughness around the local linear fit. This method is tested on self-affine curves with known fractal dimensions: the Weierstrass function and the fractional Brownian motion. The estimates of D are then compared with the ones obtained from methods of the literature.

This work is organized as follows. In the next section we present the mathematical aspects of the connection between the generalized roughness and the Minkowski-Bouligand dimension. Both the cases of profiles and surfaces are considered. The Weierstrass function and the fractional Brownian motion are briefly presented in section 3. In section 4 we define the new numerical algorithm based on the generalized roughness for the evaluation of D. The results obtained for the two standard profiles of section 3 are also discussed in this section. A summary of ours results is presented in the last section.

2 Generalized Roughness and Fractal Dimension

In this section, we present a proof of the relation $D = d - H$ for arbitrary graphs of functions of one (profiles, $d = 2$) or two (surfaces, $d = 3$) real variables. Although this relation was for long known for exact self-affine fractals [10], and can also be derived from scaling arguments for statistical self-affine profiles as the fractional Brownian motion [2, 11], there is no rigorous proof, at least at our knowledge, that it works in the general case. We should point out that for the general case of statistical fractals, even the notion of fractal dimension can be viewed as a first approximation to the microscopic structure, and a multifractal analysis [2, 5, 6, 12] is necessary. Our proof relays in the fact the fractal dimension as well as the roughness exponent, measures how far a fractal interface is from any smooth function that one use to approximate it.

Throughout this work the term interface refers to the graph G_f of a function f of a real variable x, or two real variables x, y, continuous on its domain of definition. To uniformize notation we call the variable $\mathbf{x}$, meaning x or (x, y) depending on the dimension d we are, so the interface is the graph of $f(\mathbf{x})$. Without loss of generality, we fix the domain of definition of f as the interval $[0, 1]$ or the square $[0, 1] \times [0, 1]$ and call this domain $\mathcal{D}$. We are interested in the evaluation of the fractal dimension of the graph G_f. We shall investigate the Minkowski-Bouligand dimension [13], which is equivalent to the box dimension[10]. Although these definitions of dimension characterize the local fractal properties of profiles, they can differ from the Hausdorff dimension [2, 4, 6, 14]. In this work we do not evaluate the Hausdorff dimension. Upper bounds for the Hausdorff dimension of graphs of functions, as well as a very nice discussion about fractal dimensions, can be found in Falconer [14].

Tricot et al.[15] have used the concept of generalized covers in order to demonstrate the equivalence of box and Minkowski-Bouligand dimension. They have shown that for covers by suitable geometric figures (satisfying some non-distortion relations) the fractal dimension is given by

$$D = \lim_{\epsilon \to 0} \left[d - \frac{\ln |U(\epsilon)|_d}{\ln \epsilon} \right] , \tag{2}$$

where $|U(\epsilon)|_d$ is the d-dimensional volume of the union of figures $U(\epsilon)$ covering the graph G_f. Equation (2) gives the same dimension for all generalized covers. For example, the figures are balls of radius ϵ centered on $(\mathbf{x}, f(\mathbf{x}))$ in the Minkowski-Bouligand cover and d-dimensional cubes in the box-dimension evaluation. The numerical algorithms for $d = 2$ resulting from the Minkowski-Bouligand and box dimensions have been analized in Dubuc et al. [16]. The estimates obtained for D were poor when these algorithms have been applied in profiles with known dimensions. It has been found that the basic problem was the "thickness" of the covers. In order to develop better numerical algorithms, Dubuc et al.[16] have introduced two alternating ways of evaluation of fractal dimension: the *horizontal structuring element method* and the *variation method*. Both of them are based on the idea of building covers with intervals instead of geometrical figures. The first uses horizontal segments and the second vertical ones. The idea was extended to n dimensional graphs by Dubuc & Tricot [17].

For further use, let us present briefly the variation method for $d = 2$ [15, 16] and $d = 3$ [18]. Since we are interested in fractal graphs, we can assume that f is nowhere or almost nowhere differentiable. The fractal dimension of the graph is related to the ϵ-oscillation $v(\mathbf{x}, \epsilon)$ of the function f. The ϵ-variation of f in the region $\mathcal{D}$ is defined by

$$V(\epsilon, f) = \int_{\mathcal{D}} v(\mathbf{x}, \epsilon)\, d\mathbf{x} , \tag{3}$$

$$v(\mathbf{x}, \epsilon) = \sup_{\mathbf{x}' \in R_\epsilon(\mathbf{x})} f(\mathbf{x}') - \inf_{\mathbf{x}' \in R_\epsilon(\mathbf{x})} f(\mathbf{x}') . \tag{4}$$

Here $R_\epsilon(\mathbf{x})$ is a rectangle of side ϵ centered at $\mathbf{x}$ or its intersection with $\mathcal{D}$ and is called an ϵ- neighborhood of $\mathbf{x}$. The ϵ-oscillation is also equivalent to the *maximum range* of f in $R_\epsilon(\mathbf{x})$ as defined in [14]. For $d = 2$, the union of vertical segments $S(x, \epsilon) = \{x\} \times [\inf_{x' \in R_\epsilon(x)} f(x'), \sup_{x' \in R_\epsilon(x)} f(x')]$ is a cover for G_f and has, by definition, area equal to the ϵ-variation of f, $|\mathcal{U}(\epsilon)|_2 = V(\epsilon, f)$. It can be shown [15, 17] that $V(\epsilon, f)$ leads to the fractal dimension in any dimension. As long as f is not a constant, we have

$$D = \lim_{\epsilon \to 0} \left[d - \frac{\ln V(\epsilon, f)}{\ln \epsilon} \right] . \tag{5}$$

Intuitively this defines the fractal dimension as the scaling exponent of the area or volume occupied by the graph G_f. Essentially, $V(\epsilon, f)$ measures the difference between the area (volume) under the upper approximation of f at scale ϵ and the lower approximation of f at the same scale.

The idea we introduce now, is to define a finer cover in $d = 2$ by measuring a local oscillation of the function f by a local roughness, defined by

$$w^2(x,\epsilon) = \frac{1}{2\epsilon}\int_{x-\epsilon}^{x+\epsilon} [f(x') - \langle f(x')\rangle_\epsilon]^2 \, dx' \ , \tag{6}$$

$$\langle f(x)\rangle_\epsilon = \frac{1}{2\epsilon}\int_{x-\epsilon}^{x+\epsilon} f(x') \, dx' \ . \tag{7}$$

Note that $\langle f(x)\rangle_\epsilon$ is a local average of the function f. Actually this local average function may be replaced by any smooth interpolating function. It can be replaced by, for example, the local RMS straight line in order to have a better local fit. A sufficient condition is that

$$0 < w(x,\epsilon) \leq v(x,\epsilon) \qquad \text{for all } x \text{ and all } \epsilon > 0 \ . \tag{8}$$

We can see that $w(x,\epsilon) > 0$ holds for any smooth fitting if we assume that the function f is *nowhere differentiable*. The condition $w(x,\epsilon) \leq v(x,\epsilon)$ means that this cover is finer than the one due to the ϵ oscillation and therefore fits the graph G_f better. For the average of the function $\langle f(x)\rangle_\epsilon$ the condition (8) is easily verified from equations (4) and (6). Now we demonstrate that for oscillations $w(x,\epsilon)$ satisfying (8) we have

$$\lim_{\epsilon\to 0} \frac{\ln W(f,\epsilon)}{\ln \epsilon} = \lim_{\epsilon\to 0} \frac{\ln V(f,\epsilon)}{\ln \epsilon} \ , \tag{9}$$

where we have defined the average roughness $W(f,\epsilon) = \int_0^1 w(x,\epsilon)\, dx$. For every fixed $\epsilon > 0$, $w(x,\epsilon)$ is a continuous strictly positive function of x on the interval $[0,1]$. Therefore we can find a δ_ϵ such that $w(x,\epsilon) \geq \delta_\epsilon > 0$ for all x. On the other hand, $v(x,\epsilon)$ is also continuous with regard to ϵ and x on compact sets of the plane. Then we can choose ϵ' such that $v(x,\epsilon') \leq \delta_\epsilon$ for all x. Therefore we have

$$V(f,\epsilon') \leq W(f,\epsilon) \leq V(f,\epsilon) \ . \tag{10}$$

Taking the logarithm, dividing by $\ln\epsilon$, considering the limit as ϵ', $\epsilon \to 0$ and using equation (5) we obtain that the fractal dimension D is given by

$$D = 2 - \lim_{\epsilon\to 0} \frac{\ln W(f,\epsilon)}{\ln \epsilon} \ . \tag{11}$$

For self-affine curves we have that $W(f,\epsilon) \sim \epsilon^H$, where H is the Hurst exponent, implying that $D = 2 - H$.

In higher dimension, the proof is compleatly equivalent, we only need to replace x by $\mathbf{x}$, and integrals over $[0,1]$ by integrals over $\mathcal{D}$, all the compactness arguments still work. Therefore we have

$$D = d - \lim_{\epsilon\to 0} \frac{\ln W(f,\epsilon)}{\ln \epsilon} \ . \tag{12}$$

This equation establishes a formal connection between roughness and fractal dimension in all dimensions. For self-affine interfaces we have a relation between the fractal dimension and the roughness exponent, namely $D = d - H$.

3 Two Standard Self-Affine Profiles

For further use, let us introduce briefly two standard self-affine profiles. The first one is the real cosine Weierstrass function (WF) $f_H(x)$, defined as

$$f_H(x) = \sum_{n=-\infty}^{\infty} b^{-nH} \left[1 - \cos(b^n x)\right] \quad , \tag{13}$$

where $b > 1$ and $0 < H < 1$. Note that the WF has a scaling behavior with different scaling ratios, since $f_H(bx) = b^H f(x)$.

The second one is the trace $B_H(t)$ of the fractional Brownian motion (FBM). $B_H(t)$ is a single-valued function of one real variable (t) and its increments $\Delta B_H(\Delta T) = B_H(t_2) - B_H(t_1)$ have a Gaussian distribution. The variance of this distribution is given by

$$\langle \Delta B_H^2(\Delta t)\rangle \sim \Delta t^{2H} \quad , \tag{14}$$

with $0 < H < 1$, $\Delta t = |t_2 - t_1|$ and $\langle \ldots \rangle$ denotes the ensemble average. There are several algorithms to generate the FBM. We use an algorithm described in Feder [2], page 174,equation (9.25). In Fig. 1 it is shown some typical traces. The FBM is a statistical self-affine object because its scaling properties. If Δt is changed by the factor r, the increments ΔB must be changed by a different factor r^H, since

$$\langle \Delta B_H^2(r\Delta t)\rangle \sim r^{2H} \langle \Delta B_H^2(\Delta t)\rangle \quad . \tag{15}$$

The values of the fractal dimension depends on the way one measures it [2]. In order to obtain the local box dimension we can use a scaling argument. We must cover the trace of the FBM and the graph of WF with small boxes. Then it is easy to obtain that $D = 2 - H$ for both the profiles. A detailed evaluation of D for the WF and FBM can be found in textbooks [2, 6]. We must also observe that a multivariate WF in the continuum limit is equivalent to the multivariate FBM [19].

4 Numerical Algorithm for the Evaluation of D

It is straightforward to develop numerical algorithms to evaluate the dimension D or the roughness exponent by using (11). We will describe the method for $d = 2$ but its extension for $d = 3$ is simple. Suppose that we have $N + 1$ points $f(\frac{n}{N})$ $(n = 0, 1, 2 \ldots N)$ of the digitized profile. In this method, called the roughness around the RMS straight line (SLR), we consider a ϵ neighborhood of each point $x = n/N$ of the digitized profile. First we evaluate the best straight

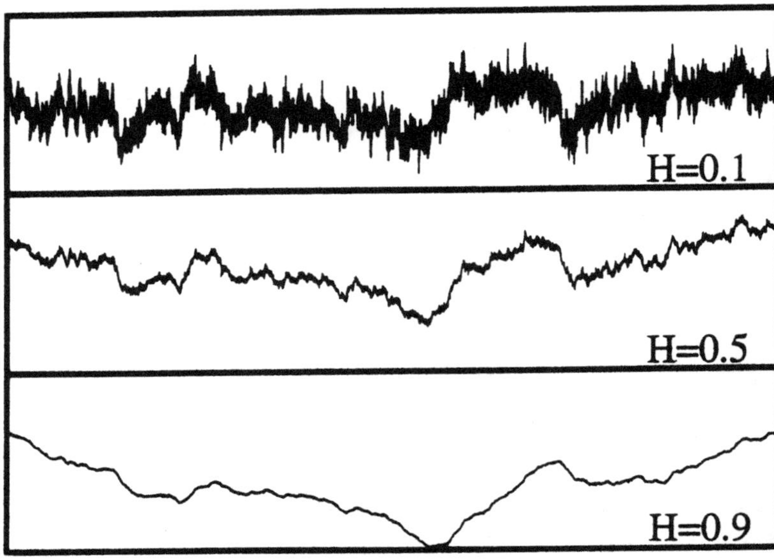

Fig. 1. The fractional Brownian motion trace with $n = 8$ and $M = 700$ for (a) $H = 0.9$; (b) $H = 0.5$; and (c) $H = 0.1$.

line $a_x(n/N) + b_x$ in this interval, with the coefficients a_x and b_x evaluated by the root mean square method. The local roughness is then given by

$$w^2(x,\epsilon) = \frac{1}{\epsilon}\sum_{n\in\epsilon}\left[f(n/N) - \left(a_x\frac{n}{N} + b_x\right)\right]^2 . \tag{16}$$

Observe that this is equivalent to (6). Finally, we obtain the ϵ dependent roughness as

$$W(f,\epsilon) = \frac{1}{N+1}\sum_{x=0}^{1} w(x,\epsilon) . \tag{17}$$

This procedure is repeated for several scales ϵ and the fractal dimension D is given by the slope of the $\ln[\frac{W(f,\epsilon)}{\epsilon^2}] \times \ln(1/\epsilon)$ plot. Due to discreteness, the log-log plot does not give an accurate estimate of D for all scales ϵ. We have defined a smallest scale ϵ_0 and consider all scales of the form $\epsilon_k = c^k\epsilon_0$ ($\epsilon_k < N$), where $c > 1$ is a constant. Then we evaluate the local slope D_i of the graph by taking a fixed number of successive points starting at k_i. When D_i is almost constant we have a reliable estimate of D. We consider the most frequent value of D_i as the best estimate for the fractal dimension. The error bar is estimated by considering the values near D_i that are significantly frequent.

In order to test the robustness and the efficiency of this algorithm, we evaluate the D dimension of mathematical objects with well known fractal dimension. The first object is the Weierstrass function (WF), defined in (13). The second object is the trace $B_H(t)$ of the fractional Brownian motion (FBM), already presented in the preceding section. The numerical values of D for the WF are shown in Table 1 . The exact fractal dimension is given in first column; the estimates D_{VAR} obtained by using the variation method [16] are shown in the second column; the third column shows the results obtained by the traditional roughness, i. e., the roughness around the mean height (D_{MHR}); the fourth column shows the evaluation of the fractal dimensions obtained by our method (D_{SLR}). The error of the last digit is shown between parentheses.

Table 1. Fractal dimensions D of profiles generated by the Weierstrass function with $b = 2.1$ in the interval $[0.60, 0.61]$; the studied profiles have been digitized with 20000 points; D is the exact value and D_{VAR} is the estimate of the variation method; D_{MHR} and D_{SLR} are the results obtained from the roughness around the local average height and around the local RMS straight line respectively. The error in the last digit is indicated between parenthesis.

D	D_{VAR}	D_{MHR}	D_{SLR}
1.9	1.83(1)	1.898(1)	1.901(1)
1.7	1.655(5)	1.702(2)	1.709(1)
1.5	1.475(1)	1.500(6)	1.515(1)
1.3	1.281(1)	1.292(6)	1.319(1)
1.1	1.096(1)	1.101(3)	1.121(1)

The results for the fractional Brownian motion are shown in Table 2, which is similar to the preceding table. Note that the estimates of the fractal dimension are worst than the ones obtained for the WF.

Table 2. Fractal dimensions D of profiles generated by the fractional Brownian motion with $n = 8$ and $M = 700$; the studied profiles have 20000 points; D, D_{VAR}, D_{MHR} and D_{SLR} are defined as in the preceding table.

D	D_{VAR}	D_{MHR}	D_{SLR}
1.9	1.80(2)	1.88(1)	1.87(1)
1.7	1.68(2)	1.725(5)	1.715(5)
1.5	1.52(2)	1.535(5)	1.525(5)
1.3	1.34(2)	1.36(2)	1.320(5)
1.1	1.22(4)	1.24(4)	1.105(5)

The $\ln[W(f,\epsilon)/\epsilon^2] \times \ln[1/\epsilon]$ plots of the FBM for three values of H are depicted in Fig. 2. Similar plots have obtained for the WF.

From these plots we evaluate the slope D_i of several intervals of scales ϵ. The $D_i \times i$ plots for the WF and the FBM are shown in Fig. 3 and in Fig. 4, respectively.

Let us discuss now the results obtained for the WF. We have very good straight lines for the three methods in $\ln[W(f,\epsilon)/\epsilon^2] \times \ln[1/\epsilon]$ plots. In fact,

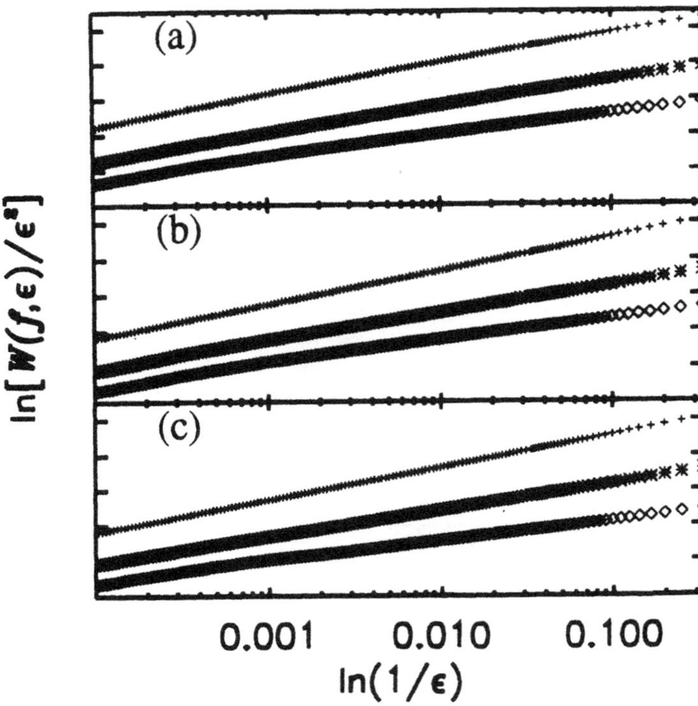

Fig. 2. The $\ln[W(f,\epsilon)/\epsilon^2] \times \ln(1/\epsilon)$ plot of the fractional Brownian motion obtained by (a) the variation method, (b) the MHR method and (c) the SLR method for $H = 0.1$ (+), $H = 0.5$ (∗) and $H = 0.9$ (◇).

the fitting is much more better than the ones for the FBM. We must observe, however, that exist some differences between the three methods. The results of the variation method at low H (high roughness) only stabilizes for large scales in a value below the expected one. For the other two methods the results are almost stable for all scales and in good agreement with the theoretical value of D. When the roughness is small ($H \sim 0.9$), the variation method is more stable and the three methods give good results for the fractal dimension.

The results obtained by the three methods for the FBM are not so good as the ones obtained for the WF. For high roughness, the MHR and SLR methods give us a better estimation of D than the variation one. In the cases that the roughness is small, all the methods are not stable for large scales, mainly for the VAR and MHR methods. For this cases, the straight line roughness method is the only one that gives a reliable estimate of the fractal dimension.

The above results can be explained as follows. Figure 5 shows the coverings of the trace of the FBM with an intermediate value of the parameter H. The scales of the plots are the same for the three methods. We can see that the covering of SLR method is the finest and the one of the variation method is very crude. Independent of the test function, the VAR method does not give a good estimate of D for low H because the curve is so rough (see Fig. 1c) implying that the

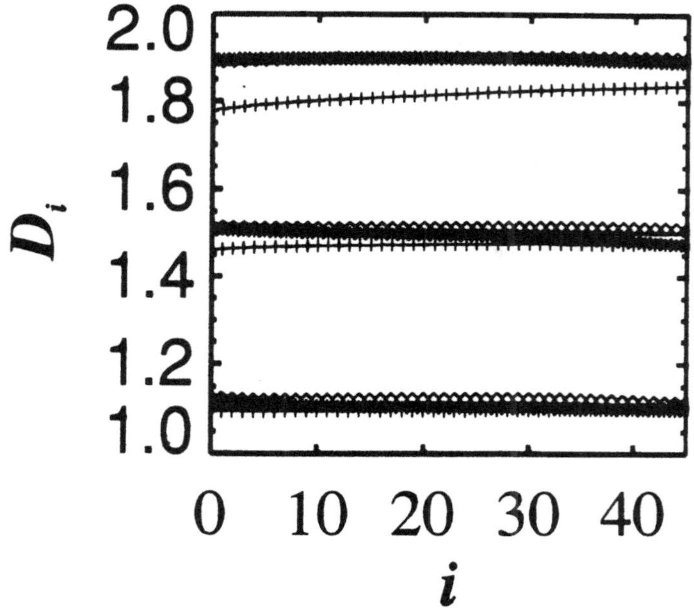

Fig. 3. The graphs of the fractal dimension $D_i \times i$ of the Weierstrass function for three different values of the parameter H. The exact values are represented by solid lines. Shown also are the estimates obtained by the variation (+), the MHR (∗) and the SLR (⋄) methods.

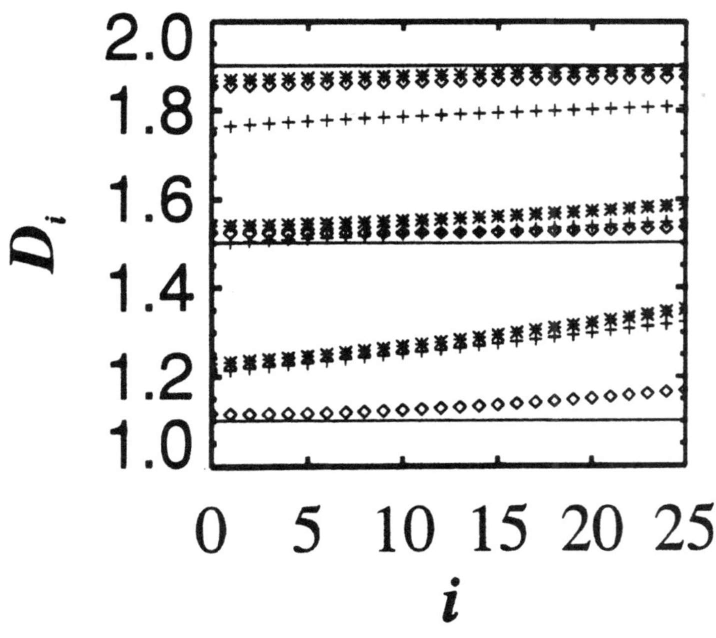

Fig. 4. Figure similar to Fig. 3 for the fractional Brownian motion.

cover is particularly crude. For the WF with $H \sim 0.9$ the profile is quite smooth and the three methods work well. On the other hand, for the FBM trace in the same range of H, the profile is also quite smooth but has some valleys and hills (see Fig. 1a). It means that we have a roughness of large scales, which are not taken in account by the VAR and MHR methods.

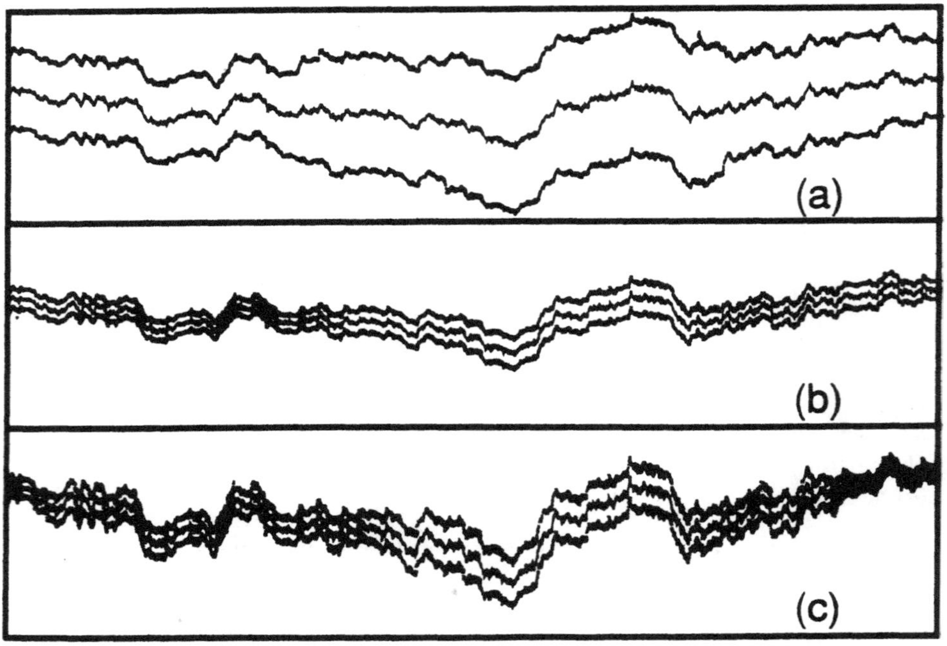

Fig. 5. Covering of the FBM trace with $H = 0.5$ by the (a) VAR; (b) MHR; and (c) SLR methods. The middle curves represent the traces, the top (bottom) curves are obtained by adding (subtracting) the local roughness $w(x, \epsilon)$ to the coordinates of the traces. For the VAR method the roughness is given by Eq. (4). The traces are digitalized with 5,000 points and $\epsilon = 1,000$.

All results shown here were obtained for a particular set of parameters described in the captions of the tables. However, we have checked the robustness of the methods by considering the WF and the FBM with other sets of parameters and found similar results. The estimates of D were obtained by visiting each point of the 20,000 ones of the digitized profile. Moreover, the results do not change when the profiles are digitized with less points. For instance, we have studied profiles with 5,000 and 2,000 points and observed a difference only in the last significative digit in the case of 2,000 points.

5 Summary

In conclusion, we have shown that the roughness is closely related to the fractal dimension (Minkowski-Bouligand or box dimensions) of profiles ($d = 2$) and surfaces ($d = 3$) nowhere differentiable. In particular, we have demonstrated a relation between the fractal dimension and the roughness exponent of self-affine fractals, namely $D = d - H$. This has allowed us to present a new numerical algorithm for the evaluation of the fractal dimension or the roughness exponent. This algorithm, based on generalized roughness, is more robust and reliable than the standard ones. We should pointed out that this algorithms is especially useful if one wants to observe the roughness over profiles or surfaces which have an arbitrary shape. This kind of shape is found, for example, in fractures. Finally, let us observe that the numerical algorithms for surfaces ($d = 3$) are equal to the ones for profiles ($d = 2$), except that we must now evaluate the roughness around the best plane in a ϵ neighborhood.

Acknowledgments

The authors wish to thank P. Mendes and A. T. Bernardes for useful discussions.

References

[1] B. B. Mandelbrot : "The Fractal Geometry of Nature" (Freeman, San Francisco, 1982).
[2] J. Feder, "Fractals" (Plenum, New York, 1988).
[3] P. Meakin, J. of Coll. and Inter. Science **134**, 235 (1990).
[4] B. B. Mandelbrot in "Fractals in Physics", edited by L. Pietronero and E. Tosatti (North-Holland, Amsterdam 1986), p. 3.
[5] "Random Fluctuations and Pattern Growth", edited by H. E. Stanley and N. Ostrowsky (Kluwer, Dordrecht 1988).
[6] T. Vicsek, "Fractal Growth Phenomena" (World Scientific, Singapore, 1989).
[7] F. Family, Physica A **168**, 561 (1990).
[8] H. E. Stanley, S. V. Buldyrev, A. L. Goldberger, Z. D. Goldberger, S. Havlin, R. N. Mantegna, S. M. Ossadnik, C.-K. Peng and M. Simons, Physica A **205** 214 (1994).
[9] K. J. Måløy, A. Hansen, E. L. Hinrichsen and S. Roux, Phys. Rev. Lett. **68** 213 (1992).
[10] B. B. Mandelbrot, Phys. Scr. **32**, 257 (1985).
[11] R. F. Voss, Physica D **38**, 362 (1989).
[12] "Dynamics of Fractal Surfaces", edited by F. Family and T. Vicsek (World Scientific, Singapore 1991).
[13] G. Bouligand, Bull. Sci. Math. **2**, 185 (1929).
[14] K. Falconer, "Fractal Geometry" (John Wiley & Sons, Chichester, 1990).
[15] C. Tricot, J. F. Quiniou, D. Wehbi, C. Roques-Carmes and B. Dubuc, Rev. Phys. Appl. **23**, 111 (1988).
[16] B. Dubuc, J. F. Quiniou, C. Roques-Carmes, C. Tricot and S. W. Zucker, Phys. Rev. A **39**, 1500 (1989).

[17] B. Dubuc and C. Tricot, C. R. Acad. Sci., Paris **306**, 531 (1988).
[18] B. Dubuc, S. W. Zucker, C. Tricot, J. F. Quinou and D. Wehbi, Proc. R. Soc. Lond. A **425**, 113 (1989).
[19] M. Ausloos and D. H. Berman, Proc. R. Soc. Lond. A **400**, 331 (1985).

Competition between two kinds of entities in a Diffusion–Limited Aggregration process

N. Vandewalle[1] *and M. Ausloos*[2]

S.U.P.R.A.S., Institut de Physique, B5 University of Liège, Sart Tilman
B-4000 LIEGE, Belgium
[1] E-mail address: vandewal@gw.unipc.ulg.ac.be
[2] E-mail address: u2150ma@bliulg11.bitnet

Abstract: An extra degree of freedom (the "spin") is added in the well–known Diffusion–Limited Aggregation growth model. A physically relevant coupling energy between aggregating spins in presence of an external magnetic field is defined. This model generates a kinetic growth competition between two kinds of entities and leads to a wide variety of kinetic processes and morphologies distributed in the "phase diagram" of the two growth parameters (coupling and field). Out of the cluster, the motion of the spins is still Brownian. The process leads to cluster fractal structures with fractal dimensions varying from 1.68 ± 0.02 to 1.99 ± 0.01 depending on the coupling parameters. Some physical ideas are presented in order to describe the new kinetic processes. Beside geometrical properties of the clusters, physical properties were also measured. For finite size clusters, a "transition" in the magnetization occurs at fixed $(\beta J, \beta H)_c$ values. This "transition" shows that a species can dominate the other in finite clusters. Moreover, the fractal dimension of the clusters drastically drops at the same critical values. These behaviours are interestingly new and unexpected.

1 Introduction

Since the last decade, kinetic growth models have received great attention because of their universality and the natural processes that they can generate [1]. Growth models are studied in many domains of science like gelation and colloids [1], percolation [2], or just crystal growth [3]. The more studied model is certainly the "Diffusion–Limited Aggregation" (DLA) model introduced in 1981 by T. A. Witten and L. M. Sander [4]. It generates aggregation of identical particules to a cluster through a Brownian motion.

DLA provides a basis for understanding a large range of natural pattern formation phenomena: electrodeposition problems [5], viscous fingering [6], dielectric breakdown [7], thin film growth [8],.... Although DLA is certainly the most studied of the growth models, many questions about kinetic growth remain open.

Usually, natural systems are constitued by entities taking different states. E.g. copolymers are macromolecules made of two kinds of monomers [9]. Some bacterian cells (like *salmonella*) can have some "on" or "off" gene [10]. It is thus of interest to generalize the kinetic growth models in order to study the kinetic of a growing competition between physically different entities. Recently, a variant of DLA was imagined where geometrically different entities aggregate (particules with different sizes) [11]. The model presented here is quite different.

In statistical physics, a simple way of representing physical entities taking a set of values is through a magnetic or a "spin" analogy. Here, we restrict our discussion to spins which can take only two states: up and down (or +1 and −1). In previous work, we have introduced the Magnetic Eden Model (MEM) which adds an extra degree of freedom (the "spin") to the classical Eden model A [12]. The competition between two kinds of entities in an Eden process leads to a wide variety of cluster–types morphologies varying from compact to lacunar clusters. In the diagram of the growth parameters (see below), the compact structures varies also from smooth faceted clusters to rough faceted clusters.

It is interesting to consider the generalization of DLA in a similar way. Here, we propose to introduce an internal degree of freedom (the spin) in DLA. The model simulates e.g. electrodeposition of two species of atoms. The geometrical (fractal) and the physical properties of the generated patterns are studied.

2 The model

The "Magnetic Diffusion–Limited Aggregation" (MDLA) model is defined by the aggregation of spins moving toward a cluster through a brownian motion as in DLA. On a two–dimensional square lattice with lattice spacing a, the growth rule is illustrated in Fig.1 and defined by the
following steps:

(i) On a site of the lattice: a spin is dropped. It is called the "seed" of the growth.
(ii) A diffusing up or down spin is dropped on a circle of radius $r_{\max}+5a$ centered on the seed site ($r_{\max}$ is the largest distance between a cluster site and the seed site).
(iii) Because we consider only short range interactions, the motion of the diffusing spin is still Brownian (as DLA) when it moves toward the cluster. During the motion, the spin randomly can turn up or down at each step. If the spin moves outside a circle of radius $3r_{\max}$ centered on the seed site, it is removed from the process and one goes back to step (ii). If the spin reaches a neighbour site of the perimeter, the diffusion is controlled by the local "magnetic" configurations and one goes to step
(iv) Probabilities of jumping to the (four) specified neighbour sites of the perimeter are defined to be proportionnal to $\exp(-\Delta\beta E)$ where $\Delta\beta E$ is the local gain of the Ising energy between the initial and final state of the cluster before and

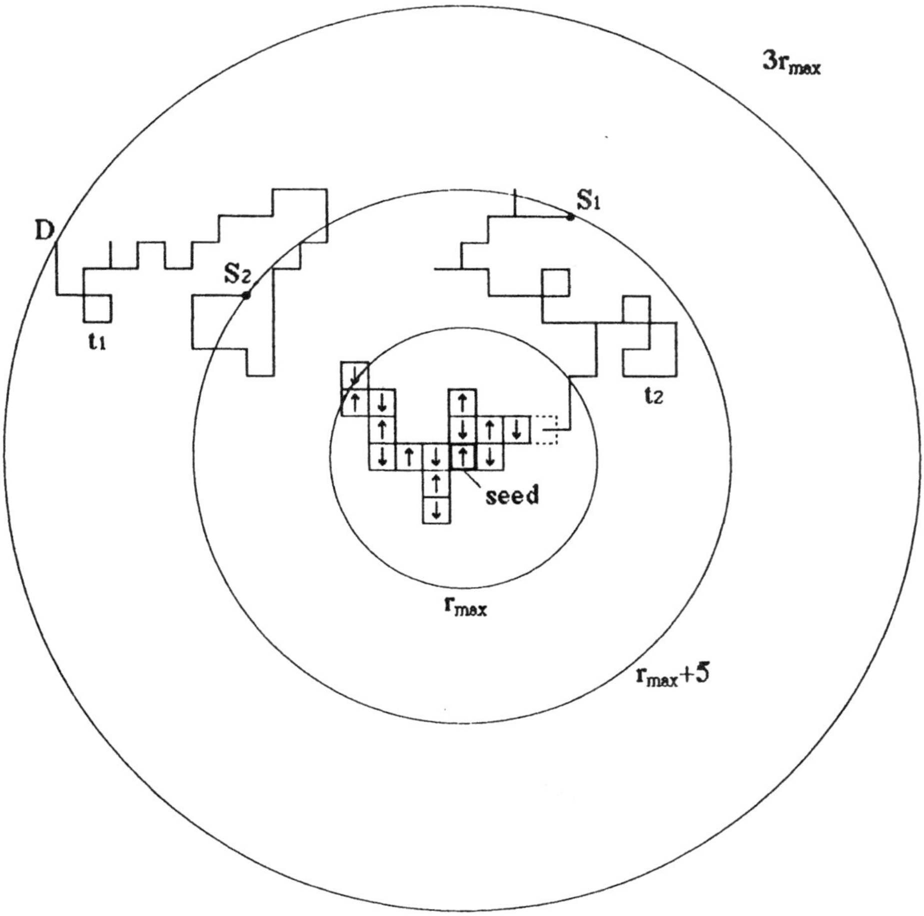

Fig. 1. Magnetically Controlled Diffusion–Limited Aggregation rule.

after spin addition. The dimensionless Ising total energy βE is defined by

$$\beta E = -\frac{\beta J}{2} \sum_{\langle i,j \rangle} \sigma_i \sigma_j - \beta H \sum_i \sigma_i$$

where the first summation occurs on the nearest neighbours only. The first term describes a short range interaction (coupling) between nearest neighbour spins. The second term defines a dimensionless energy of the orientation of the spins in a magnetic field.

We precise that the probabilities of the 8 possible configurations for each jump are renormalised and one specified configuration among them is choosen through a random number generator. If the spin jumps on a perimeter site, it sticks immediately on the cluster, is frozen and the next diffusing spin is launched: one goes to step (ii). The process is irreversible. However, if the spin

jumps toward a neighbouring site of the perimeter, one goes to step (iv). If the moving spin jumps toward an other empty site, one goes to step (iii).
(v) The procedure is repeated for a desired number of particules to be frozen on the cluster.

The MDLA model has two major differences with the Witten and Sander model. First, the introduced extra–degree of freedom is original. Notice that eventhough J and H are "energies", MDLA is still a purely kinetic growth model because the diffusion and growth are driven by the probabilities $\exp(\beta J)$ and $\exp(\beta H)$.

This lead to the second difference, on the neighbouring of the perimeter, sticking and diffusing probabilities are richer than DLA [13] because 17 possible perimeter site configurations are available on the two–dimensionnal square lattice and distributed among 7 sticking probability levels (noted P_1 to P_7). The configurations and the associate probability levels as function of βJ and βH are shown in Fig. 2. Equaling the 7 probability levels, we are led to 12 linear relations between βJ and βH, i.e. $\beta J = 0$ and $\beta H = n\beta J/2$ with n integer varying from -5 to 5. (The MEM was determined by 16 linear relations [12].) These boundaries determine 24 regions in the plane (βJ, βH) where the growth processes on the perimeter differ from each other. Along the vertical axis, the spins are decoupled ($\beta J = 0$) and the model generates simply a DLA process. The processes and morphologies are symmetrical with respect to the $\beta H = 0$ axis. In the upper half diagram exhibited in Fig. 3, the regions are labelled AF and F for

$\beta J < 0$ and $\beta J > 0$ respectively and numbered from 1 to 6 when the field is increased.

3 Morphologies and geometric properties of MDLA clusters

Fig. 4 exhibits typical clusters of 3000 spins generated in various regions of the diagram. In the whole ferromagnetic interaction (F1 to F6) regions of the diagram, the structure of MDLA clusters is DLA–like but is more diffuse and rather less side–branched than the DLA one. In the AF1 region, the same cluster morphology is observed in Fig. 4. In the AF2 region, the dendritic structure is still conserved but it results in an important thickening of the branches as shown in Fig. 4. In AF3 region, compact structures are generated as shown in Fig. 4. However, these clusters are provided with internal lacunes and channels as shown. In AF4 to AF6 regions, "Eden Tree"–like structures [14] are generated as shown in Fig. 4. The antiferromagnetic order dominates the magnetic field in AF1 to AF3 regions but in AF4 to AF6 regions, the field is enough to order the majority of spins in the up direction. The antiferromagnetic order in a branch of a cluster grown in the AF2 region is shown on Fig. 5.

We have also simulated 80 clusters of 3000 spins at definite ($\beta J, \beta H$) parameter values in each region of the diagram. The fractal dimensions are evaluated

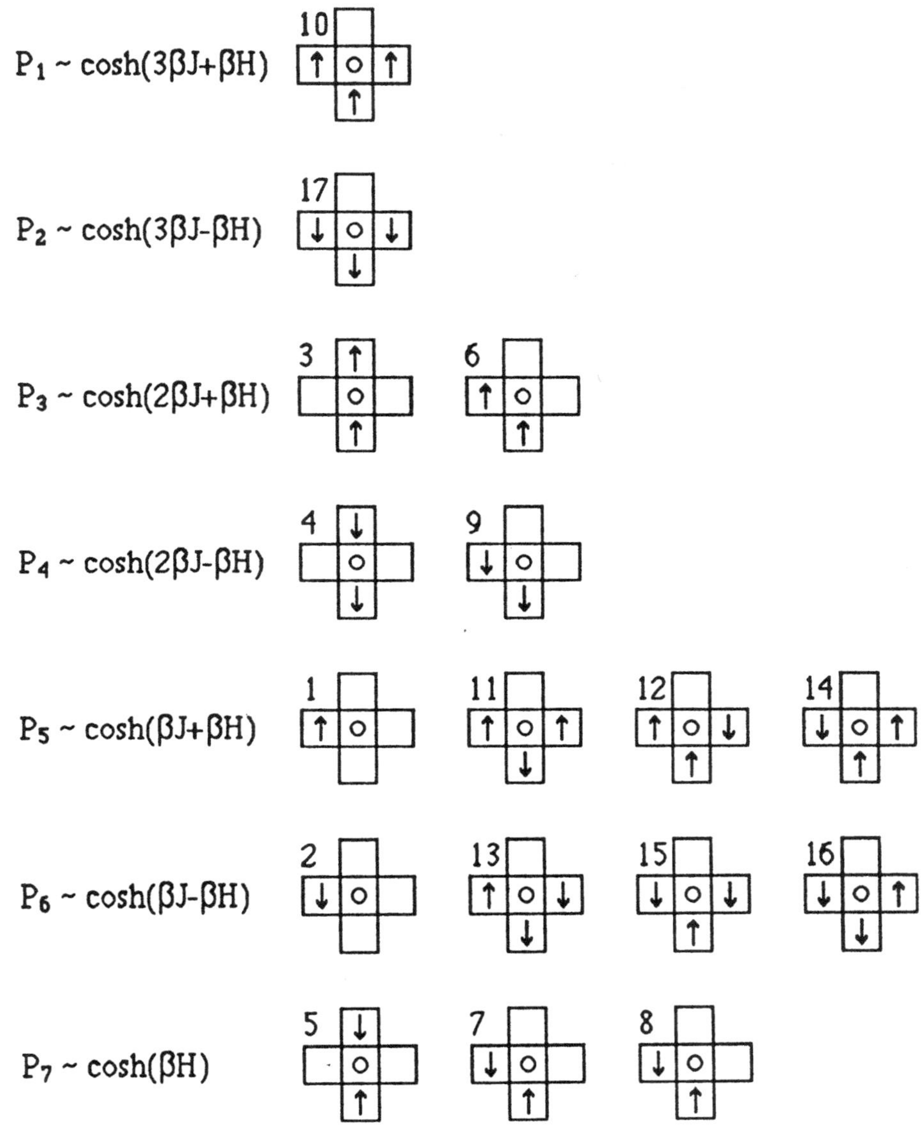

Fig. 2. Probability of growth (sticking) and configurations in the 2D Magnetically Controlled Diffusion-Limited Aggregation.

by the radius of gyration method. As in Witten and Sander simulations, the number of spins which has been used seems to be sufficient to give a good estimate of D_f here also. Fractal dimensions for each region and parameter values are given in Table 1. We have found $D_f = 1.71 \pm 0.01$ for decoupled spins. This is in good agreement of course with usually reported fractal dimension of DLA clusters [4].

As mentioned, the growth processes are determined by the probabilities P_1

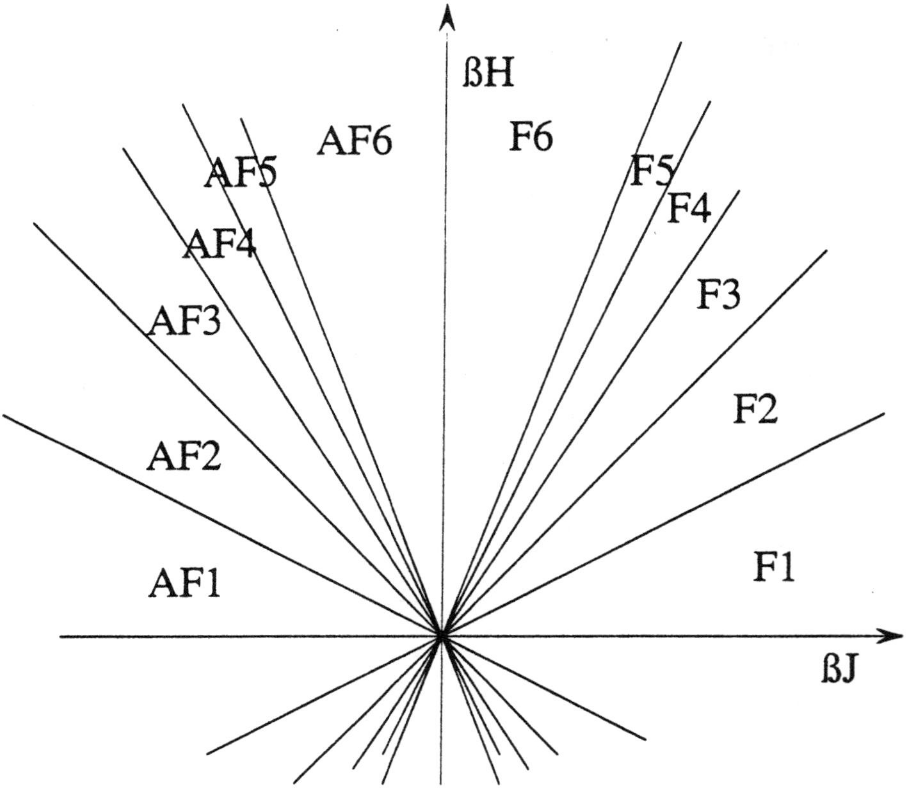

Fig. 3. Upper-half plane of the phase diagram of Magnetically Controlled Diffusion-Limited Aggregation.

Table 1.

region	βJ	βH	D_f
DLA	$= 0$	$\forall H$	1.71 ± 0.01
F1	4	1	1.70 ± 0.01
F2	4	3	1.69 ± 0.01
F3	4	5	1.71 ± 0.01
F4	4	7	1.70 ± 0.01
F5	4	9	1.71 ± 0.02
F6	4	11	1.72 ± 0.01
AF1	-4	1	1.68 ± 0.02
AF2	-4	3	1.70 ± 0.01
AF3	-4	5	1.80 ± 0.02
AF4	-4	7	1.95 ± 0.01
AF5	-4	9	1.98 ± 0.02
AF6	-4	11	1.99 ± 0.01

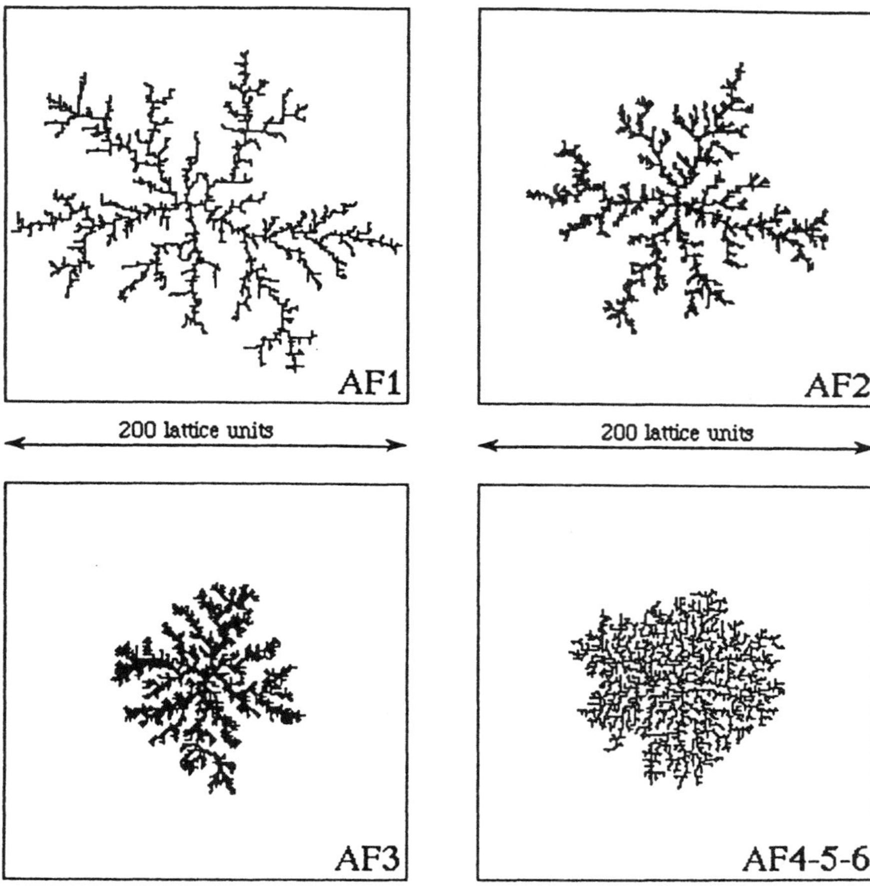

Fig. 4. Typical clusters of 3000 spins grown from an up spin as seed in various regions of the diagram of Fig. 3.

to P_7. From these, some physical processes can be derived and discussed below. Any magnetic field tends to enhance the effect of the positive coupling between spins because the field favours the growth of one spin species. In the fjords and channels, the motion is still brownian and on the neighbouring of the perimeter, the probability of sticking to the cluster is high. Furthemore the hot perimeter sites close to the launching circle are more favoured than internal ones [15, 16]. This easily explains the structure of MDLA clusters in the ferromagnetic part.

However, in the antiferromagnetic part, the coupling and field effects are in greater conflict. This leads in this part of the phase diagram to a wider variety of processes and morphologies. When the field is not high enough to break the antiferromagnetic order (in the AF1 region), the coupling effect dominates the others: the process is closely–related to that which occurs in the ferromagnetic regions (F1 to F6).

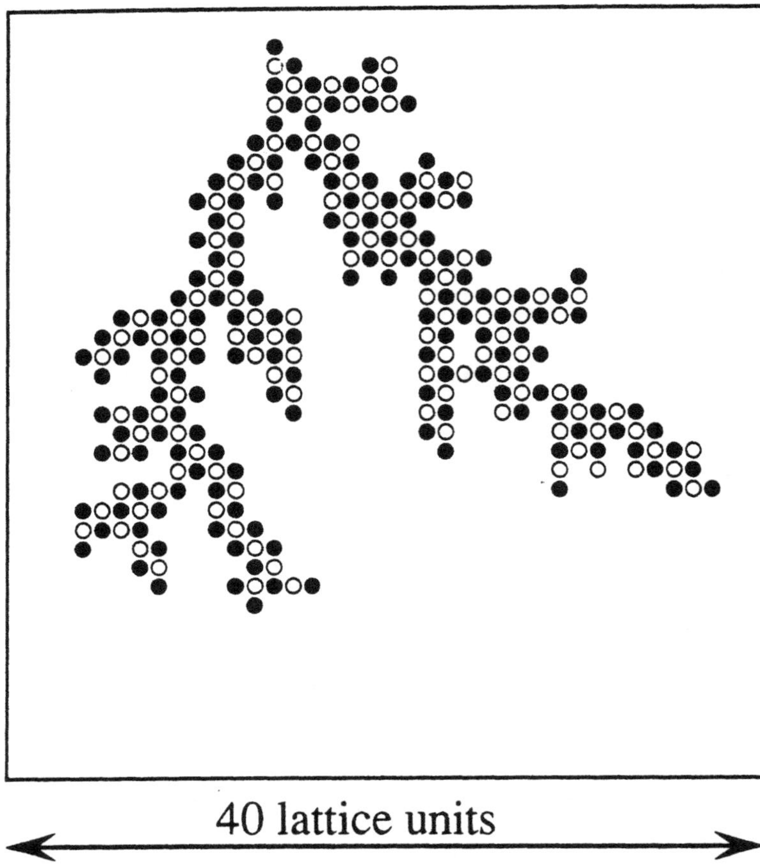

Fig. 5. A branch of a Magnetically Controlled Diffusion–Limited Aggregation cluster grown in the AF2 region. The black and white dots represents down and up spins respectively.

In AF2, the probabilities of gluing a spin on the cluster are different from AF1: the probability of sticking on the tip–perimeter sites is smaller than jumping to diffusing sites. Thus internal perimeter sites are more favoured by the AF2 growth process than tip ones. It results a thickening of the branches.

In the AF3 region, the probability of sticking close to a spin of the same species oriented in the field is small. The diffusing spins wander a long time before freezing on the cluster. It is as if the whole surface is visited. Thus this leads to compact clusters. The D_f is 1.80 ± 0.02 which is obviously a process and a value quite different from ordinary DLA. The computing times are here very long: decreasing βJ but staying in this region, we note very low sticking probabilities.

In AF4 to AF6 regions, the field βH dominates the coupling βJ, the sticking probabilities are also smaller than diffusing ones. But sticking on the tip perimeter sites is more favourable than sticking on the other perimeter sites. This process is similar to Dielectric Breakdown [7, 17] and Eden Tree [14] models.

4 Physical properties of MDLA clusters

The magnetic analogy allows us to define the magnetization M of the growing clusters. M is defined by the difference between the number of up and down spins in the growing clusters. A non-zero magnetization means that a species dominates the other in the formed pattern. In this section, we restrict our simulations and discussion to particular cases where the external magnetic field is absent and the coupling between spins is positive (on the horizontal line of Fig. 3).

Fig. 6 shows the magnetization M of clusters of 3000 spins as a function of the coupling between spins. The clusters have grown from an up-spin and in

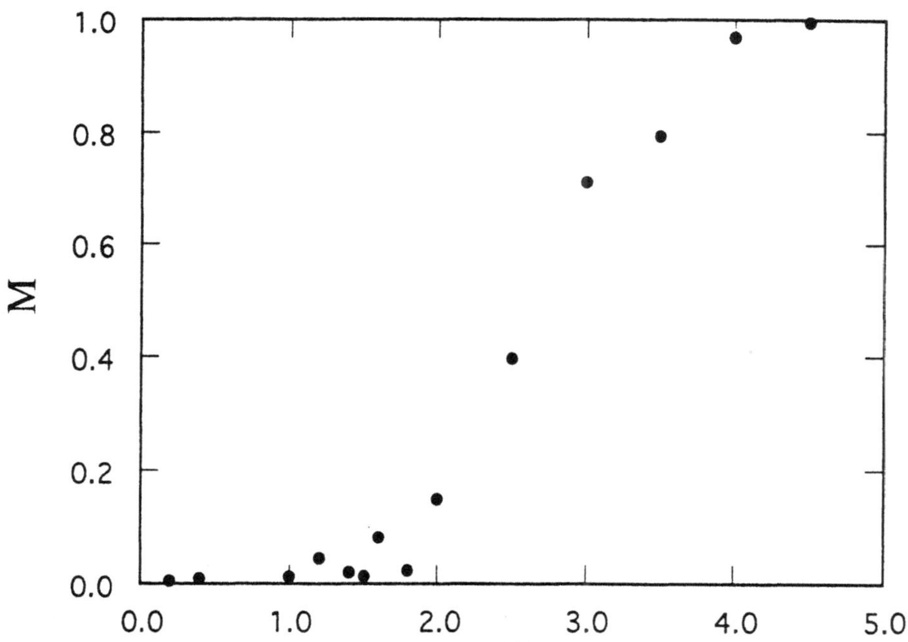

Fig. 6. Magnetization of MDLA clusters of 3000 spins grown without field from an up-spin as seed as function of the coupling βJ.

absence of an external field. Each dot represents the average magnetization over 80 clusters. The magnetization M is seen to increase rapidly from 0.0 to +1.0 around $\beta J_c = 1.5 \pm 0.1$. For a fixed mass S of the clusters, there exists a critical value $(\beta J)_c$ above which the spin species of the seed dominates the other spin species. The same behaviour with M ranging from 0.0 to -1.0 is found for cluster grown from a down spin as seed (not shown here).

Moreover, the fractal dimension D_f of the clusters were measured for the clusters simulated here above. D_f was determined by the gyration radius method. Fig. 7 shows D_f as a function of the coupling ranging from 0.0 to 5.0. D_f is seen

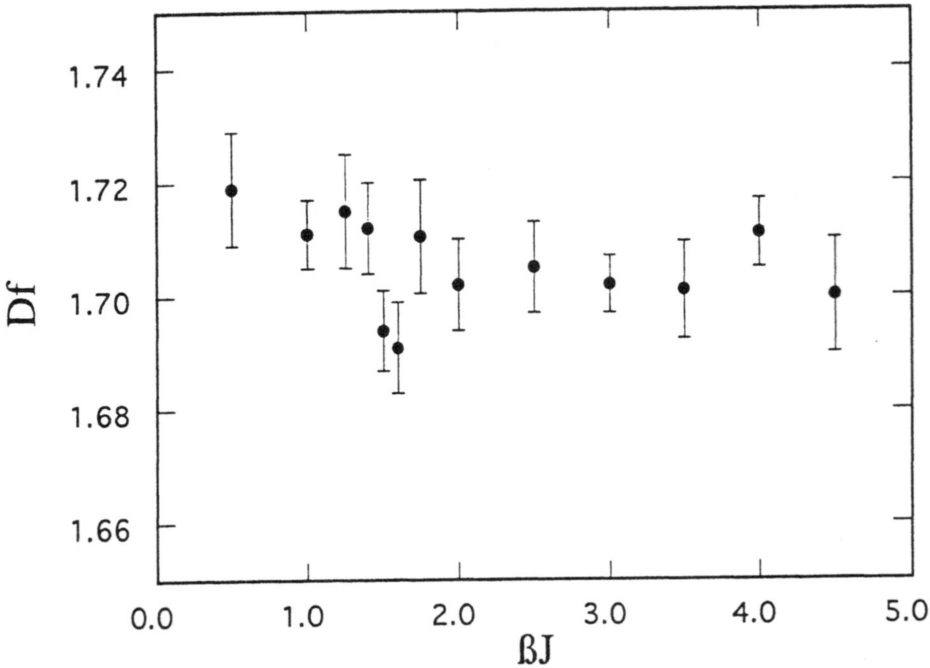

Fig. 7. Fractal dimension Df of MDLA clusters of 3000 spins grown without field from an up–spin as seed as function of the coupling βJ.

to decrease slowly and continuously from 1.71 ± 0.01 to 1.702 ± 0.006; but around βJ_c, D_f drops rapidly to 1.691 ± 0.008. Fig. 8 shows three clusters of 3000 spins generated around and on βJ_c. The morphologies are not seen to vary drastically because the Df dropping is slight. One should note that the cluster generated at $\beta J = 1.5$ is more extended that the two others (it reaches the sides of the box of 200 lattice units size). This extension corresponds to a lower D_f. The origin of this dropping of D_f stays unclear: the theoretical models [19, 20] cannot explain such a behaviour at this time. The question *"Why D_f of the aggregates is dropping when one spin species began to dominate the other spin species?"* remains also open. This particular case ($\beta H = 0$), illustrates also that the growth processes are more complex than simply described by the diagram of Fig. 3. It is interesting to note that such a behaviour was also found in the MEM cluster where we have measured the magnetization the fractal dimension of one species (the species of the seed) [18].

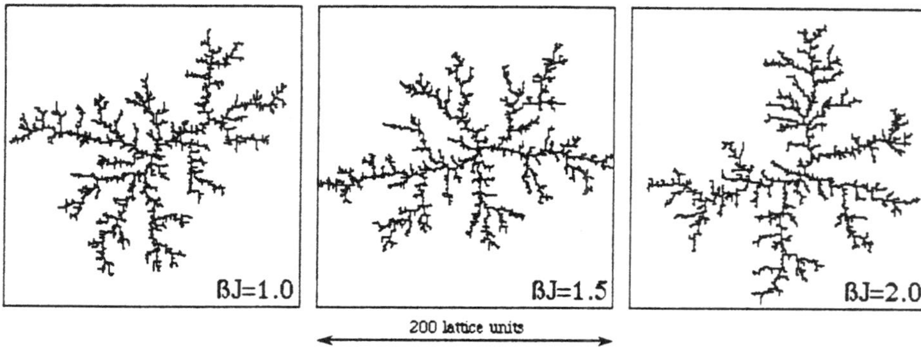

Fig. 8. Typical clusters grown from an up spin as seed without field around and on the critical coupling $\beta J_c = 1.5$.

5 Conclusion

In conclusion, the magnetic DLA (MDLA) model simulates the aggregation of particules with one internal degree of freedom taking here, as an example, two states, i.e. up and down. This generalization of DLA is physically and chemically relevant and leads to new competing growth processes and morphologies distributed here in 24 regions of a phase diagram of the growth parameters. The model allows one to give a systematic framework for many, sometimes apparently uncontrolled, extensions of kinetic growth models.

The wide variety of magnetic growth processes and cluster types was not immediately expected, and is thus of great scientific interest. The fractal dimension D_f takes different values ranging between 1.68 ± 0.02 and 1.99 ± 0.01 (nearly the dimension of the euclidian space which is 2) in the phase diagram. This connects geometrical fractals and thermodynamic fields. It is clear that D_f continuously varies with the parameters as in the MEM [12]. Further simulations must explain how D_f varies from a DLA process to a MDLA process. Boundary lines should be examined but this leads to 7 different cases which are not the scope of the present report.

The magnetization M of the finite–size clusters shows that above a sufficient coupling value, a spin species dominates the other in the growing clusters. Without external field, the dominating spin species is the seed one. More importantly, the fractal dimension D_f of the aggregates drops at the same critical values. The origin of these transition behaviors remains unclear but one should note that the model is like a spreading and diffusing Ising droplet model. The transition behavior and its origin must still be explored by standard statistical physics methods.

The basic idea clearly opens the way to many new models and considerations, and to many applications in various fields.

Acknowledgements

Part of this work is the continuation of some work financially supported through the SU/02/013 contract of the Impluse Program of the Belgium Prime Minister Service for Scientific Policy (SPPS). N.V. thanks the Belgium Institute for Encouragement on Scientific Research in Industry and Agriculture (IRSIA, Brussels).

References

[1] H. J. Hermann, Phys. Rep. **136**, 153 (1986)
[2] A. Bunde, H. J. Hermann, A. Margolina and H. E. Stanley, Phys. Rev. Lett. **55**, 653 (1985)
[3] R. F. Xiao, J. I. D. Alexander and F. Rosenberger, Phys. Rev. A **38**, 2447 (1988)
[4] T. A. Witten and L. M. Sander, Phys. Rev. Lett. 47, 1400 (1981) ; see also T. A. Witten and L. M. Sander, Phys. Rev. B **27**, 5686 (1983)
[5] M. Matsushita, M. Sano, Y. Hayakawa, H. Honjo and Y. Sawada, Phys. Rev. Lett. **53**, 286 (1984)
[6] J. Nittmann, Physica **140A**, 124 (1986)
[7] L. Niemeyer, L. Pietronero and H. J. Wiesmann, Phys. Rev. Lett. **52**, 1033 (1984)
[8] Z. Xin and W. Zi–Quin, Z. Phys. B **73**, 129 (1988)
[9] L. H. Sperling, in Introduction to Physical Polymer Science, (New York, John Wiley & Sons INC., 2d. 1992), p. 41
[10] M. Silverman and M. Simon, in *Mobile genetic elements.* ed. J. A. Shapiro (Orlando, Academic Press, 1983), p. 537
[11] P. Ossadnik, C.-H. Lam and L. M. Sander, Phys. Rev. E **49**, R1788 (1994)
[12] M. Ausloos, N. Vandewalle and R. Cloots , Europhys. Lett. **24** , 629-634 (1993)
[13] The Diffusion–Controlled Aggregation model introduced by Meakin adds only one sticking probability $p \in [0, 1]$ to DLA and this does not imply any internal degree of freedom. P. Meakin, Phys. Rev. A **27**, 604-607 (1983); ibid. Phys. Rev. A **27**, 1495 (1983)
[14] D. Dhar and R. Ramaswamy, Phys. Rev. Lett. **54**, 1346 (1985)
[15] M. Plischke and Z. Racz, Phys. Rev. Lett. **53**, 415 (1984)
[16] L. Pietronero, A. Erzan and C. Evertsz, Phys. Rev. Lett. **61**, 861 (1988)
[17] Y. Sawada, S. Ohta, M. Yamazaki and H. Honjo, Phys. Rev. A **26**, 3557 (1982)
[18] N. Vandewalle and M. Ausloos, submitted for publication. (1994) *Lacunarity, Fractal and Magnetic Transition behaviors in a generalized Eden growth process*

Diffusion models of internal dynamics of proteins

Michał Kurzyński

Institute of Physics, Adam Mickiewicz University,
Poznań, Poland

Abstract. Formulation of a truly advanced statistical theory of biochemical processes needs simple but adequate models of phenomena underlying microscopic dynamics of biomolecules, in particular enzymatic proteins. A synthetic picture of microscopic dynamics of proteins emerging from the recent studies is outlined and two classes of theoretical models of slow conformational (activated) dynamics within protein native state, both of diffusion type, are described. In the first class, referred symbolically to as *Protein-Glass*, the dynamics is represented by diffusion of structural deffects in liquid-like region of protein of an effective dimension between 1 and 2, to be be approximated by various fractal lattices. In the second, *Protein-Machine* class of models the conformational dynamics is treated as a relative motion of solid-like fragments of secondary structure, also of the nature of quasicontinuous diffusion. It is presumably a rule that it is the process of conformational relaxation, and not the details of chemical mechanism, that affects the rate of biochemical processes. Under this assumption a particular Protein-Machine model is applied for constructing a theory of single enzymatic reaction. The important result obtained is that the reaction pathways close to and far from the chemical equilibrium can differ. A possibility is indicated of direct coupling among several reactions taking place at the same multienzyme complex.

1 Introduction

Any statistical theory of physical processes has to refer to simple models of phenomena underlying microscopic dynamics. The quality of models in use is, however, different in different fields of research. The theory of solid state, successfully exploiting a variety of models named after prominent physicists of 20th century from Debye, though Heisenberg, to Anderson, is an exemplary case. The theory of biochemical phenomena seems, unfortunately, to be the opposite extreme. Until recently, biochemists acted as if following the rule stating that what is not known, simply does not exist. The transition state theory, commonly used in interpretation of biochemical reactions [1], excluded the existence of any microscopic dynamics of biomolecules (in particular enzymatic proteins) more complex than fast vibrations about a single well-defined equilibrium conformation.

This picture, adapted directly from chemistry of low-molecular weight compounds, has eventually proved untrue. Many experiments performed in the 1980s using various techniques demonstrate that within the protein native state, apart from usual vibrational dynamics, a rich interconformational (activated) dynamics exists in the whole range of time scales from 10^{-11} s to 10^{5} s or more [2, 3]. The slowness of this dynamics makes any conventional theory of chemical reactions inapplicable for description of biomolecular processes [4]. A consequence is a challenge to physicists-theoreticians to construct a truly advanced statistical theory of biochemical processes based on still simple, but realistic models of microscopic dynamics of proteins. In the present paper we describe two classes of such models, both of diffusion type, symbolically referred to as *Protein-Glass* and *Protein-Machine*. A particular realization of the latter is applied for constructing a theory of single enzymatic reaction and to draw some general conclusions.

2 Protein dynamics

Let us start with an outline of the main features of the picture of protein dynamics emerging from recent studies [3]. Proteins are linear polymers of amino acids (Figure 1). The fundamental structural unit of protein is a *domain*. It consists, on the average, of one hundred amino acids, thus has about 5×10^{3} internal degrees of freedom. These can be identified with covalent bond lengths and angles as well as dihedral angles of rotations about the bonds. It is the ability to such rotations (limited only to some degree by steric hindrances), combined with the possibility of hydrogen bond breaking and reforming, that makes the landscape of the potential energy of internal degrees of freedom extremely complex. A general feature of this landscape is the presence of an astronomical number (of an order of 10^{100}) of local minima separated by higher or lower energy barriers of non-covalent nature.

As in the stereochemistry of low-molecular weight organic compounds, the regions of microstates surrounding the local minima can be referred to as protein stable *conformations*. In the first approximation, internal dynamics of protein comprises *vibrations* within particular conformations and *conformational transitions*. The former are damped harmonic oscillations subjected accidentally to stochastic perturbations, whereas the latter are purely stochastic activated processes described by a set of master equations. The spectrum of vibrational periods ranges from 10^{-14} s (weakly damped localized N-H or C-H stretching modes observed spectroscopically) to 10^{-11} s (overdamped collective modes involving the whole domains, studied numerically). In physiological temperatures, the spectrum of relaxation times for conformational transitions begins at 10^{-11} s (local side chain rotations or hydrogen bond rearrangements on the protein surface).

A characteristic feature of biologically active proteins in physiological conditions is their well-defined spatially folded structure. Protein folding is a process

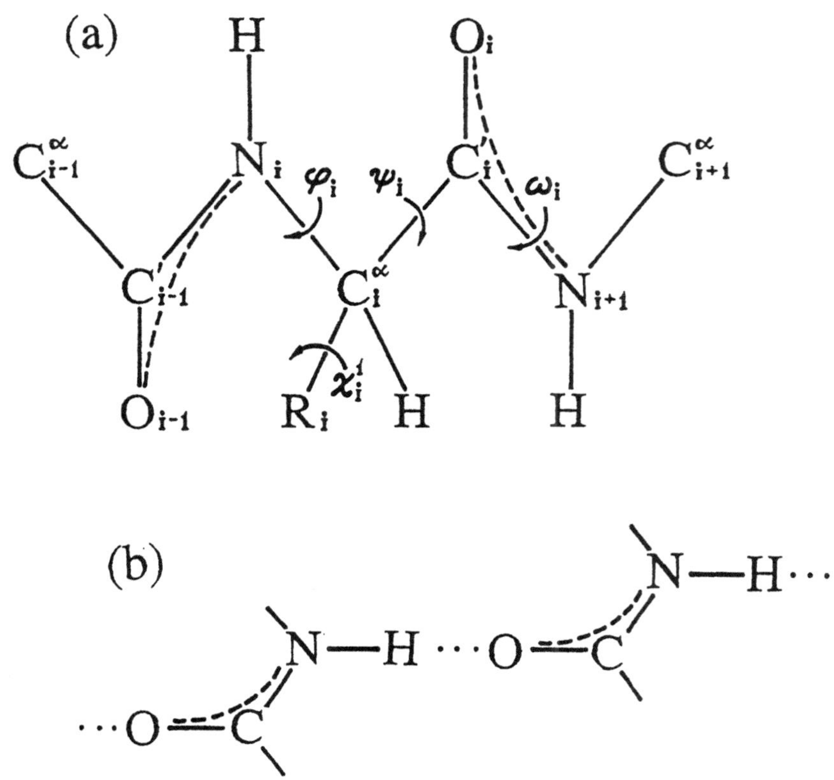

Fig. 1. Chemical structure of protein. (a) The linear main chain. Subsequent amino acids with central carbon atoms C_i^α and characteristic side chains R_i are connected by covalent amide bonds O–C'–N–H in the planar *trans* configuration. This configuration is stabilized by delocalization of bonding π-electrons, which is indicated with the help of broken lines. Notation of dihedral angles is shown. (b) Hydrogen bonds between amide groups of the main chain.

of discontinuous phase-transition type, independent for each domain, thus all particular conformations in physiological conditions are to be divided uniquely between the *native* (folded) and *unfolded* state. The mean waiting-time of folding is within the range $10^{-1} - 10^2$ s and the mean waiting-time of spontaneous unfolding should be longer than that as many times as the equilibrium concentration of the native state is higher than the equilibrium concentration of the unfolded state. A careful estimation provides for this time a value within the range $10^3 - 10^{11}$ s and this is the upper limit of the relaxation time spectrum of protein conformational transitions.

As we have stated in Introduction, a great achievement of the physics of proteins in the past two decades is demonstration of the existence of a rich interconformational (activated) dynamics within the native state of protein. The time scale of this dynamics observed with the help of various techniques is given in Figure 2. Conformational transitions take place not in the whole bulk of the domain, but are limited to liquid-like regions surrounding solid-like fragments of secondary structure (α-helices or β-pleated sheets) surviving, in physiological conditions, the transition to the unfolded state. It is not easy to infer the actual nature of conformational dynamics from experimental data. Figure 2 indicates that conformational transitions occur within the whole time scale from 10^{-11} s to 10^{5} s or more, but we do not know how numerous is, in fact, the population of conformations composing the native state. From this point of view numerical simulation results appear important, pointing directly to the existence of the whole quasi-continuum of short-lived conformations within the native state [5, 6]. There is no reason to doubt that also longer-lived conformations form some kind of a quasi-continuum. In Figure 3, this view is illustrated as the outline of the spectra of reciprocal relaxation times of protein conformational dynamics, separately in the unfolded and the native state.

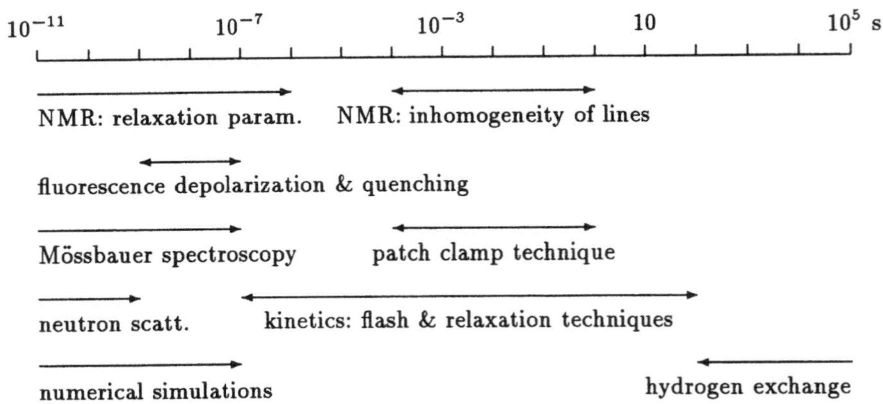

Fig. 2. Time scale of conformational transitions within the protein native state observed with the help of various experimental techniques (after Ref. 3). Time period 10^{-11} s at one edge of the spectrum characterizes localized conformational transitions on the protein surface. Time period 10^{5} s at the other edge is a (rather underestimated) value of the waiting-time for spontaneous unfolding of the protein in physiological conditions. Note that typical reciprocal turnover number of enzymatic reactions, 10^{-3} s, is exactly in the middle of the scale.

3 Theoretical models of conformational dynamics

Because the experiments at hand cannot elucidate the nature of conformational dynamics in detail, the problem is to some extent left open to speculation. In two classes of models provided hitherto by literature the speculative element

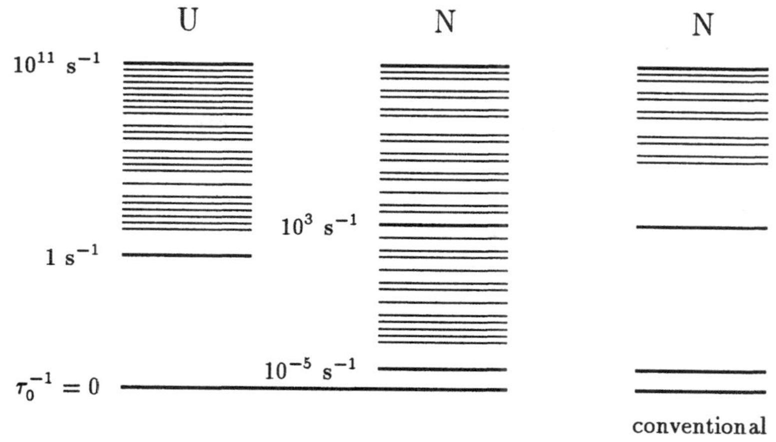

Fig. 3. Fig. 3. Schematic spectra of reciprocal relaxation times of conformational dynamics within the unfolded (U) and the native (N) state of protein domain in physiological conditions. The ground zero level (infite relaxation time) corresponds to the sum of probabilities of all the conformational states preserved. Indicated are rates of spontaneous folding ($\sim 1\,s^{-1}$), unfolding ($\sim 10^{-5}\,s^{-1}$), and the typical rate of enzymatic reaction ($\sim 10^{3}s^{-1}$). Our discussion implies that biochemical processes are not separated in time scale from the processes of conformational relaxation. This view contradicts the conventional belief claiming that the spectrum of reciprocal relaxation times within the protein native state looks rather like the scheme to the most right.

seems to be kept within reasonable limits. We shall refer to them symbolically as *Protein-Glass* and *Protein-Machine*. In essence, the question concerns the form of the relaxation time spectrum of a set of conformational master equations. As mentioned in the previous section, this spectrum is not expected to have a well-defined time scale separation. The simplest way to tackle such problems is to assume that the dynamics of a system looks alike in every time scale, i.e., the spectrum of relaxation times has a self-similarity symmetry. This assumption is the core of any Protein-Glass model (Figure 4(a)). An alternative makes the Protein-Machine class of models in which the variety of conformations composing the native state is supposed to be labelled with only a few "mechanical" variables. The reciprocal relaxation time spectrum is then a sum of several more or less equidistant subspectra (Figure 4(b)). Let us say a few more words about each class of models.

Protein-Glass. Time scaling, considered to be a generic property of glassy materials, originates either from a hierarchy of barrier heights in the potential energy landscape, or from a hierarchy of "bottlenecks"in the network joining conformations between which direct transitions take place [7]. A hierarchy of interconformational barrier heights was proposed ten years ago by Frauenfelder and coworkers in order to combine the results of various experiments concerning

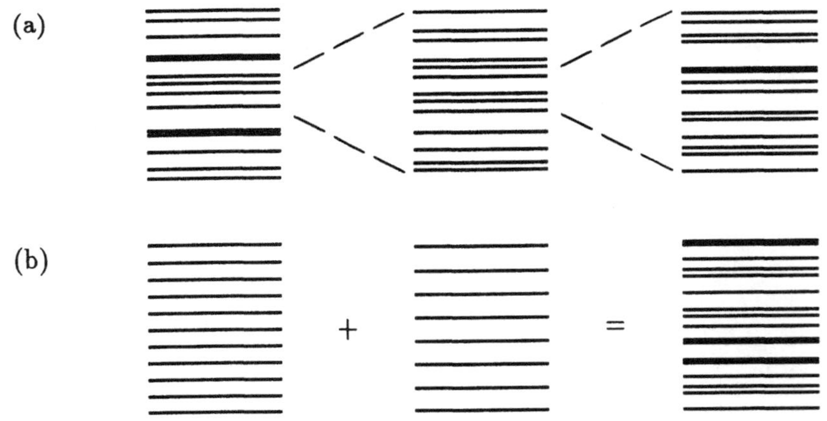

Fig. 4. Schematic spectra of reciprocal relaxation times of conformational dynamics within the native state of protein. (a) *Protein-Glass* model: spectrum looks approximately alike in several subsequent time scales. (b) *Protein-Machine* model: spectrum consists of a few more or less equidistant sub-spectra.

the process of ligand binding to myoglobin [2]. The reasonable mathematical realization of such a hierarchy in the context of its application to proteins are spin glasses [8]. The mathematical realization, on the other hand, of hierarchical networks are lattices with the effective dimension between 1 and 2, e.g. geometrical fractals (Figure 5(a)) or the percolation clusters (Figure 5 (b)). The latter have been already applied to describe dynamics of protein [9]. The process of diffusion on fractal lattices simulates structural defect motions in the liquid-like regions between solid-like fragments of secondary structure.

Protein-Machine model was proposed more than twenty ago by a group of Moscow biophysicists assembled around Chernavsky [10] and Blumenfeld [11]. Speculative at the origin, the model becames presently better and better experimentally justified [12]. In the simplest case, the mechanical variables can be identified with angles describing the mutual orientation of rigid fragments of secondary structure (α-helices or β-pleated sheets). In the continuum limit, subsequent conformational transitions along a given mechanical coordinate are to be approximated by diffusion in an effective potential, parabolic at the simplest [13-15]. It is worth noting that identical mathematical equations describe overdamped collective vibrations of domains, moreover, numerical analysis indicates that also they take the form of mutual motions of relatively rigid fragments of secondary structure.

Each of the presented classes of models may be true to some extent but, in the author's opinion, the Protein-Machine type approach seems to be more universal. The Protein-Glass models behave unrealistically both in the limit of

(a)

(b)

Fig. 5. Examples of hierarchical lattices with the fractal dimension between 1 and 2. (a) Sierpiński gasket. (b) Percolation cluster.

very short and very long times, and in practice should be restricted only to a few levels of hierarchy [2]. However, a finite hierarchy can occur also in the Protein-Machine models, if one identifies mechanical elements, first, with some chosen side chains, next, with the fragments of secondary structure and, in the end, with the whole domains. Important is also that there is no essential difference between conformational transitions along the mechanical coordinates and overdamped collective vibrations, and that the cross-over between both types of motion is more or less a matter of convention.

4 Protein-Machine model of a single enzymatic reaction

Conventional biochemistry, like ordinary chemistry of low-molecular weight compounds, does not distinguish, in general, between the *kinetic* and the *chemical* mechanism of reaction [1]. Thus, the reaction involving a single covalent step

$$\mathrm{R} \longleftrightarrow \mathrm{P}$$

is modelled by the three-step kinetics of Haldane, given in Figure 6(a), with only two association-dissociation steps added.

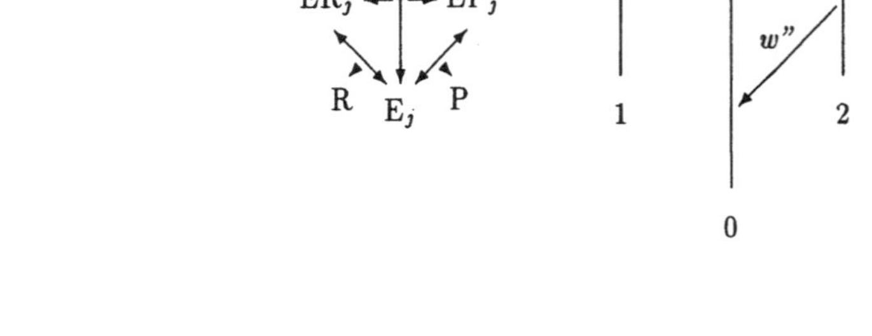

Fig. 6. Single enzymatic reaction. (a) Conventional kinetics. (b) Actual kinetics involving a quasi-continuous set of conformations of the enzyme or its complexes labelled with indices i and j. (c) Protein-Machine model. Dynamics of conformational transitions within each of the three chemical species E, ER and EP is approximated by diffusion along a mechanical coordinate. Perpendicular chemical transitions are in general reversible; the arrows indicate only the signs of the corresponding components of local reaction rates (3).

A typical rate of enzymatic reaction is $10^3\,s^{-1}$. In the conventional biochemistry this reaction is assumed to be well separated in the time scale from other processes. But at present we know that this is not true: conformational relaxation within the native state of protein is as slow, if no slower, as the reaction (Figures 2 and 3). Certainly, some conformational transitions are independent of the reaction but others must affect the latter in a way. In consequence, to describe the actual kinetic mechanism of enzymatic reaction one has to treat conformational (non-covalent) transitions on an equal footing with the chemical (covalent) transformations. The resultant scheme for the enzymatic reaction involving a single covalent step is infinitely more complex than the conventional scheme [4]. It is shown in Figure 6(b).

It is obvious that, on the phenomenological level, a scheme like that in Figure 6(b) is operationally useless as one cannot distinguish (and prepare) so many species, and no detailed investigation of the reaction time course can verify it. As in the conventional approach, the process should be described phenomenologically in terms of equations for concentrations of a few *observable* species: E, ER and EP. The problem of formulating reasonable approximations of those equations is highly non-trivial. In the following we present a partial solution to this problem, based on a particular Protein-Machine model of conformational dynamics [12, 15].

In the Protein-Machine model considered, dynamics of conformational transitions within each of the three chemical species is approximated by diffusion along a mechanical coordinate x in some effective parabolic potential. Chemical transitions are perpendicular to this coordinate, thus the complete dynamics of enzyme is described by a set of three coupled partial differential equations for probability densities $p_i(x,t)$ ($i = 0, 1, 2$ for E, ER and EP, respectively):

$$\frac{\partial}{\partial t}p_i = -\frac{\partial}{\partial x}j_i + w_i, \tag{1}$$

with diffusion fluxes

$$j_i = -\gamma\left[\frac{1}{2}\frac{\partial}{\partial x} + (x - x_i)\right]p_i, \tag{2}$$

and local reaction rates

$$w_0 = -w' + w'', \qquad w_1 = -w + w', \qquad w_2 = w - w'' \tag{3}$$

(compare Figure 6(c)). The dimensionless mechanical variable x, proportional to $\sqrt{kT}$, is normalized in such a way that its equilibrium dispersion is 1/2; x_i's are positions of minima of the conformational potential in individual chemical species, and the parameter γ has a meaning of the reciprocal relaxation time of the mean value of the mechanical variable. We assume that the chemical transitions are localized in narrow regions of the values of this variable:

$$w(x) \propto \kappa\delta(x), \qquad w'(x) \propto \kappa'\delta(x - x'), \qquad w''(x) \propto \kappa''\delta(x - x'') \tag{4}$$

(the reaction is *gated* by conformational dynamics). We assume also that local chemical transitions are much faster than conformational diffusion:

$$\kappa, \kappa', \kappa'' \gg \gamma. \tag{5}$$

This assumption is quite natural if we recall that conformational relaxation within the native state of enzyme is as fast as the very reaction so that it is this relaxation that determines the resultant reaction rate and not the details of chemical mechanism. An important consequence of this assumption is a different path of the reaction close to and far from the equilibrium.

Close to the thermodynamic equilibrium the reciprocal chemical relaxation time is given by the formula:

$$\tau_{\mathrm{eq}}^{-1} = \frac{\gamma}{2\sqrt{\pi}} \sum_{i=0}^{2} (1 - C_i^{\mathrm{eq}}) \sqrt{\Delta G_i^{\ddagger}/kT} \exp(-\Delta G_i^{\ddagger}/kT), \tag{6}$$

where C_i^{eq}, $i = 0, 1, 2$, are equilibrium molar fractions of chemical species E, ER and EP, respectively, and $\Delta G_i^{\ddagger}$ are values of the free energy that have to be reached within particular species in order to reach the nearest gate from the equilibrium conformation x_i (see Figure 7).

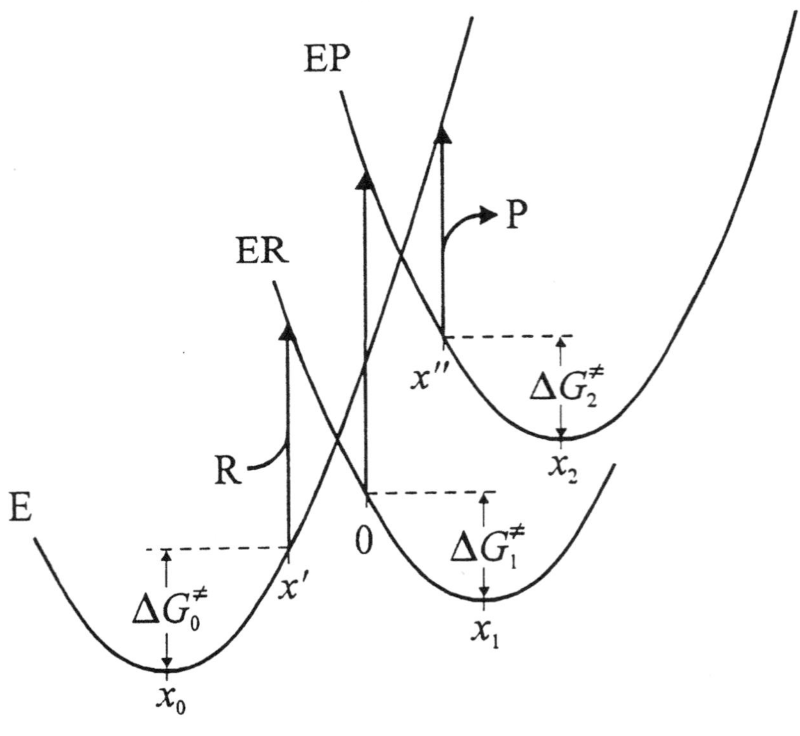

Fig. 7. Protein-Machine model of a single enzymatic reaction (compare Figure 6(c)). Three different conformational potentials with minima at x_0, x_1 and x_2 correspond to individual chemical states E, ER and EP, respectively, of the enzyme. Three chemical transitions are localized at the points 0, x' and x''.

Equation (6) has the meaning of the average reciprocal time of diffusion uphill the conformational potential from its minimum, provided, however, that $\Delta G_i^\ddagger \gg kT$. Quite generally, time of difusion from x' to x'' in the potential x^2 (in kT units) is given by the formula

$$\tau(x' \to x'') = 2\gamma^{-1} \int_{x'}^{x''} dy\, e^{y^2} \int_{-\infty}^{y} dx\, e^{-x^2}. \tag{7}$$

In Figure 8 diffusion times uphill and downhill the potential are plotted vs the distance x in the logarithmic scale. It is seen that diffusion uphill can be several orders of magnitude slower than diffusion downhill, taking place in the time of the order of γ^{-1}.

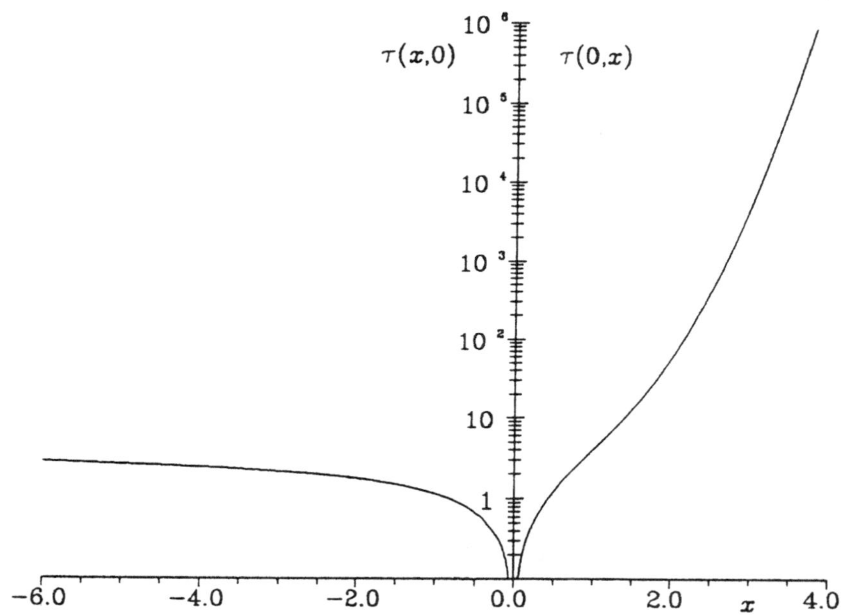

Fig. 8. Diffusion time uphill, $\tau(0 \to x)$, and downhill, $\tau(x \to 0)$, the potential x^2. Time is counted in characteristic diffusion time units γ^{-1}. Dimensionless coordinate x is proportional to $\sqrt{kT}$.

On examining Figure 7 we find that after transition at the point x' from the state E + R to ER, the enzyme can either equilibrate within ER or pass directly to the next state EP, or even E + R, with the process of equilibration omitted. As a consequence, in the case discussed, direct reactions between *each*, in general, pair of kinetic states are possible [4]. The chemical relaxation time (6) determines rate constants occurring in such a generalized scheme.

No partial equilibration within any kinetic species is necessery if the enzymatic reaction proceeds far from the chemical equilibrium. In steady state conditions with the concentration of reactant kept constant, [R] = const, and with the constant removal of product, [P] = 0, the rate of the reaction is described (as in the conventional approach) by the Michaelis-Menten law:

$$[\dot{\mathrm{P}}] = \frac{k_c[\mathrm{R}]}{K_m + [\mathrm{R}]}, \tag{8}$$

with the reciprocal turnover number

$$\begin{aligned} k_c^{-1} = {} & \tau_0(x'' \rightarrow x') + \tau_1(x' \rightarrow 0) + \tau_2(0 \rightarrow x'') \\ & + (C_1^{\mathrm{eq}}/C_2^{\mathrm{eq}})\,[\tau_2(0 \rightarrow x'') + \tau_2(x'' \rightarrow 0)], \end{aligned} \tag{9}$$

and the apparent dissociation constant

$$\begin{aligned} K_m = {} & k_c[R]^{\mathrm{eq}}\,\{(C_0^{\mathrm{eq}}/C_1^{\mathrm{eq}})\,[\tau_1(x' \rightarrow 0) + \tau_1(0 \rightarrow x')] \\ & + (C_0^{\mathrm{eq}}/C_2^{\mathrm{eq}})\,[\tau_2(0 \rightarrow x'') + \tau_2(x'' \rightarrow 0)]\} \end{aligned} \tag{10}$$

τ_i denoting diffusion time in the potential of i-th species with the minimum at x_i. Note independence of k_c of the chemical relaxation time near equilibrium τ_{eq}, Equation (6).

The theory presented gives a rule of the optimum action of enzyme: turnover number k_c is maximum, if conformational diffusion within free enzyme E is downhill, and the transition points $x = 0$, x', x'' lie not very far from each other. In that case $k_c \simeq \gamma \gg \tau_{\mathrm{eq}}^{-1}$. Of course the backward reaction along the same path in steady state conditions should be much slower, but the latter reaction, in Protein-Machine model, can proceed along another path it finds more convenient.

5 Concluding remarks

Great progress in the studies of protein dynamics in the 1980s propels an essential alteration in our understanding of the enzymatic catalysis phenomenon. The simple classical statement: *"enzymes accelerate reactions by decreasing the free energy of activation"* represents only half of the truth. We do not deny enzymatic reactions to proceed through gates of relatively low free energy. These gates can have either the character of transient packing defects giving free space for substrate motions, or that of local polarization fluctuations facilitating displacement of charges during the very covalent act. However, because of the slowness of conformational relaxation *it is usually not the process of gate crossing that limits the reaction rate, but the process of gate opening.*

In Figure 9, the presumably universal statistically independent unit of biochemical processes, a *supramolecular multienzyme complex* [16], is outlined. From the dynamical point of view essential is a distinction within its body between solid-like fragments of secondary structure (α-helices or β-pleated sheets) and liquid-like surrounding regions, either nonpolar (domain interiors, lipid membrane environment), or polar ones ("channels" between domains, water environment). The complex, of the size of approximately 30 nm, is too large to be described in terms of *microscopic* mechanics of individual atoms, and too inhomogeneous to be described in terms of *macroscopic* thermodynamics. The proper language is the *mesoscopic* theory of stochastic processes. Both considered classes of models of conformational dynamics use such a language. Models of Protein-Glass type treat the dynamics of conformational transitions as a quasicontinuous diffusion of structural defects through the liquid-like medium. Alternatively, models of Protein-Machine type treat this dynamics as a relative motion of solid-like elements, also of the nature of quasicontinuous diffusion. Slow diffusion dynamics controls all, in principle, reactions that take place in localized catalytic centers.

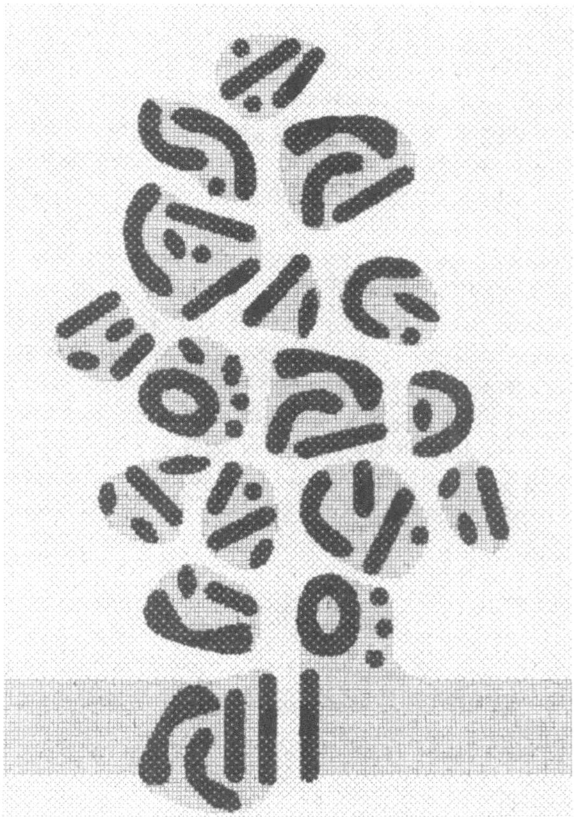

Fig. 9. Schematic cross-section of the universal statistically independent unit of biochemical processes, a supramolecular multienzyme complex. Heavily shaded are solid-like fragments of secondary structures, medium shaded are nonpolar liquid-like regions, and weakly shaded are polar liquid-like regions. Black are individual catalytic centers usualy localized at two neighbouring solid-like elements.

The picture presented contradicts the conventional outlook, following which the complex consists of several component macromolecules, each occurring only in a few *discrete* chemical or conformational states. The corresponding dynamics is also a stochastic process, however, not of a quasicontinuous diffusion type, but the jump one, of all-or-nothing type. Both close to and far from the total thermodynamic equilibrium, kinetic transformations take place only after reaching partial equilibrium within individual kinetic states.

Slow diffusive dynamics, on the contrary, results in different pathways of enzymatic reaction close to and far from the equilibrium. Close to the equilibrium, equations of conventional kinetics are valid, possibly corrected for direct component reactions. In the steady state conditions far from the equilibrium, the thermodynamic variables proper for description of chemical processes are not *concentrations* of chemical species, but rather quantities characterizing conformational nonequilibrium of the enzyme, for instance mean values of *mechanical variables*. The turnover number of enzyme is not related to reaction rate constants close to the equilibrium. This was suggested already twenty years ago by Blumenfeld [11], whose ideas influencened the author of this paper very strongly. Note, that the new approach changes the interpretation but leaves the phenomenology essentially unaltered, so only evident demonstration of the lack of this relation could be the experimentum crucis directly proving the conventional interpretation wrong.

The conformational nonequilibrium should play an extremely important role in the process of coupling of *several* reactions taking place at the same multienzyme complex. In processes of this type the biological free energy transduction is carried out at the cellural level. In the conventional mechanism of *chemical* coupling the complex is needed only for keeping the appropiately high concentration of intermediates. The conformational nonequilibrium implies the possibility of different, *mechanical* coupling of reactions, upon which energy released in the center of one reaction is directly transferred to the center of another reaction. In this field, theoretical studies are, unfortunately, at the beginning stage.

Acknowledgements

This study has been supported in part by the Alexander von Humboldt Foundation and in part by the Polish State Committee for Scientific Research within project 2 P302 061 04.

References

[1] A. Fersht, Enzyme Structure and Mechanism, 2nd edn., Freeman, New York (1985)
[2] H. Frauenfelder, S. G. Sligar and P. G. Wolynes, Science **254**, 1598 (1991)
[3] M. Kurzyński, Biophys. Chem., review article submitted (1994)
[4] M. Kurzyński, FEBS Lett. **328**, 221 (1993)
[5] R. Elber and M. Karplus, Science **235**, 318 (1987)

[6] N. Gō and T. Noguti, Chemica Scripta A **29**, 151 (1989)
[7] A. Blumen and H. Schnörer, Angew. Chem. Int. Ed. **29**, 113 (1990)
[8] D. L. Stein, Physica Scripta **34**, 92 (1986)
[9] W. Doster, W. Schirmacher and M. Settles, Biophys. J. **57**, 681 (1990)
[10] D. S. Chernavsky, Yu. I. Khurgin and S. E. Shnol, Biophysics (USSR) **32**, 775 (1987)
[11] L. A. Blumenfeld, Problems of Biological Physics, Springer Series in Synergetics vol. 7, Berlin (1981)
[12] M. Kurzyński, FEBS Lett., submitted (1994)
[13] K. V. Shaitan and A. B. Rubin, Biophysics (USSR) **27**, 386 (1982)
[14] N. Agmon and J. J. Hopfield, J. Chem. Phys. **78**, 6947; **79**, 2042 (1983)
[15] M. Kurzyński, J. Chem. Phys. **100**, No. 13 (1994), in print.
[16] P. A. Srere, Annu. Rev. Biochem. **56**, 89 (1987)

IV–th MAX BORN SYMPOSIUM : SOLID STATE PHYSICS I

KARPACZ, 28–30 SEPTEMBER 1993

ORGANIZING COMMITTEE

Prof. Z. M. Galasiewicz, Dr. A. Strycharska, Dr. J. Jędrzejewski, Dr. Z. Petru, Dr. Cz. Oleksy

LECTURES

1. Helmut R. Brand (Bayreuth)
   Phase dynamics–the analogue of hydrodynamics for pattern forming systems

2. Wilhelm Brenig (Garching)
   Optical phonons and delocalization in hydrogen overlayers: lattice dynamics without displacement coordinates

3. Jerzy Czerwonko (Wrocław)
   Two–dimensional electron gas in the magnetic field: a toy model

4. Peter Entel (Duisburg)
   Electron–phonon interaction and strong correlations in high-temperature superconductors

5. Zygmunt Galasiewicz (Wrocław)
   The life and scientific work of Max Born

6. Dale A. Huckaby (Fort Worth)
   Cubic versus hexagonal structure of noble gas crystals and ice

7. A. G. Naumovets (Kiev)
   Surface diffusion and phase transitions in electropositive adsorbed layers on metals

8. Władysława Nawrocka (Wrocław)
Shape transitions in finite fermion systems

9. Tadeusz Paszkiewicz (Wrocław)
Phonon images of crystals

10. Andrzej Pękalski (Wrocław)
Diffusion of oxygen in YBCO type systems

11. Andrzej Radosz (Wrocław)
Anyon superconductivity at non–zero temperatures

12. E. Shikhovtseva (Ufa)
One–dimensional high conductive polymer state. A possible creation mechanism

13. Łukasz A. Turski (Warszawa)
Simple diffusion or simply diffusion?

14. Wolfgang Weller (Leipzig)
Field theory of the fractional quantum hall effect

15. Elżbieta Zipper (Katowice)
On the possibilty of orbital ferromagnetism in mesoscopic systems

FSC
www.fsc.org
MIX
Papier | Fördert gute Waldnutzung
FSC® C083411

Kolibri 360 GmbH
Ferdinand-Jühlke-Straße 7
99095 Erfurt, Deutschland
produktsicherheit@kolibri360.de